THE
DIGITAL
REPUBLIC

THE
DIGITAL
REPUBLIC

On Freedom and Democracy
in the 21st Century

JAMIE SUSSKIND

PEGASUS BOOKS

NEW YORK LONDON

THE DIGITAL REPUBLIC

Pegasus Books, Ltd.
148 West 37th Street, 13th Floor
New York, NY 10018

First Pegasus Books cloth edition July 2022

ISBN: 978-1-64313-901-2

10 9 8 7 6 5 4 3 2 1

Printed in the United States of America
Distributed by Simon & Schuster
www.pegasusbooks.com

For Joanna

Give thy kings law – leave not uncurbed the great.
John Keats

Contents

Preface   xiii

PART ONE: THE AGE OF THE DIGITAL REPUBLIC

Introduction: Unaccountable Power   3
1   The Indignant Spirit   19
2   Thinking, Old and New   25

PART TWO: THE HOUSE OF POWER

3   Autocrats of Information   35
4   Data's Dominion   39
5   Masters of Perception   45
6   Republic of Reason   51
7   The Automation of Deliberation   57

PART THREE: THE DIGITAL IS POLITICAL

8   The Morality of Code   65
9   The Computational Ideology   71
10 Technology and Domination   79

PART FOUR: THE MARKETPLACE OF IDEALS

11 The Market's Place   85
12 Selfie   93
13 Ethics Washing   99
14 The Consent Trap   105

PART FIVE: THE GHOST OF GOVERNANCE

15 Making Our Own Laws   115
16 The Mild West   119
17 Private Order   127

PART SIX: FOUNDATIONS OF THE DIGITAL REPUBLIC

18 Four Principles   135
19 Technology and Democracy   145
20 Deliberative Mini-Publics   153
21 Republican Rights   163
22 Republic of Standards   169

PART SEVEN: COUNTERPOWER

23 Tech Tribunals   177
24 Collective Enforcement   185
25 Certified Republic   191
26 Responsible Adults   197
27 Republican Internationalism   203

PART EIGHT: OPENNESS

28 A New Inspectorate   213
29 Zones of Darkness   217
30 Transparency about Transparency   221
31 A Duty of Openness   225

PART NINE: GIANTS, DATA AND ALGORITHMS

32 Antitrust, Awakened   235
33 Republican Antitrust   243
34 Beyond Privacy   249
35 Acceptable Algorithms   257

PART TEN: GOVERNING SOCIAL MEDIA

36 The Battlefield of Ideas   267
37 Toasters with Pictures   277
38 A System of Free Expression   285
39 Governing Social Media   293

Conclusion: The Digital Republic   301

Acknowledgements   307
Notes   311
Bibliography   387
Index   435

Preface

Not long ago, the tech industry was widely admired and the internet was regarded as a tonic for freedom and democracy. Not anymore. Every day, the headlines blaze with reports of racist algorithms, data leaks and social media platforms festering with falsehood and hate. Politicians denounce the tech giants in extravagant terms. Regulators crack their knuckles ominously. In the boardrooms of Silicon Valley, lawyers and lobbyists are limbering up for the fight of their lives.

What went wrong?

It is tempting to point the finger at a few big companies and the people who run them. Indeed, the story of digital technology is often told as a kind of Shakespearean tragedy, propelled by the flaws of its leading characters. But what if human failings are only a small part of the story? What if the problems at the heart of the tech industry are much bigger than any individual or company?

This book aims to persuade you that the challenges presented by digital technology are not the fault of a few bad apples. They are the result of our shared failure to govern technology properly, a failure derived from decades of muddled ideas and wishful thinking. To reclaim the promise of digital technology, and protect the things that matter most to us, we will need to do more than wag our fingers or wring our hands. The task is more fundamental. We will have to change the way we think about technology, ourselves and each other.

This book suggests how.

THE
DIGITAL
REPUBLIC

PART I

THE AGE OF THE DIGITAL REPUBLIC

We cannot solve our problems with the same thinking we
used when we created them.
Albert Einstein (apocryphal)

Introduction:
Unaccountable Power

This book is about how freedom and democracy can survive, and even flourish, in a world transformed by digital technology. It is for those who are excited by digital innovation, but concerned that we may be unprepared for the future that is coming into view.

The central challenge can be captured in two words: unaccountable power. In the early days of the commercial internet, scholars discovered that, in cyberspace, computer code operated as a kind of 'law'.[1] Not law as we know it – public rules decided by legislators and judges – but a different kind of law, embedded in the tech itself. Whenever we use an app, platform, smartphone or computer, we have no choice but to follow the strict rules that are coded into these technologies. Some rules are commonplace, like the rule that *you cannot access this system without the correct password*. Hence the lad who lost more than $200 million because he couldn't remember the password to his virtual currency wallet.[2] Other rules are more controversial. In late 2020, Twitter made it impossible for users to share a controversial *New York Post* article containing allegations of corruption about Joe Biden's son, on the basis that it violated the platform's rules against sharing hacked material.[3] As more and more of our actions, interactions and transactions are mediated through digital technology, those who write code increasingly write the rules by which the rest of us live. Software engineers are becoming social engineers.[4]

Code carries a different kind of power, too: the power to affect how we perceive the world. Every time we use a search engine, digital assistant, news app, social media platform or the like, we let others subtly shape our outlook. Digital systems propel issues to the top of the public agenda, or make them disappear. They frame the way we see ourselves and each other. They influence our norms and customs; what we regard as true or false, real or fake, right or wrong. This form of power is more subtle than the ability to write hard-edged rules. It operates on hearts and minds. But it is no less potent for that.

Another form of power lies in the capacity of digital technologies to gather data. It is predicted that the world will have 175 zettabytes of data by 2025. If you downloaded that onto DVDs (remember DVDs?), the resulting stack of discs would stretch round the planet 222 times.[5] More and more of our thoughts, feelings, movements, purchases and utterances are captured and analysed by systems working silently around us. With each passing year, they get better at identifying our tastes, fears and habits. This leaves us increasingly exposed to influence and anxious about the data trail we leave behind.

Despite what is sometimes claimed, digital systems are not morally neutral or objective. They are laden with biases, prejudices and priorities. Every algorithm used to determine job or credit applications, for instance, is necessarily engineered according to a particular set of values. The same is true of other algorithms too. In 2021, Facebook users who watched a video featuring black men received an automated prompt asking if they would like to 'keep seeing videos about Primates'.[6] This error was probably the result of poor-quality training data. But it was not an anomaly. Technology doesn't merely shunt us around. It has the power to shape the moral character of society, for better or worse.

In short: technologies exert power; that power is growing; and it is entrusted to those who write code. The tricky thing is that tech doesn't often look obviously political, at least as that term is usually understood. Digital power doesn't reside in a palace or parliament. It operates outside the traditional channels of high politics. This presents a danger. If we continue down our current path, liberty could be stifled, and democracy undermined, by diffuse technical forces that cannot be attributed to any single corporation or person.[7]

***

How have the advanced democracies of the world reacted to the rise of this new and strange form of power in their midst? Until recently, with a mix of confusion and inertia, particularly in the US, where action is needed most.

Take the most recent presidential election. Before the campaign, the New Yorker magazine ran an article headlined Can Mark Zuckerberg Fix Facebook Before It Breaks Democracy?[8] Then, once the campaign was underway, Joe Biden started a petition asking Facebook to keep 'paid misinformation' from influencing the election.[9] And on Capitol Hill, Speaker of the House of Representatives Nancy Pelosi plaintively asked advertisers to tell technology companies to reduce online misinformation.[10]

Stories like these are so common that they have lost their capacity to shock. But in a healthy political system, public officials should not have to plead with corporations to protect the integrity of the democratic system. It shouldn't be for Mr Zuckerberg, or advertising executives, or any business for that matter, to decide the fate of a great democracy. Must we simply hope that the tech companies will do the right thing with the power at their disposal? Or should we be asking a different question: why are they given that choice at all?

In the face of growing criticism, the tech industry has sought to reassure the world that market competition will compel firms to act in the interests of consumers. After all, if businesses don't give the people what they want, they will lose out to their competitors – right? This argument is seductive but ultimately wrong. This book argues that instead of reining in the might of tech corporations, the market often gives them more power. Instead of curtailing bad behaviour, the market encourages it. And instead of empowering citizens, the market strips us of our individual and collective agency.

Self-regulation has been suggested as an alternative way to hold the tech industry to account. But when tech lobbyists speak of self-regulation, they are not describing it as it is understood by professionals like doctors. Unlike in medicine, there are no mandatory ethical qualifications for working as a software engineer or technology executive. There is no enforceable industry code of conduct. There is no obligatory certification. There is no duty to put the public ahead of profit. There are few consequences for serious moral failings; no real fear of being suspended or struck off. Recent years have seen an explosion of AI ethics charters and the like, filled with well-meaning generalities about the responsible use of powerful computers. But without consequences for breaching them, these charters are just toothless statements of aspiration. The tech industry is basically saying: trust us. But blind trust is not how we govern doctors, lawyers, bankers, pilots or anyone else in unelected positions of social responsibility. Tech is the exception, and it's not clear why.

What about laws? Not the 'law' of code but the actual laws that are supposed to protect us from the predations of the powerful? Contrary to myth, digital technology is not unregulated. It is governed by many overlapping legal regimes. The trouble is that today's laws don't protect us in the way that we need protecting. They leave individuals to fend for themselves. They let tech companies get away with serious misconduct. Some even shield platforms from legal liabilities to which they would otherwise be exposed.

The growth in technology's power has not been matched by corresponding growth in legal responsibility. But if you think about it, there is no good reason why business corporations should be able to wield great influence over the rest of us simply because they design and control digital technologies. Freedom, wrote the Roman jurist Cicero more than twenty centuries ago, does not mean 'having a just master'. It means having no master at all.[11] Cicero and his followers believed that important decisions about our lives – what we believe, what we say and do, how we govern ourselves – must be our own, not handed over to powerful others. But the more power we delegate to those who write code, the less free we remain to plot our own course.

History tells us that unaccountable power of any kind is a disease that eats away at society. It erodes the bonds of community that hold us together. It undermines the capacity of democratic institutions. It tarnishes our dignity and diminishes our liberty. And often it does so *inadvertently*, without anyone intending for it to happen. Political decay is rarely the work of a dastardly magnate steepling his fingers in a volcanic lair.

It has been too easy, in recent years, to roll our eyes at Mark Zuckerberg and the other Silicon Valley bigwigs whose mistakes always seem so obvious in hindsight. In the final analysis, however, the problem isn't Mark Zuckerberg; it's the *idea* of Mark Zuckerberg. It's that he, and all the other Zuckerbergs out there and still to come, will be able to make decisions about our collective fate with impunity. Some Zuckerbergs will be wise, other Zuckerbergs will be knaves. But in a world of increasingly powerful Zuckerbergs, where does that leave the rest of us? Passive and impotent in the face of forces we cannot understand, still less control.

All this may sound a little excessive in the early 2020s. After all, most consumer tech feels empowering, and the world isn't falling apart quite yet, despite some of the more sensational reporting. But that is no cause for comfort or complacency. We have to look ahead. As the rate of technological change accelerates, technology's power will grow faster than our rickety social systems are able to adapt. Our laws are ageing. Our institutions are crumbling. Our governing ideas are muddled and outdated. And into this turmoil we are introducing systems so powerful that they would shake the foundations of even the most well-ordered societies.

No wonder we feel out of control – we are.

***

Let's start by framing the problem properly. The current laws are not the natural order of things. They are human-made creations. We can undo or remake them however we choose. The choice has never been between *regulation* and *deregulation*. The real question is: *what kind of regulation is best?*[12]

If you're interested in the practical steps that can be taken to govern the tech industry more effectively, then this book should be of value to you. It offers lots of concrete proposals. New legal standards. New public bodies and institutions. New duties on platforms. New rights for ordinary citizens. New regulators with powers of audit, inspection, certification and enforcement. New codes of conduct for people in the tech industry. And along the way, it considers questions like:

- What is 'ethics washing' and why is it a problem?
- Is it possible to 'democratise' digital technology?
- What kinds of rules and standards should govern important algorithms?
- Should powerful figures in the tech industry be regulated, like doctors or lawyers?
- Why are 'terms and conditions' so useless?
- Is antitrust law fit for purpose?
- What rules should govern the use and abuse of personal data?
- Should technology be governed globally, or by states or regions?
- Can we regulate social media without stifling freedom of speech?

Many countries around the world are trying to answer these questions. The last few years have seen a flurry of legislative and regulatory initiatives, with the EU at the head of the pack. There is also a growing body of academic research on the governance of digital technology, lots of which can be found in the endnotes of this book. But I believe these policy questions can't be answered – at least, not in a coherent or sustainable way – without first tackling some even more fundamental issues:

- Why do we regulate things at all?
- What is the role of law and regulation in a free and democratic society?
- What is our purpose in regulating digital technology? What are we trying to achieve?

These are big (and, I think, exciting) questions. They are questions for philosophers, not just lawyers or politicians. And this book suggests, in outline, one big answer to them: a guide to why we govern things, how we should govern digital technology and what we should aim to achieve by doing so. The idea is ancient in origin – drawn from some of the oldest traditions in Western political culture – but modern in application. I call it *digital republicanism*.

***

The term *republicanism* can be confusing, so it's first worth explaining what, in this context, it doesn't mean. It's not a reference to the modern Republican Party in the US, though obviously the early party saw itself as belonging to the republican tradition. Nor is it a reference to the distinction between a republic and a democracy sometimes attributed to James Madison (as when people say, *America is a republic, not a democracy*). Nor, finally, is republicanism about executing kings and queens, although the guillotine is part of republicanism's history. No, the republicanism described in this book is an ancient way of thinking about *power* and *freedom*. It cannot be claimed either by left or right. It is not the exclusive preserve of any culture or era. It has come to the fore at many of history's turning points – wars and revolutions, declarations and constitutions – and found expression in the Roman Republic, the English and American revolutions and many other political movements besides.

In essence, to be a republican is to oppose social structures that enable one group to exercise unaccountable power, also known as *domination*, over others.[13] In the past, republicans struggled against the domination of kings, emperors, conquerors, priests, landlords and bosses. They did so not by complaining about the cruelty or ignorance of their oppressors, but by reforming the social structures that allowed their oppressors to dominate them in the first place. Republicans – I stress, in the sense of the word used in this book – oppose the very idea of empire, not just bad emperors. They reject the institution of absolute monarchy, not just the flaws of particular kings. They fight for tenants' rights, not just for more beneficent landlords. They demand legal protections at work, not just kinder bosses. In the incarnation introduced in this book, they object to the *idea* of someone with Mark Zuckerberg's power, not Mr Zuckerberg himself.

In the Roman Republic, the greatest threat to liberty was believed to lie in *imperium*: unaccountable power in the hands of the state. Centuries later, that threat has not gone away. The state remains a formidable concentration of power, and digital technology is set to supercharge that power. But republicanism also warns against *dominium*: unaccountable power in the hands of private individuals and corporations.[14] James Madison understood this, urging of the need 'not only to guard the society against the oppression of its rulers; but to guard one part of the society against the injustice of the other part'.[15]

For the *digital* republican, I argue, the law's purpose should be to keep the awesome power of digital technology from escaping acceptable bounds of control, and to ensure that tech is not allowed (by design or accident) to undermine the values of a free and democratic society. That translates into four basic principles, which should guide the future of regulation in this field:

1. The law must preserve the basic institutions necessary for a free society (such as a functioning democratic and judicial system).
2. The law should reduce the unaccountable power of those who design and control digital technology, and keep that power to a minimum.
3. The law should ensure, as far as possible, that powerful technologies reflect the moral and civic values of the people who live under their power.
4. The law should restrain government too, and regulation should always be designed in a way that involves as little state intrusion as possible.

You might be thinking that these principles seem rather obvious. If so – good news – you may already be a republican. I believe that the 'lost language of freedom' in the republican philosophy does indeed reflect many of our shared ideals and intuitions.[16] Indeed, many of the proposals in this book are unashamedly based on the work of scholars and activists who might never have previously seen themselves as republicans.

In practice, however, a transition to digital republicanism would represent a sharp change in direction. In the last few decades, digital technology has mainly developed according to a rival system of thought that I call *market individualism*.[17] Market individualism is old, if not quite as old as republicanism. It has been a feature of Western political thought for centuries. It holds that social progress is primarily the result of individuals pursuing their own interests. It sees society as the product of a grand contractual bargain between each of its members; a vehicle for the pursuit of individual advantage, with no overarching pursuit of the common good. For the market individualist, the law is fundamentally hostile to liberty, such that more law generally means less freedom.

Market individualism has shaped the modern tech industry. In the name of economic orthodoxy, we have rejected laws that curtail the power of tech companies, while passing laws that protect them. In the name of liberty, we have fixated on the power of the state, while ignoring the swelling power of corporations. In the name of innovation, we have treated tech primarily as an economic phenomenon, when all the signs are that it has become a political force too. In the name of individual autonomy, we have left people to fend for themselves in the face of vast corporate power, protected only by pop-up 'terms and conditions' that few read and even fewer understand.

Often unknowingly, Silicon Valley insiders and political reformers alike have adopted the language of market individualism, deploying its concepts and precepts as if they were the only way of imagining and ordering the world. But another way is possible. Unless we start to think differently about technology, we will keep making the same mistakes.

So this isn't just a book about tech policy. It tries to offer a vision for a freer and more democratic society. It imagines a new type of society – a digital republic – starting with its historical foundations, then the philosophical scaffolding and finally the architecture of laws and institutions.

Even while writing this book, there has been a burst of legislative initiatives across the world. Some of these proposals are republican in spirit if not in name.[18] It remains to be seen what will become of them. But in the UK, the US and the EU at least, the ideas remain in flux. So this is not a book about specific laws proposed by the EU, or by other governments thinking along the same lines. Such a book would be out-of-date within weeks. Instead it seeks to offer a more durable guide for those who want to understand how we have governed technology until now, and what we might do better in the future.

***

To manage expectations, if you hanker after hardcore analysis of technical protocols, or lust for line-by-line study of the General Data Protection Regulation, you are going be disappointed. Although it tries to bring some of the best ideas from academia to a wider audience, this is not a textbook. It offers a sketch of a different future, not a blueprint (though further detail can often be found in the endnotes). Nor is this a work of political strategy. The grim reality is that in most democracies, debates about tech regulation are beset by political gridlock. Achieving meaningful change is going to be fiendishly hard. My modest hope is that if we can make the arguments for reform a little clearer, that might remove at least one obstacle from the path of progress.

I also recognise that most of the proposals in this book would not be appropriate in every country – only those with a functioning democratic system and the rule of law. On the thorny question of how technology ought to be governed in nondemocratic countries, I defer to others.

Some readers may wonder why more time is not dedicated to *global* governance initiatives. My view is that global governance of technology is probably more likely to emerge from a patchwork of national efforts rather than a top-down mission to impose a world law.[19] But even if that's wrong, I suggest that there are sensible reasons for trying to govern technology at the national or regional level first, even imperfectly (chapter twenty-seven).

Finally, there are many books out there for those who want to know how to better protect their online identity, manage their personal data or protect their children from online harm. This is not one of those books. Instead, this book respectfully departs from the market individualist assumption that we – you and I, as individuals – must be the first and last line of defence when it comes to protecting our digital freedom. The truth is that there are serious limits to what we can do by ourselves. As well as asking *what can I do to protect myself?* the republican asks, *what can we do to protect ourselves and each other?*

***

As you read this book, you may have legitimate doubts about what is being proposed. Wouldn't that do more harm than good? That's all very well in theory, but would it work in practice? That sounds expensive; who's going to pay for it? This wouldn't be a book about political change if it didn't provoke reactions like these. *Res technica, res publica* as the Romans might have said. The digital is political.

There are certainly reasons to temper our resolve with caution. Any new system of governance will have to reckon with the fact that innovation moves faster than legislation. Public authorities are usually playing catch-up.[20] Bedraggled regulators are often outgunned inside and outside the courtroom. As K. Rahman Sabeel notes, complex industries like digital technology put regulators at an 'epistemic disadvantage' because the industries themselves hoard the information needed to come up with new laws.[21]

Then there's the perennial risk that lobbyists and special interests will 'capture' the regulatory process and turn it to their own advantage.[22] Regulators, inspectors, judges and the like could get too close to those on whom they are supposed to be keeping an eye – perhaps because they come from within the industry themselves, or (worse) because they want juicy jobs in the private sector after they leave public office.[23] Not so long ago, Facebook's top public relations guy, Nick Clegg, was the Deputy Prime Minister of the United Kingdom. In 2020, the director of content standards at Ofcom, the UK's putative regulator of social media, also joined Facebook – perhaps to advise it on regulations he had himself helped to write. The 'revolving door' is a particular problem in industries like tech in which there is a relatively small pool of experts.[24]

Another worry is that new laws and regulations become political footballs kicked around by politicians looking for a quick win for their side, rather than lasting reform. That risk is particularly acute when it comes to governing social media platforms (part ten), which have become fiercely contested fields of political struggle in themselves.

An even thornier challenge is that regulation can lock in the advantages of big players while making it harder for new entrants to get a foothold.[25] Sometimes this is simply because bigger companies have more influence on the legislative process, and the laws are tailored to their needs. But it can also be a side-effect. It's easier for companies with deep pockets and armies of lawyers to deal with 'regulatory thickets' than it is for smaller ones on tighter budgets.[26] Indeed, one criticism of the EU's flagship data governance regime is that younger companies have found it harder to comply than more established ones.[27]

These challenges are formidable but not new. Some are as old as the idea of governance itself. They should not be seen as arguments against change. They should be seen as reasons to find new and superior modes of governance: better resourced, more insulated against corruption, nimbler and more flexible, with fresh locks on the 'revolving door' and due attention to the needs of smaller firms.[28] It is easy to list the difficulties associated with good governance, but they don't amount to an argument for doing nothing. They're an argument for doing better.

***

This book kicks off with a brief introduction to republicanism, and how it differs from its great intellectual rival, market individualism. Then the rest of the first half is essentially diagnostic. It makes five main points.

The first is that digital technologies can exert real power. They contain rules that the rest of us have to follow. They condition our behaviour, often without us realising. They frame our perception of the world, determining which information reaches us and in what form. They subject us to near-constant scrutiny. They set the rules of public deliberation, deciding what the rules are, when they are enforced and when they don't apply. These forms of power are still in their infancy. They will grow as technologies become more capable.

Secondly, technology is not neutral, objective or apolitical. Digital systems are soaked through with biases and prejudices.

Third, under the current system, digital technology has been ordered primarily according to the logic of the market economy. This brings benefits, like economic efficiency. But it also has drawbacks. Instead of restraining the might of corporations, the market empowers them. Instead of curtailing the worst instincts of those in the industry, the market encourages them. Instead of empowering citizens, the market strips them of individual and collective agency.

Fourth, there is nothing natural or inevitable about the current system. It is, in large part, the product of a custom-made legal regime that prioritises private ordering over public safeguarding. This regime can be changed.

Finally, the entire system of technological development – both the ways we develop new technologies *and* the ways we have tried to regulate them – have been excessively conditioned by the ideology of market individualism.

The second half of the book is about what we should do differently. It lays out the philosophy of digital republicanism, and then a prospectus for a new system of republican governance. There are chapters on governing data, algorithms, antitrust, social media and much else besides. The book is easiest to follow if read from start to finish, but you can also jump between parts without too much difficulty.

Finally, a personal note.

My generation (I am 32) is the last that can remember a time before the commercial internet. We were too young for 1990s cyber-utopianism, but came of age in the early years of online platforms. Facebook launched when we were teenagers. Social life and social media became inseparable. Smartphones came out in our early twenties, and distinctions that had been clear to our parents – between online and offline, real space and cyberspace – began to melt away. We could see that the world was changing, but that seemingly obvious reality was nowhere to be found in our textbooks. In my first book, *Future Politics* (2018), I argued that we were not yet ready – intellectually, philosophically or morally – for the world we were creating. I believed then, as I do now, that we were living through the first tremors of a great convulsion that would have profound and irreversible consequences for the way we live together.

One of my anxieties in writing *The Digital Republic*, however, is that it seeks to change a system that is very appealing to millions, even billions, of people. 'No old regime is merely oppressive,' writes Michael Walzer; 'it is attractive, too, else the escape from it would be much easier than it is.'[29] Most of us regard technological development with awe and optimism, and that's a good thing. The future is set to be exciting, if nothing else. But I reject the notion, and you should too, that we can only enjoy the wonders of digital technology if we submit to the unaccountable power of those who design and control it.

A different system is possible, and this book is about how to build it: a digital republic in which human and technological flourishing go hand in hand.

ONE

The Indignant Spirit

When we try to imagine the future, we use the same parts of our brain as when we remember the past.[1] We have no memories from the future, so the building blocks of imagination are necessarily made from the raw material of experience. Thus when we think about our next visit to a favourite restaurant, we are really remembering great meals we ate there before.

What is true of the human brain is also true of political cultures. The line between the past and the future, between memory and prophecy, is often less clear than it seems. We are not the first generation to grapple with the rapid rise of a new form of power. And history is filled with legends of the many throwing off the yoke of the few. Behind many of these tales lies the idea of republicanism. But what is republicanism? What can we learn from its long and chequered history? And what does it have to teach us about the power of digital technology? Before we get to the tech, these questions are the subject of the next two chapters.

***

In the Roman Republic, more than two and a half thousand years ago, every citizen was expected to be *sui juris* – his or her own master.[2] But the Romans, like the ancient Greeks, understood that if everyone merely pursued their own interests with no regard for others, there would be anarchy, not liberty.[3] They faced a paradox: how could people live together while still maintaining mastery over themselves? Their answer was the self-governing republic. The term 'republic' comes from the Latin *res publica*, which means the people's thing or affair.[4] It is also sometimes translated as 'commonwealth'.[5] The idea of the republic came to mean a society in which the state was free from the domination of foreign powers, and citizens were free from the domination of other powerful members of society, and the state itself.

The Roman Republic was a highly imperfect system. But one of its strengths was that citizens were expected to participate in collective life, and to cultivate public awareness, empathy and vigilance.[6] It lasted 500 years. After it collapsed, a millennium followed in which republicanism was just a fleeting memory.[7] Europe fell under the dominion of kings, emperors, clergymen and warlords.

A republican renaissance stirred in the eleventh and twelfth centuries. The city-states of northern and central Italy began to replace their lordly rulers with magistrates known as *podestà*, who governed for limited periods through networks of citizen councils.[8] These were not democracies, to be sure, but rulers could be held accountable at the ballot box or in the courthouse.[9] Gradually, Italians came to refer to their system of self-rule as *res publica*.[10]

The spirit of republicanism spread to England in the late sixteenth century, around the time that the great chroniclers of republican Rome – Cicero, Sallust, Livy, Tacitus – were translated into English.[11] England proved fertile terrain for republican ideas. The English believed they were the heirs to an unwritten constitution that endowed them with liberties that no one could take away. What the Romans had called a liber, the English called a freeman.[12] But republicanism was mere fantasy in a country that was still ruled by a hereditary monarch.

Matters came to a head in the seventeenth century. The Stuart kings insisted that they had unlimited powers of taxation, conscription and criminal punishment. Republicans, by contrast, argued that the king's powers were constrained by ancient tradition and the common law.[13] In 1628, dissidents bitterly condemned the king's taxes and raged that his subjects 'have of late been imprisoned without any cause showed'.[14] These rumblings turned into rebellion, and eventually the king was captured, prosecuted and beheaded. His days of ruling 'in accordance with his own caprice', as the poet-polemicist John Milton put it, were over.[15]

England had become a republic, at least in name.

What began in England as a political revolt against a particular king grew into an intellectual revolt against the idea of kingship itself.[16] Inspired by the Romans (if not by the rather dictatorial way that England was governed in practice after the regicide), English republicans argued that any ruler unchecked by the law was effectively a tyrant.[17] In Eikonoklastes (1649), Milton thundered that the people's wellbeing could never be entrusted to the 'the gift and favour of a single person'.[18]

The English republic did not last long, but its aftershocks rippled through the centuries. In England, the common law remains a bulwark against unaccountable power.[19] In 2019, when the UK government declared that powers derived from the crown entitled it simply to shut Parliament, the Supreme Court disagreed, citing precedent from 1611: 'the King hath no prerogative, but that which the law of the land allows him'.[20]

***

In Northern Europe, republican ideas spread fast in the seventeenth and eighteenth centuries. The Free States of Poland and Switzerland styled themselves as heirs to the Roman tradition. The mighty

Dutch republic developed a religious conception of republicanism, inspired by the Jewish Commonwealth of the ancient Hebrews: 'a *respublica* of God's people'.[21]

Then, in 1765, the UK Parliament passed a law requiring its colonial subjects in America to use embossed paper for certain documents, and to pay a tax for the privilege of doing so. The ensuing outrage meant that the Stamp Act was repealed within a year. But Parliament still claimed that it had 'full power and authority to make laws and statutes' to govern its American colonies 'in all cases whatsoever'.[22] This was reckless. Not only were the English saying that Americans would be unrepresented in the body that made their laws, but there would be no checks or balances on Parliament's power.[23]

This did not go down well in America, to put it mildly. The revolutionary backlash, which led to independence, launched the greatest republican experiment yet seen in human affairs. Alexander Hamilton recognised the special place that the US had assumed for itself in history:[24]

> It seems to have been reserved to the people of this country, by their conduct and example, to decide the important question, whether societies of men are really capable or not, of establishing good government from reflection and choice, or whether they are forever destined to depend, for their political constitutions, on accident and force.

Whatever their gripes at the time, the British could hardly accuse the Americans of being unreasonable. After all, the logic of the American revolution was essentially the same as the English revolution a century before: no one should have to live at the mercy of an unaccountable ruler, benevolent or otherwise.[25] It was not acceptable to rely on the goodwill of the powerful. As Joseph Priestley put it in 1769, by 'the same power' as the English could compel Americans 'to pay *one penny*, they may compel them to pay the *last penny* they have'.[26] A few years later, France joined the ranks of the world's republics.

***

This potted history would suggest that republicanism is only concerned with the unaccountable power of kings and conquerors. But the republican philosophy actually opposes all forms of unaccountable power. It has been the language of workers resisting oppression since at least the plebeian secessions of the Roman peasants.[27] It has been the language of feminists, like Mary Wollstonecraft, who, in 1792, demanded to know 'Who made man the exclusive judge?'[28] It has been the language of abolitionists like Frederick Douglass, who opposed slavery not because of ill treatment he received, but 'the consideration of my being a slave at all'. The problem was not the cruelty of slave-masters; it was the institution of slavery itself.[29]

Today, the philosophy of republicanism holds that freedom is lost wherever there is a systematic imbalance of power between two parties, leaving one in a position to interfere arbitrarily in the life and affairs of the other. Consider the abused daughter who only escapes a beating when her father is too tired to raise his hand. Or the undocumented worker who must fawn and flatter his boss to keep his job. To the republican mind, the question is not whether the father will be too tired to thrash his daughter, or whether the boss would ever really fire his worker. What matters is that one human being is utterly at the mercy of another. Freedom that depends on the goodwill of the powerful is not real freedom at all.[30] When the Minnesota police officer Derek Chauvin, who had been the subject of seventeen previous misconduct complaints, knelt on the neck of George Floyd until he stopped breathing, he showed the world the face of unaccountable power.[31] It does not matter, from the republican perspective, that most police officers would never behave like Chauvin, or even that Chauvin himself was subsequently sent to prison. He worked within a system that gave him power and let him exercise it without suitable control.

***

The republican project has been trundling on for more than 2,000 years. But republican reality has often failed to match the promise of its theory. Like most Western philosophies, at some time or other it has been used as a facade for authoritarian rule, or for the hoarding of power by (usually white male) elites.[32] At the heart of the republican project, however, is not a policy, or even a principle, so much as a frame of mind. Adam Ferguson spoke of the 'indignant spirit' of the republican citizen.[33] In the years to come, we shall have to find new indignance about the unaccountable power of digital technology. And republicanism can be revived and reformed to confront this new form of power. But how is republicanism different from today's governing ideas? That's the subject of the next chapter.

TWO

Thinking, Old and New

Flaubert wrote that the best form of government is one that is exhausted, because 'in dying it makes way for another'. As we will see, the system we use to govern digital technology is dying, if not already dead. But what will come after it? We need a legal regime that is different from ours, but faithful to our instincts and traditions. We start not with a set of policy solutions – those come later in the book – but with a different way of thinking.

***

Until now, the problems raised by technology and many of our responses to those problems have been confined, often unthinkingly, within a set of ideological commitments to which I refer as *market individualism*.[1]

In its pure form, the philosophy of market individualism regards the individual pursuit of self-interest as the ultimate source of political order and legitimacy. It expects humans to make progress by bargaining and trading with each other, primarily according to norms of competition. The government's role in this model is minimal: to protect the market and maintain the safety of the realm, and otherwise to stay out of things as much as possible.

Market individualism is probably the defining political philosophy of the modern age. It has certainly been an ordering principle of Anglo-American civilisation for more than a century. But it has its critics. Some see it as a shorthand for corporate greed. Others claim that it will soon be surpassed by the state-capitalist model favoured in China. The argument in this book is more modest. It is not that market individualism should be dismantled in its entirety. But in the sphere of digital technology, it has failed on its own terms and should now be replaced with something better.

The alternative I suggest is republicanism. To begin with, it differs from market individualism in two important ways.

First, republicans think differently about *freedom*. The distinction is best explained in an analogy. Imagine a king so mighty that he can meddle in the lives of his subjects however he pleases. At his command, they will be thrown into dungeons or put to the sword. His powers of taxation and conscription are unlimited. If another man's property takes his fancy, the king need only snap his fingers. Men live or die at his mercy. His word is the law.

Just because the king has these powers, however, does not mean that he will exercise them. He might be nasty, but he might also be benign. Periods pass in which he exercises his power with restraint.

Would you say this is a free society?

The market individualist answers this question: 'It depends'.[2] When the king behaves like a tyrant, then the people are unfree.[3] But when he is in a good mood and lets the people get on with their lives, then yes, strictly speaking, there is freedom.

The republican offers a different answer. To be truly free is to live beyond the arbitrary power of anyone else. So long as the king has the *capacity* to govern as a tyrant, and the choice is his alone, then his is not a free nation. To live under a dictator, even a benevolent one, is to live in bondage. A slave with a kind master is still a slave.

The difference can be summed up in this way. The market individualist minds mainly about *interference*, holding that we are free whenever we are not being physically or legally coerced by someone else.[4] The republican shares this concern but also cares about *domination*: we are free only when no one is given the capacity to coerce us without accountability. For the republican, unaccountable power is a problem in itself.

The application of this principle to digital technology is simple enough. We should be concerned about the growing power of digital technology *even when it is being used in ways that don't bother us*. For as long as a technology company or government could simply change its mind and use its power to impinge on our rights and liberties, we are unfree. To borrow a metaphor from the English civil war, trusting blindly in the power of others is like putting your head in the mouth of a wolf and hoping it will not bite.[5] That's true even if the wolf is fluffy and cute most of the time.

Republicans and market individualists also disagree about the nature of *democracy*.

Both traditions see self-rule as important. But for the market individualist, the purpose of democracy is fairly limited. Democracy is like any other form of contractual arrangement. It is a vessel for personal advancement *through* community, rather than for the advancement *of the* community as a whole (although that may be a pleasant side effect). When market individualists make arguments that appeal to the interests of others, it is understood that they do so mainly out of self-interest – a desire to reach a bargain – not public duty.

Republicans, however, do not see democracy merely as the aggregation of the private preferences of individuals. Republican democracies expect their members to behave as citizens, not just consumers; to acknowledge the common good as well as private desires; to deliberate in good faith, at peace with the idea that someone might change their minds. This last point is important, and it explains why digital republicans might be wary of *virality* on social media – promoting speech according to how many clicks or likes it gets (chapter thirty-six). Virality seems democratic: what's wrong with directing more attention to the content that the most people enjoy? The republican answer is that the point of politics is not just to tot up people's pre-existing opinions, but to allow them to be formed and shaped through public debate. That's not what virality does. Instead of challenging majority views, it reinforces them. It fosters a thumbs-up/thumbs-down culture in which minority perspectives and inconvenient truths are consigned to algorithmic irrelevance.[6]

***

So we need to think differently about *freedom* and *democracy*. But we can go further. We can reimagine what it means to speak of *society*, *politics* and *law*. Once we do that, the path ahead becomes clearer.

Perhaps unsurprisingly, market individualists see the basic unit of social organisation as the individual. They emphasise the fundamental separateness of people: their cognitive autonomy, their various self-interests. On this view, there is no such thing as society apart from the individuals who make it up. Societies are the products of social contracts. Each one is a 'cooperative venture for the pursuit of individual advantage' or a 'scheme of mutual cooperation' for the pursuit of pre-formed desires.[7] This way of looking at the world originates in the seventeenth century.

The republican vision is much older. Like market individualists, republicans believe in the integrity and autonomy of the individual. But republicans see our natural state as being in the company of other people. We are 'creatures of community', not solitary atoms drifting through the ether.[8] As Hannah Arendt observes, for the Romans, the word for 'to live' meant the same as 'to be among men'.[9]

What may seem like a subtle disagreement over the meaning of *society* leads to a more fundamental one over the nature of *politics*. In the market individualist tradition, the default response to a social problem is to expect that the market or civil society will take care of it. That is to say: market individualists trust that economies of people acting in their personal self-interest will generally work their way towards an acceptable equilibrium.

The republican tradition is more sceptical about the idea that the uncoordinated actions of disparate people will solve social problems. There is a reason why we are not left to decide for ourselves which side of the road to drive on, or what rate of tax to pay. Some social schemes require coordination and cooperation, even coercion, not just competition. Market individualists and republicans both believe in individual freedom. The difference is that republicans believe that the preservation of individual freedom often requires us to act *together*.

Indeed, it is not just political action but the very *concept* of politics itself that divides market individualists and republicans. The market individualist inclination is to treat what happens in the economy as something 'private', separate from the 'public' sphere of politics. That's why for the market individualist, there is little injustice in the asymmetries in power that arise in the rough and tumble of the economy. Because (on this worldview) the market is not really part of politics at all.[10]

Republicanism has no time for this way of thinking. A conception of politics that is blind to every power imbalance other than the one between the government and the people is obviously incomplete. It leaves out the politics between sexes and genders, between nations, races and religions, between old and young, rich and poor – really all of the dynamics that make politics interesting and important. And, of course, it entirely omits one of the defining political relationships of our time: between those who design and control digital technologies and those who must live under the power of those technologies.

The final difference between the two traditions relates to the concept of *law*. This is important, because it concerns the proper role of the state in governing the power of digital technology.

The market individualist conception of law goes something like this: people are born free but made subject to legal restrictions on their activity. Although some laws are necessary for the preservation of safety and security, each one represents a restriction on liberty.

A justifiable restriction, perhaps, but a restriction nonetheless. Jeremy Bentham put it bluntly in the nineteenth century: 'All coercive laws,' he wrote, are 'abrogative of liberty.'[11] This perspective has some quite pungent consequences. If law and freedom are really enemies, then more law necessarily means less freedom. It is a short hop to the idea that freedom lies in the absence of government, or governance, altogether.

The republican position is more nuanced. It sees an underregulated society as one in which the strong can prey on the weak at their pleasure; where power inevitably accrues to the richest, largest and most violent. The rest are left with whatever dregs remain after the mighty have taken their fill. For the republican, law should be seen as a means of reining in the over-powerful, thereby carving out space for everyone to flourish. James Madison put it like this: 'liberty may be endangered by the abuses of liberty, as well as by the abuses of power.'[12] In the republican tradition, therefore, law and freedom are not opposites. Quite the reverse: law often makes freedom possible.[13]

At the same time, however, republicans do not believe that more law is always better. On the contrary, the tradition is obsessed with restraining the power of government. Like any other concentration of power, government can only be justified so long as it serves to reduce domination in society. If it goes further than that, it ceases to be justifiable. 'The best rule as to your laws in general,' wrote the seventeenth-century republican James Harrington, 'is that they be few.'[14]

***

Republicans and market individualists do not disagree about everything. But there is no escaping the fact that these traditions, both of the Western canon, take very different views of the world. When it comes to freedom and democracy, republicans imagine a different future from the one we have now. And when it comes to society, politics and law, that difference only grows.

If you see yourself as more of a market individualist, you are certainly in good company. As we will see, for most of modern history, market individualism has been the default setting for Anglosphere (and much of Western) politics, while republicanism has been little more than a footnote.[15] Indeed, many of our 'solutions' to the challenges of digital technology come from the same intellectual paradigm that created the problems in the first place: more marketisation, more commodification, more individualised decision-making, more competition.[16]

This book argues that it is time to try a different way of thinking and a new way of governing, before the chance disappears forever.

We will unpack the idea of digital republicanism later in the book. But our first task is to find a diagnosis of the central problem. That is the focus of the next part: digital power. Where it comes from, how it is growing, and why it matters.

PART II

THE HOUSE OF POWER

No cause is left but the most ancient of all . . . the cause of
freedom versus tyranny.
Hannah Arendt

THREE

Autocrats of Information

Electric scooters are the latest vogue in urban transportation. They wait in clusters on the pavement, ready for hire by anyone with a smartphone and a credit card. Scooters are faster than walking, easier than cycling and nimbler than cars. They turn the cityscape into a playground. Riding them is a carefree experience – but it is more controlled than it might seem. Every journey is tracked from start to finish. No matter how hard the throttle is pressed, the scooters will not go above a particular speed. They refuse to leave designated urban areas. And there is no haggling over the fare: an app deducts a precise sum depending on the length of the journey.

None of this is inherently objectionable. But scooters do offer a helpful example of the paradox of digital technologies: they offer freedom, but only in exchange for some surrender of control.[1] This is not a paradox that will ever be fully resolved. The question will always be whether the balance between freedom and control is struck in the right place.

***

Computer code has a formidable ability to control human activity – silently, automatically, precisely and without tolerating any objection.[2] And it is used to enforce a growing number of society's rules. In the past, borrowing a book meant returning it to the library before the expiry of the agreed period. Late returns would be met with a small fine and a disapproving glare from the librarian. Now, when a customer borrows a book from a lending app, late return is not an option. When the term expires, the book simply disappears from the reader's device. In the same way, a movie rented on Amazon Prime evaporates when time is up. The rules are self-enforcing.

Code is now present in almost all the occupations, transactions and interactions that make up a meaningful life. It's unavoidable. And the empire of code is no longer confined to the separate dimension of 'cyberspace'. We cannot escape by shutting a laptop or logging off. We are physically surrounded by technology, encountering hundreds or thousands of digital objects every day. As Bruce Schneier says, it used to be that things contained computers, but now 'they *are* computers with things attached to them.'[3] Soon, tens of billions of once-dumb artefacts will be connected to the internet, equipped with sensors and endowed with processing power, enabling them to interact with us and each other.[4] All these technologies will contain rules, and we will have to follow them.

Of course, some of the hype about the 'internet of things' is indeed just hype. Not everything will be digitised. My guess, for instance, is that consumers will not want 'smart toilets' that can identify them by the unique shape of their backsides (just one product currently said to be under development).[5] But even on a conservative view, the future will not permit much meaningful escape from digital technology, not if you want to live a full and rounded life. Tech's empire recognises no borders or barriers. It is a ubiquitous feature of twenty-first-century life, spreading outwards and surrounding us. Algorithms are increasingly used to determine our access to the necessities of civilised existence: work, credit, insurance, housing, welfare and much else besides. As a social force, code will eventually rival the invisible hand of the market and the great clunking fist of the state.

The border between online and offline is fading, and with it, the distinction between tech and non-tech corporations. Is a manufacturer of 'smart' medical devices a 'tech' company? Is Airbnb a landlord? Are Ford and Google in the same category because they both develop self-driving cars? The Chinese rideshare platform Didi Chuxing moved from transport into personal finance by gathering data about its passengers – where they live, who they live with, where they work, where they eat – and using it to make predictions about their financial status and proclivities. It now offers loans and insurance without customers needing to complete a single questionnaire.[6] Digital technology lets businesses shift between social functions, allowing power in one sphere of life to be carried over into others.

The nineteenth-century computer pioneer Ada Lovelace believed that a coder was an 'autocrat of information', marching at the head of 'the most harmoniously disciplined troops'.[7] Her analogy was apt. Code does not care if humans agree with it or not. It does not need their consent or compliance. It is rarely the product of a democratic process or legislative debate. It doesn't derive from any kind of social contract. It is almost impossible for ordinary people – the billions of us who are its passive subjects – to alter or appeal. Code makes no claim to legitimacy, only efficacy. Increasingly it is just a fact of life: an invisible border which marks, with hard edges, the limits of what we can and cannot do.

***

How is code's power used? Mostly for commercial purposes. Think of the workplace, for example. Businesses use digital tools to manage their workforce in ways that would have been hard to imagine a couple of decades ago. Amazon's systems chide workers who fail to pack their boxes quickly enough, and terminate those who underperform.[8] Uber's software deactivates the accounts of drivers whose ratings fall below a required level.[9] Businesses target new recruits with online advertisements then screen them using systems that scan their CVs for keywords. They 'interview' candidates with software that simulates basic text conversations[10] and programs that draw inferences from diction, tone, eyebrow activity, lip movements and chin gestures.[11] (Whether this technology works is another question. What does an unacceptable chin gesture look like? What is the appropriate level of eyebrow activity?) In some businesses, 'people analytics' could eventually replace CVs and interviews altogether. Employers will be able to draw inferences about candidates based on their habits and leisure patterns, their compulsions and addictions, their friends and social circles – all revealed by the data.[12]

Code's power is not limited, however, to business. Increasingly it is wielded in ways that are overtly political, and not just by governments. During the Covid-19 crisis, Facebook summarily took down event pages that were being used to organise anti-quarantine protests, effectively quashing the protests themselves.[13] No law required Facebook to take this step. No law prevented it from doing so either. The company identified what it perceived as a social harm and took public policy into its own hands. Some criticised Facebook for this decision, but if it had left those groups alone, it would have been criticised for that too. Decisions of this kind used to be the province of public officials. In the future, more and more will be made by technology companies. And as the next chapter shows, these companies know more about us than any government of the past.

FOUR

Data's Dominion

Imagine a huge bazaar bustling with activity. A sign by the entrance proclaims that traders come here from all over the world. Hundreds of billions of dollars change hands every year.

But this is not a normal market.

In stalls and kiosks where you'd expect to find fresh produce or homeware, only one product is for sale: people's personal data.[1] What they eat for breakfast, where they sleep, what they do for work, what they believe and care about, the kind of TV they watch, their ages and sexual preferences, their insecurities and fears – there is data here about hundreds of millions of people. Some merchants sell filthy buckets of raw data in need of a clean. Others offer neat packages, crisply tailored to the needs of their clientele. One vendor hawks a list of rape victims – seventy-nine bucks for a thousand – along with a list of victims of domestic abuse.[2] There are no law-enforcement officials in sight.

This data market actually exists but is, of course, too large for any single physical space. It thrives, particularly in the US, because there is an abundance of supply and a great deal of demand.[3] The world is increasingly reflected in data. Satellites capture the entire planet every day at a resolution so fine that 'every house, ship, car, cow, and tree on Earth is visible'.[4] On Earth, almost every human encounter with technology leaves a trail of data that can be hoovered up and sold on. Data is inexhaustible. We generate it just by existing, and more and more of it is being captured and stored. Thousands of businesses have sprung up to trade in it.[5] And business is good.

What happens to all this data? Usually, purchasers are not interested in the sordid details of individual lives, a fact that can get lost in debates about online privacy. Data's real value emerges when it is gathered in gigantic amounts to build computing systems that can find patterns and predict behaviour. For these systems, each person ceases to be an individual and becomes instead a bundle of attributes – millennial, sausage dog owner, cheese addict – a 'voodoo doll'[6] that can be chopped, changed and aggregated with the attributes of thousands of others.[7] The promise of 'big data' is that there are correlations out there that can only be seen when thousands or millions of cases are considered together. And when those are patterns are revealed, strangers can know us in ways we might not even know ourselves. This is a form of power.[8]

***

In the twentieth-century imagination, surveillance meant a secret agent peering unseen from behind the curtain of a darkened room. Some still cling to that image. Steven Pinker, for example, offers reassurance that we have been able to install surveillance cameras in 'every bar and bedroom' for a long time, but that we have not done so because 'governments' don't have 'the will and means to enforce such surveillance on obstreperous people accustomed to saying what they want'.[9] Pinker, however, is taking comfort from a world that no longer exists. Today's authorities don't have to force us to place surveillance devices in our homes. And they don't have to see us in order to watch us. We expose our lives to scrutiny whenever we interact with a phone, computer or 'smart' household device; every time we visit a website or use an app. And personal data usually finds its way to those who want it most. Thus, when police in the US wanted information about Black Lives Matter activists, they didn't need to spy on them with cameras in bars. They bought what they needed in the data market. Facebook had a trove of data about users interested in Black Lives Matter, which it sold to third-party brokers, who sold it on to law-enforcement authorities.[10] For juicy data that cannot be found on the open market, Facebook has a special portal for police to request photographs, data about advertisement clicks, applications used, friends (including deleted ones), the content of searches, deleted content and likes and pokes. Facebook provides data 88 per cent of the times it is asked.[11]

Even in the physically 'private' space of home, there is already little escape from data-gathering devices. If you used Zoom to keep in touch with your family during the Covid-19 crisis, Zoom will have sent Facebook details of where you live, when you opened the app, the model of your laptop or smartphone, and a 'unique advertiser identifier' allowing companies to target you with advertisements.[12] Zoom is not even owned by Facebook. It merely used some of its software.[13] A commercially available dataset recently revealed more than thirty devices that were using hookup or dating apps within secured areas of the Vatican city – that is, areas generally accessible only to senior members of the Catholic Church.[14] That's the kind of secret that would probably have stayed hidden in the past.

The future is not Big Brother, in the sense of one government monolith watching us all at once. It is, instead, a 'big brotherhood' of hidden, unblinking eyes, some belonging to the state but countless others belonging to private parties who watch us while remaining unseen.[15]

**

Another anxiety inherited from the twentieth century is the sense that anonymity is no longer possible; that even if we try to hide, powerful others will always know exactly who we are and where to find us. In recent years, this fear has led to concern about the spread of facial recognition systems.[16] In fact, our faces are just one way to identify and locate us. The unique identifiers in our smartphones and payment devices telegraph our presence wherever we go. Using location data from the phones of millions of people, it took minutes for journalists to track the whereabouts of the President of the United States.[17] There are systems that can monitor people's heartbeats and read their irises from afar.[18] Others use WiFi technology to identify individuals through walls. With the right tech, a person's gait can identify them as readily as a fingerprint.[19] In the future it may be possible to 'Google spacetime' to find where any of us were at a specific time and date.[20]

Taking a longer view, anxieties about being *identified* will eventually be superseded by concerns about being *analysed*. We are not as mysterious as we like to think, even if the capability of computers is sometimes overhyped. Systems are being developed to interpret our feelings and moods using the tiniest of physical cues.[21] They are said to be able to take in the sentiments of crowds in a heartbeat.[22] They can tell if we are bored or distracted from the tiny movements of our faces.[23] They can see if we're sad from the way we walk.[24] They can detect cognitive impairments from the way we poke our smartphones.[25] They can predict our mental state from the content of our social media posts.[26] Famously, Facebook 'likes' can be used to predict a person's political preferences 85 per cent of the time, their sexuality 88 per cent of the time and their race 95 per cent of the time.[27]

Using data gathered from a thousand sources, today's systems scrutinise us more closely and completely than any government agent ever could. And as we will see in the next chapter, this allows them not only to interpret our cognitive states but influence them too. Tomorrow's technologies will be even more powerful.

***

Not everyone is aware of the extent to which they are scrutinised, and it might be thought that educating people would improve their prospects for freedom. Digital literacy is certainly important, and it pays to understand the systems we use and the companies that produce them. Paradoxically, however, hyper-awareness can also have a crumpling effect on liberty. The more we are conscious of others watching us, the more we mind our own behaviour. Long before Michel Foucault wrote *Discipline and Punish* (1975), the republicans of revolutionary England argued that merely being *aware* of the power of others was enough to make people change their behaviour.[28]

Thus, if young people knew that future employers would be likely to examine their social media activity, they might avoid doing anything that could lead to an embarrassing photograph being posted on the internet. If loan-seekers were told that becoming 'friends' with a bankrupt person on Facebook might lower their *own* credit score, they would pause before admitting them to their network. If consumers knew that their conversations with digital personal assistants might be listened to by human beings, they would hesitate before asking an embarrassing question about the rash on their groin.[29] If it was widely known that substance abuse websites passed on the identities of their visitors to third parties, people might be reluctant to reach out for advice.[30]

Many employers instinctively understand the disciplinary effect of being watched. Staff told that their 'productivity score' will go down if they go to the bathroom too often might be inclined to hold it in, no matter the discomfort.[31] Couriers told that their deliveries are monitored against benchmarks – 'time to accept orders', 'travel time to restaurant', 'time at customer' – are likely to pedal a little faster.[32] Home-workers who know that their 'Hubstaff' software is tracking their mouse movements, keystrokes and web activity are unlikely to shirk or procrastinate.[33] Worse, those who know the 'Sneek' platform might photograph them every few minutes through their webcam will be understandably reluctant to nap on the job.[34] It recently emerged that Amazon received regular reports on the social media posts of its drivers, even in ostensibly 'private' online groups, and used them to monitor plans for strikes and protests.[35] If you were a driver and knew that this was going on, you'd be less likely to post online, wouldn't you?

Knowledge is not always power, and merely knowing that we are being watched can be debilitating.

***

The citizen of a free republic should be able to stand tall, look others in the eye and plan her life with confidence.[36] But people under permanent supervision will, in time, adopt behaviours that are pleasing to others rather than true to themselves. Nervous, fidgety, self-conscious – the scrutinised citizen is a meek and timid creature, afraid of her own data shadow. Of course, we will never be totally free of surveillance. But there is a quiet intensity to being watched all the time. With time, it could grow oppressive.

In the last chapter we saw the first dimension of technology's power: the ability to write rules that others must follow. Sometimes those rules are so strict that they cannot be broken. They literally force us to behave in one way or another. In this chapter we have glimpsed the second: technology gives a select few the ability to see in ways that no mortal ever could. The next chapter considers a third form of power: the ability to frame how we perceive ourselves and each other.

FIVE

Masters of Perception

Have you ever struggled to unsubscribe from a mailing list, frustrated by tiny hidden options, endless hyperlinks and baffling requests for long-lost passwords? Your inability to do what you wanted might have been caused by sloppy design, but it might also have been the result of roadblocks thrown in your path.[1] Either way, you couldn't do what you planned. And eventually you stopped trying.

Perhaps you can recall an evening in which you planned to plough through your to-do list, only to spend six hours binge-watching *The Real Housewives of Cheshire* instead. That might have been because of the quality of the drama. But it may also have been the result of careful engineering. Streaming platforms are generally built for low-level addiction. They tee up the next episode while the previous one is still playing, like the waiter who discreetly refills your glass until you are too drunk to notice.

Part of being free is being able to form our own preferences and act for our own reasons.[2] Digital technology makes this harder. We are encouraged to purchase or consume, to surrender time or attention, to offer up personal information. On one level this is to be expected. To be human is to have others place demands on us. But technology does not stick to conventional means of influence. Often it changes our behaviour through 'conditioning' rather than reason or persuasion.[3] It can operate beneath human consciousness, in a manner closer to manipulation than influence.

Of course, each small act of conditioning, taken alone, is hardly an unacceptable interference with our freedom. Netflix's algorithms do not pose a threat to our way of life. But taken together, the numberless prods, nudges and demands on our attention may eventually result in a significant compression of liberty.[4] As we have seen, technology can be used to force us to do one thing or another. But its subtler effect is to degrade our ability to decide what we want in the first place – or our will to pursue it.[5] This is a potent form of power, and it's the subject of this chapter.

***

Humans are vulnerable to cognitive influence, in part because we have only a limited capacity to gather, store and process information. Even the smartest folks can hold only a tiny fraction of the world's information in their minds. To compensate for this limitation, we build information systems that capture reality and present it in a digestible form. Languages, mathematical symbols, databases, books and newspapers are all fruits of the human effort to perceive the world more clearly. Without them, life would be even more confusing and chaotic than it already is: a formless torrent of sensations washing over us from moment to moment.

Processing information might be thought a dry and technical exercise, best left to experts in the same way that medicine is best left to doctors. But this could not be further from the truth. There is nothing more political than being able to shape how others perceive the world. The systems we build to gather, store, analyse and communicate information are as fundamental to our shared existence as any economic, legal or political institutions.[6] They filter what we know and what we believe to be worthwhile. They shape our habits and rituals. They determine what may be uttered and what must remain unsaid. They mould our interests, preferences and desires. They incubate our memes and clichés. They are the prism through which we see each other and ourselves.

For the generation that first used home computers in the 1990s, the early commercial internet lives in the memory as a jumble of personal websites, cluttered hobbyist pages and appallingly formatted message boards. The world wide web seemed like virgin territory, unknowable and ungovernable: an information superhighway without traffic laws or road markings. Enterprising individuals, however, saw value in creating order out of the chaos. Search engines were invented that could rank and sort the world's information. The old chat forums were superseded by vast, slick platforms on which we could share everything there was to know about ourselves. Systems like these were inserted into the spaces between people, and between people and information. Once ensconced there, they gained the capacity to influence our inner lives and shape our public discourse.

Take one modest recent example. The popular platform TikTok has more than a billion users, many of them young adults or children. TikTok was not designed in an obviously political way. Users post short videos of themselves, often set to music. Yet its sheer size and reach have made TikTok a potent social force. When users open the app, the first thing they see is the For You page, which promotes content likely to draw in users. This content is curated, partly by algorithms and partly by humans. In 2020, it emerged that TikTok's human curators had been instructed to block images of 'chubby' people and those with an 'abnormal body shape' from appearing on the For You page. Likewise, people with dwarfism and acromegaly were filtered from view, along with 'seniors' and those with 'ugly facial looks'. Curators were also told not to promote videos shot in 'shabby and dilapidated' surroundings.[7]

With these policies in force, TikTok subtly marginalised the unattractive and poor. The platform showed that in the digital era, a person, group or idea does not need to be blocked or censored to be made irrelevant. They need only be degraded or downrated, buried in a tide of other information or presented in a way that minimises the desire to give them attention. How a platform chooses to present and prioritise information will inevitably affect the perspective of its users. This is yet another privilege afforded to those who write code: the ability to frame our perception of the world.

Facebook once conducted a randomised controlled study during congressional elections. One group was shown a banner encouraging them to vote with an 'I voted' button. Another was shown the same banner and photographs of friends who had already clicked 'I voted'. A third group was shown nothing. The second group – those shown photos of friends who had voted – were 0.4 per cent more likely to vote. That doesn't sound impressive until you remember that the method was tested on more than 60 million people. The authors of the study claimed to have increased turnout by around 340,000 votes – a bonanza enough to swing many elections.[8]

Some of the concerns with this kind of power are well known. Being ranked too far down on Google's results pages, for instance, can mean calamity for businesses, which is why they spend tens of billions of dollars every year to try to stay high in the rankings.[9] Other worries relate to abuse and manipulation. The European Commission has found that Google's search results sometimes prioritise its own products over those of rivals. Concerns have also been raised that search engines might affect the results of elections. Before the 2016 presidential election, the top two search suggestions on Yahoo! for 'hillary clinton is. . .' were:

'. . .a liar'

'. . .a criminal'.

On Bing they were:

'. . .a filthy liar'
'. . .a murderess'.

Google, by contrast, prompted searches for:

'. . .winning'
'. . .awesome'.[10]

Why might it matter that these results differed so significantly? One reason is negativity bias: unpleasant things tend to capture and hold our attention more intensely than neutral or positive ones. If undecided voters search for the name of a political candidate and are presented with three positive search suggestions (e.g. 'Kamala Harris is. . .funny' or 'Kamala Harris is. . .clever') and just one negative one (e.g. 'Kamala Harris is. . .a disgrace'), then they will disproportionately click on the negative result, which could have a knock-on effect on their voting intention. At scale, the concern is that search engines might 'shift opinions dramatically' simply by 'varying the number of negative search suggestions' for each candidate.[11] This could be deliberate or inadvertent. Either way, it is a significant capability. In the right circumstances it could be more influential than any advertisement campaign or policy announcement. When Google tweaked its algorithm in a well-intentioned effort to shunt extremist publishers down its news feed, it inadvertently choked off traffic to public service websites that *reported* on extremist content. One saw its traffic drop 63 per cent.[12]

In the future, 'search' will look less like typing words into a box and more like an ongoing conversation with a digital personal assistant. But that won't change the underlying dynamic: every choice to provide a user with one piece of information is a choice to deny her a different piece of information. This is not a criticism; it is simply the way that information technologies work.

***

In 2018, the (then) CEO of Twitter, Jack Dorsey, came before Congress and sombrely informed the American people that the platform's algorithms had been 'unfairly filtering' 600,000 accounts, including some members of Congress, from the platform's search auto-complete and 'latest results'. This was a significant admission. It confirmed longstanding anxieties that the platform might be manipulating the political process without anyone intending it, or even noticing.[13]

Though his confession drew praise, Mr Dorsey's appearance also demonstrated how submissive we have become in the face of technology's power. It was sobering to see America's most senior politicians listen quietly while a young business executive explained how his platform had dented American democracy. When he said he would fix the problem, all they could do was hope that he meant it.

It is not helpful, for our purposes, to worry too much about individual corporations or CEOs. The point is structural. The corporations that sort and order the world's information now decide, in significant part, what goes on society's agenda. They may use that power in ways we like or dislike. But either way, it is a major political responsibility, and it is freighted with risk.

In a few decades, a relatively small number of corporations have assumed the power to frame how the rest of us see the world. And – this is the central point – while we may agree or disagree with certain of their framing choices, we rarely have a say in them. This is unsatisfactory. The structure and health of the information environment should not be treated as a corporate concern. It affects all of us. It is what the Romans would have called a *res publica*, the people's thing or affair.[14]

The next two chapters concern a further source of power for those who design and control digital technology: the power to influence democracy itself. The political system is increasingly at the mercy of the ringmasters of the great theatres of digital debate.

SIX

Republic of Reason

The people of Britain are not known for excessive displays of emotion. When something dramatic happens, the British way is to take a sip of tea, think of the Queen and never speak of the matter again. Every once in a while, however, the country is bestirred from what Her Majesty has called its 'quiet, good-humoured resolve' into a frenzy of terrible rage.[1]

One such incident took place in 2020.

Dominic Cummings was a principal aide to the prime minister and one of the architects of the UK's strict Covid-19 lockdown, which forbade people even from attending the funerals of their loved ones. It was discovered that, at the height of the lockdown, Mr Cummings took a 500-mile road trip in breach of the rules he had helped to write. Along the way, he visited a pretty tourist town, coincidentally on his wife's birthday, which he maintained (some thought rather unconvincingly) was a medical expedition to 'test' his 'eyesight'.

Enraged by the hypocrisy, the British public took to social media and demanded his resignation. On Twitter the scandal soon had its own hashtag: #CummingsGate. But seasoned observers noticed something strange. Despite dominating the national conversation, the word 'Cummings' itself did not appear on Twitter's trending topics. It was as if the issue was being hidden from public sight.

Why?

The answer had nothing to do with high politics. 'CummingsGate' was censored because the word 'cumming' was on Twitter's roster of forbidden pornographic terms. A similar mistake occurred when Facebook challenged posts referring to 'the Hoe' — not a term of misogynist abuse, but a reference to Plymouth Hoe, a scenic spot on the southwest coast of England.[2] This type of error is known in computer science as the Scunthorpe Problem, named after the English town whose name conceals a very naughty word. See also — for educational purposes only — the quaint English settlements of Penistone, Cockermouth and Rimswell.

The Cummings affair demonstrated, in microcosm, all the comedy, tragedy and farce of online political discourse: the hum of intrigue, the sugar rush of viral outrage, the mountainous difficulty of moderating at scale, the whiff of conspiracy. It reminded us what we already know: that technology and democracy are now so interconnected that it is hard to tell where one ends and the other begins. Those who design and control social media platforms set the rules of debate: what may be said, who may say it and how.

***

There are many online spaces for citizens to engage in public discussion. On Facebook platforms alone, people post more than 100 billion times a day — more than 13 posts for every person on the planet.[3] Platforms often claim that they are not in the business of *creating* the content that appears on their platforms, and that's usually true. But it doesn't mean they are neutral in public debate. Far from it. Instead of making arguments themselves, they rank, sort and order the ideas of others. They censor. They block. They boost. They silence. They decide who is seen and who remains hidden.

The simplest form of platform power is the ability to say no.[4] With a click, any user – even a president – can be banned from a platform forever. Nearly 90 per cent of terms of service allow social media platforms to remove personal accounts without notice or appeal.[5] Often the banished deserve their fate: Facebook removes billions of fake accounts every year, as well as countless fraudsters, predators and crooks.[6] Donald Trump almost certainly deserved it when it happened to him. However, there but for the grace of Mark Zuckerberg go all of us. You or I could find ourselves shut out from a prominent forum for reasons we don't fully understand. No law requires platforms to explain or justify their decisions. Even if you accept that they do not deliberately favour certain political views over others, because of the scale at which they operate, they sometimes make mistakes.[7]

Sometimes platforms' decisions about who not to exclude can be just as controversial as their decisions to throw users out. Facebook, for example, has been accused of letting powerful politicians in India breach its rules on fake news.[8]

Platforms block content as well as people. Turkey requires Twitter to remove tweets that are critical of the country's president.[9] Facebook removes 'blasphemous' content on behalf of the government of Pakistan and anti-royal content on behalf of the government of Thailand.[10] TikTok censors content mentioning Tibetan independence and Tiananmen Square.[11] The EU prods platforms to take down hate speech.[12] Often, what seems like pure corporate power is a mix of commercial and government forces, interacting in ways that are opaque to ordinary users.[13]

Sometimes platforms remove material because it is unlawful. But often they go further. In Turkey, TikTok banned depictions of alcohol consumption even though drinking alcohol is legal there. It also prohibited depictions of homosexuality to a far greater extent than required by local law.[14] Ravelry, a platform used by nearly 10 million knitting and crocheting enthusiasts, forbids (lawful) images that are supportive of Donald Trump. A photograph of knitwear emblazoned with 'Keep America Great' will be removed by the site's moderators. But garments protesting Mr Trump's 'grab 'em by the pussy' remark (known as 'pussy hats') are actively encouraged.[15]

When social media platforms permit or remove certain forms of legal expression, they make contestable judgements about the nature of free expression. Instagram's community guidelines, for instance, contain a general prohibition on the display of female nipples. Rihanna, Miley Cyrus and Chrissy Teigen have all had photographs removed for breaching this rule. But if their snaps had been of breastfeeding or post-mastectomy scarring, then Instagram would have allowed them. And if the offending areas had been pasted over with images of male nipples, that would also have been acceptable. You may agree or disagree with Instagram's approach to this issue, but either way, the platform's 500 million users have little choice but to abide by its judgment about what flesh should be seen and what should remain hidden.[16] The same goes for political speech more generally. During the Covid-19 crisis, Twitter removed tweets from the Presidents of Brazil and Venezuela that endorsed quack remedies for the virus.[17] Twitter suspended Donald Trump from its platform permanently. Some supported these moves while others were uncomfortable about a private company censoring heads of state. It has been reasonably pointed out that other controversial political figures, like Iran's Ayatollah Khamenei – who believes that gender equality is a Zionist plot to corrupt women[18] – are still merrily tweeting away.

***

Political advertising is one of the most important and sensitive forms of speech. In the US, platforms essentially decide for themselves what should be allowed. Twitter and TikTok have banned political advertisements altogether. Snapchat permits them but only if they are fact-checked. Google allows them without any fact-checking at all.[19] Facebook allows political advertisements and will fact-check them too, unless they are made by politicians, in which case they are left alone. In 2019, Donald Trump launched a Facebook advertisement that falsely claimed that Joe Biden had offered Ukrainian officials a bribe. The clip was edited to give the misleading impression that Mr Biden had confessed to this non-existent crime. It was seen by more than four million people.[20]

When Mark Zuckerberg launched Facebook in his sweaty college dorm, he probably never imagined that he might eventually have to regulate the falsehoods of the President of the United States. But we might reasonably ask whether *any* private company should be making that kind of decision, on a regular basis, all by itself. Political advertisements on the television are subject to robust regulation. Why should online advertisements be treated differently?

'Which kinds of speech should be permitted and which should be prohibited?' is a big philosophical question. As we will see, democratic societies offer different answers. The strange reality of the current era, however, is that philosophy and tradition matter less than the opinions of the lawyers and business executives who make the decisions. Frustratingly, social media platforms often seem to be making up their ramshackle philosophies as they go along. Mr Dorsey, Twitter's founder, says that his platform was founded 'without a plan' and 'built incentives into the app that encouraged users and media outlets to write tweets . . . that appealed to sensationalism instead of accuracy'.[21] In 2021, the *Wall Street Journal* revealed that Facebook had 'whitelisted' certain celebrities and politicians, allowing them to violate the platform's rules without facing sanctions. The Brazilian footballer Neymar, for instance, was apparently permitted to post nude pictures of a woman who had accused him of rape.[22]

In 1922, the authoritarian jurist Carl Schmitt wrote, 'Sovereign is he who decides on the exception'.[23] The platforms decide the rules and they decide the exceptions, often in the heat of a scandal. But knee-jerk crisis management is no way to regulate the deliberation of a free people.

SEVEN

The Automation of Deliberation

On the social media platform Reddit, a user recently posted a moving account of suicidal feelings they had had in the past:

> I think the thing that helped me most was probably my parents. I had a very good relationship with them and they were always there to support me no matter what happened. There have been numerous times in my life where I felt like killing myself but because of them, I never did it. They knew about my feelings and they were extremely supportive. I am not sure how much of an impact this had on me but it definitely helped.
>
> I think the other thing that really helped me was my teachers. I had some fantastic teachers in high school and college, they were always there to help me out when I needed it most. I don't think I would be alive if it wasn't for them.

It is very difficult to explain but there have been a few other things that I think have helped me. One of them was my friends, they were always there for me.

They did not really know about my depression or suicidal thoughts but whenever I was feeling down, they were there for me and would lift up my mood. If it wasn't for them, I probably wouldn't be alive.

This post was filled with humanity. Other users responded to it warmly. But they had no idea that its anonymous author, thegentlemetre, was not a human at all, but a text-generation machine, a 'bot' engineered to mimic human speech.[1]

This chapter is about the growing capacity of digital systems to intervene in human affairs through speech – a special kind of power in societies where political order rests on the free exchange of views between citizens.

When the Washington Post columnist Jamal Khashoggi disappeared in 2018, the Western media widely suspected that he had been murdered at the instigation of Mohammed bin Salman, the Crown Prince of Saudi Arabia. Yet Arabic-language social media told a different story. In one day, the phrase 'we all have trust in Mohammed bin Salman' appeared in 250,000 tweets, along with 60,000 tweets saying, 'We have to stand by our leader' and 100,000 messages imploring Saudis to 'Unfollow enemies of the nation'. These messages were unmissable. And the vast majority appear to have been generated by nonhuman chatbots, unleashed to do the bidding of the crown prince's supporters.[2]

Chatbots are software programs capable of using natural language. They respond to human speech by inferring the appropriate response from large repositories of human language. They are not intelligent in any meaningful sense, and for now, they are generally limited to crude propaganda. But the Khashoggi case shows that message discipline alone can dominate public discourse. Remember the commotion about the 'caravan' of Central American migrants heading north before the 2018 US midterm elections? It is estimated that 60 per cent of the online chatter was initiated by chatbots.[3]

Nonhuman systems are already used to present the news,[4] answer helplines, make marketing calls[5] and give basic legal advice.[6] IBM claims to have developed a system that 'can debate humans on complex topics'.[7] The system used to create thegentlemetre was initially said to be too dangerous to release for wider use.[8] Trained on almost a billion words of human writing, this system, the latest version of which is known as GPT-3, is said to be able to answer questions, fix poor grammar and compose entirely original stories. It is now commercially available. Systems like it will soon be available for advertisement, debate, banter and flirtation. They may, in time, be paired with technologies that can read human emotions and respond to them with faces and voices engineered to be attractive and persuasive.[9]

There is already concern about false but photorealistic videos of real people, so-called 'deepfakes'.[10] So far, deepfakes have mainly been used to superimpose the faces of celebrities onto the bodies of pornographic actors. But they have also been used to impersonate politicians.[11] In the future, they could feasibly be used to depict a fake act of corruption on the part of a president, or a fake announcement on the part of a police chief.[12] The main risk posed by deepfakes is not merely that people will believe in falsehoods. It's that the very notion of truth will decline in value. People will not know who or what they can trust.

Social media platforms are gearing up for the next generation of challenges, but again, they are largely being left to themselves, as if the challenges were merely technical in nature. But the question of how to deal with sophisticated nonhuman speech is one on which reasonable citizens might have different views. Some will wish to see such 'voices' outlawed or hidden; others will see them as legitimate and even helpful forms of free expression. The future of democracy depends on the balance that is struck. Right now, the power to decide lies in private hands.

The bigger picture is that we have entrusted the most sacred treasure of the republic – the democratic exchange of ideas – to business corporations with few safeguards or systems of oversight. Is it wise to allow them to decide with unfettered discretion what the rules are, when and how to enforce them and when they don't apply, without the rest of us having a say?

For years this issue was addressed – or, rather, deflected – by the response that platform moderation was best left to the companies that developed the technologies in the first place. Obviously, at a technical level, the platforms know their systems better than anyone else. But it is wrong to suppose that the work they do is anything other than deeply and inescapably political. Even Mr Zuckerberg appears to agree with this now: 'I don't think private companies should make so many decisions alone when they touch on fundamental democratic values.'[13]

***

In this part we have looked at four forms of digital power: the power to write rules that others must follow, the power to scrutinise others, the power to frame how others see the world and the power to shape democratic deliberation. This is not power in the traditional sense, the kind found in legislatures and courtrooms and armies. It is new and different, emanating from ownership and control of digital technology.

What should we make of this power? After all, most technological systems are designed to make people's lives better, not worse. And those who design and control digital technologies are often highly intelligent, well educated and well intentioned. But from a republican perspective, that isn't what matters. It is inherently risky to entrust great power to a narrow and unaccountable slice of the population.

The challenge runs even deeper, however. The technology industry likes to present itself as a domain of unfettered rationality; a place where decisions are governed by data and logic rather than the fickle whims of human prejudice. This lack of self-awareness would be amusing if it wasn't dangerous. It depoliticises the power held by those who work with digital technology. It makes that power seem natural, inevitable and unthreatening.

The next part of this book holds a moral mirror up to the technology industry. It argues that those who wield power through digital technology will always have to make hard choices – just like politicians, judges or anyone else with great social responsibility. In short: the digital is political.

PART III

THE DIGITAL IS POLITICAL

Science today should be entrusted to men of spiritual
understanding, to prophets and saints. But instead it's been
left to chess-players.
Vasily Grossman

EIGHT

The Morality of Code

Not long ago, I gave a presentation to a class of mid-career public policy graduate students. During my talk, I noted that Google had a history of returning problematic results to search queries, including autofill responses which ended queries like 'Why do Jews...' with '...love money so much?' In the discussion afterwards, a student put up his hand. He explained that he had worked as an engineer at Google, and that the company worked hard to ensure that its algorithms were neutral. If racist or unpleasant results came back, that was simply because lots of people had found those results useful. All the algorithm did was promote the websites that got the most clicks and links.

This defence, though well intentioned and probably factually correct, was a prime example of what I call *the neutrality fallacy* – the flawed idea, widely held in Silicon Valley, that *an algorithm is just if it treats everyone the same*.[1] The flaw lies in the fact that justice often requires us to treat people differently, not the same. That's why we implement special safeguards for children, or target social resources at deprived communities where they are needed most. This student was not necessarily wrong to say that Google's algorithm was neutral. His mistake was to assume that neutrality was a defence against a charge of injustice. I asked him whether Google ought to take steps to *reduce* inequity by adjusting its algorithm so that racist results were given *less* prominence rather than more (something Google has since started to do). The student looked horrified. That was the work of a social engineer, not a software engineer. It was not what he had prepared for when he learned to code. He didn't want that responsibility.

The neutrality fallacy is endemic in Silicon Valley. It can be seen in the design of many prominent social networks. Reddit, for instance, makes posts more or less visible depending on the votes of the community. Moderators do not often intervene. They maintain neutrality. But because Reddit is dominated by white men, and often by an aggressive 'geek culture', the result is an atmosphere that can be poisonous for women and people of colour.[2] Neutrality towards abuse, harassment and extremism means supremacy for abusers, harassers and extremists. Neutrality and justice are not the same thing.

In recent years an exciting crop of commentators – Joy Buolamwini, Meredith Broussard, Ruha Benjamin, Safiya Noble and others – have begun to expose the injustices that lurk within even the most sophisticated digital systems. This chapter draws on some of their ideas.

***

Programmers and engineers pride themselves on their rationality. On their behalf, it is often claimed that digital technologies can offer a scientific and objective basis for the ordering of society and its resources. This is a myth. The technologies of power are rarely neutral in their operation, and even if they were, neutrality is usually a poor guide to justice.

Consider two smartphone apps.

One, developed for the residents of Hong Kong, aggregates and shares data about the movements of the police in the city. Its aim is to allow pro-democracy demonstrators to evade tear gas, pellets and batons as they assemble to protest. A second app, encouraged by the law of Saudi Arabia, gives men the ability to monitor and restrict the movement of their wives. Which of these apps should American technology companies make available on their platforms?

As it turns out, Apple has banned the police-monitoring app from its App Store, along with more than sixty other apps used to circumvent Chinese internet censorship. The Saudi wife-monitoring app, meanwhile, is freely available for download within the Kingdom of Saudi Arabia.[3]

Technology companies make moral decisions of this kind on a regular basis. They do so with considerable freedom and often in secret. It is easy to be misled by the chrome and glass. Beneath the surface, many consumer technologies are a seething mass of values, biases and ideologies. Some reflect the explicit moral preferences of their manufacturers. Other times, the politics of a given technology are hidden or accidental – the unintended product of design rather than ideology. And sometimes it's just about commercial expediency: whatever sells best.

***

A few years ago, Amazon developed (but apparently never used) a recruitment system to find correlations between (a) the most successful employees at Amazon and (b) the contents of their CVs. Once it had identified those correlations, the system was supposed to screen job applicants to decide whether they should progress to the next stage. The idea was that those with CVs that contained similar attributes to those of Amazon's successful employees would get through; those whose CVs did not would be rejected. This probably seemed like a logical approach, embodying the kind of scientific detachment for which the tech industry is often praised. But there was a problem. Amazon had historically been a male-dominated company, and so its system ended up concluding that a strong indicator of likely success at Amazon was *being a man*. Thus, CVs that contained the words 'women's soccer team' or the names of all-women colleges would be sent to the bottom of the pile.[4] The designers of this system did not set out to discriminate against women, but the data embodied a legacy of injustice and the system amplified it.

Systems like these often end up relying on proxies: attributes that appear neutral but which are correlated with characteristics like race, sex or age. Crudely, if you combine data about a person's residence, purchasing history, social media connections and music tastes, then you may well end up with a profile of their race, sex or age even if that's not what you set out to find.[5] In the US healthcare system, algorithms used to allocate medical resources will often mistakenly assign the same risk level to seriously ill black patients as less sick white patients. Why? Because the system uses patients' health costs as a proxy for their health needs. Because fewer healthcare dollars are spent on black patients than white ones with the same levels of need, the system erroneously concludes that black patients are healthier than white ones when they suffer from the same conditions.[6] Likewise, software in digital cameras routinely warns photographers that Asian subjects are 'blinking',[7] and many facial recognition systems struggle to identify black people, because they were 'trained' on datasets that disproportionately feature white faces.[8]

These kinds of computing systems are usually referred to as 'machine learning' or 'artificial intelligence'.[9] The term 'learning', like 'artificial intelligence', implies consciousness, awareness and creativity. But these words can be misleading. They give the false impression that these systems think like we do – or, indeed, that they think at all. (I use the term machine learning in this book under protest, as it has become the common descriptor.)

The risk with animistic metaphors like 'learning' and 'intelligence' is that they encourage us to treat computers as having *mind-like* faculties. But there is nothing inherent in the design of computer systems to make them do the right thing when presented with a social problem (or, indeed, to choose between competing visions of the 'right thing'). It depends on how they have been programmed. The name Analytical Engine, given to one of the earliest computers, better captures the mechanistic way in which even the most advanced systems still work. They have achieved superiority to human performance in a number of fields, not because they replicate the human mind but because they are so different from it. Machine learning systems work by processing huge quantities of data and detecting the patterns that lie within. Some look for a specific correlation, like oncological systems that scan images of freckles, compare their appearance against images of cancerous lesions and thereby diagnose which freckles are likely to be cancerous. Others acquire 'skills' like game-playing, asset-trading, fraud-detecting, product-recommending, language-translating, speech-mimicking, document-analysing and car-driving − all by chomping through data that reflects how those skills are performed. However they do it, it's not like the human brain.

A machine learning system was recently used to analyse a very large body of text from the internet. It looked for words that are closely associated. It found, for instance, that musical instruments are more commonly associated with pleasant terms than weapons. No surprise there. But it also found that European American names like Harry, Josh or Roger are 'significantly' more associated with pleasant terms than African American names like Leroy, Lamont or Tyrone. And men are more associated with words like *executive, management, professional* and *corporate*, while women are more associated with *home, children* and *family*.[10] Studies like these show society its reflection. They reveal the injustices embedded in our language. But when we 'train' systems to read or speak using datasets like these, we reproduce and amplify injustices rather than rectify them.

The Romans had a word, *iniuria*, to describe a violation of the 'equality of respect' that citizenship demanded. To demean or deny the individuality of another citizen was a wrong recognised in law (slaves were another matter).[11] The Greek word *isothymia* described the basic demand for each human to be recognised as of equal worth.[12] To be treated with dignity is to be seen as a full person and a unique individual, with an identity that cannot be disaggregated into generalisations about its various parts. To violate that dignity is a form of injustice.

***

In the long run, digital innovation should make life easier, safer and more prosperous. But it would be a mistake to assume technical proficiency and social progress go hand in hand. The power of digital technology is not abstract. It accrues to some groups in society, while others are left marginalised or even oppressed. Every digital innovation comes with a set of presumptions and assumptions. The world is being remade by those who write code, according to their (implicit or explicit) political preferences, and usually without recourse to the traditions or opinions of those who have to live with the consequences. The next chapter considers a common, and problematic, way of thinking in Silicon Valley – a new ideology that has slipped unseen into our midst.

NINE

The Computational Ideology

In 2020, because of the pandemic, British schoolchildren were unable to sit public examinations, and so were given grades based on their past work. For lazy but bright students who had planned to idle all year before cramming furiously in the last two months, this policy will have been as welcome as a hog roast at a Bar Mitzvah. They were to be judged on their past performance, not their future excellence. Then, to make matters worse, the Department for Education adjusted students' grades using a simple algorithm that took into account the performance of *previous* students at the school from which each pupil came. This had the (apparently unintended) effect of downgrading 39 per cent of results, with poorer pupils seeing the greatest downward adjustment because schools in less prosperous areas had previously underperformed.[1] Three days later, young people took to the streets of London with placards reading 'Fuck the Algorithm'.[2]

In a different context, an entrepreneur in the lending business recently declared, 'all data is credit data, we just don't know how to use it yet.'[3] What he meant was that the more you know about a person, the easier it is to predict whether they will repay their loans (just as the British government had assumed that if you know what school a pupil goes to, you can be confident enough about their likely academic performance to adjust their grades).

The exam fiasco and the statement that 'all data is credit data' share an implicit political philosophy – one that sees society as a dataset, humans as data points and the goal of social organisation as the pursuit of optimisation and efficiency. Call it *the computational ideology.* The computational ideology is widespread in the tech industry and in government use of digital technology. It masquerades as apolitical, or neutral, or objective, or scientific, but in reality it is deeply political, both because of what it treats as important and what it ignores.

***

Today's humans are able, if they wish, to treat political and moral questions as exercises in the analysis of very large numbers.[4] Even the great mathematicians and actuaries of the last century had nothing like our quantity of data or our capacity to process it. The whole of life is increasingly seen through the prism of data.

In *The Human Condition* (1958), Hannah Arendt cautioned against looking for humanity in the broad sweep of 'everyday behaviour' rather than in 'rare deeds' and seminal events.[5] Yet one need only look at the rise in algorithmic modelling to decide who gets important social goods like jobs, insurance, credit and housing to see how widespread computational thinking has become. It assumes that there is order in human affairs even if it is not always visible to the human eye; that ranking and classifying human beings is not only technically impressive but socially useful; and that hierarchy and segmentation are more desirable than equality and solidarity.

The computational ideology is growing in influence, not because it is found in philosophy textbooks but because it increasingly underpins practice in the real world. You can agree or disagree with the computational ideology, but you can't pretend it's not political. Like any public philosophy, it ought to be rigorously examined, not just in business schools and boardrooms, but in the arena of public opinion.

There are at least three reasons to be concerned about the computational ideology.

First of all, to treat people as mere data points in a larger algorithmic exercise is to risk violating the principle that *every person counts*. When we stand before a judge, an employer, a parole officer or a mortgage lender, we want to be seen for who *we* really are, rather than as mere bundles of shared traits. It may be true that an older worker is more likely to suffer cognitive impairment than a younger one, but shouldn't a mature job applicant be judged on her own mental acuity, not a stereotype? For an algorithm, it matters little if a system gets one answer wrong once if it is right the other 10,000 times. But for the person affected, it can be devastating.

Secondly, the computational ideology is hard to reconcile with the idea that *people have free will and are capable of change*. Machine learning systems necessarily assume that our past is a good guide to our future. Often that is right: someone who smoked in the past is likely to smoke in the future, or at least more likely than someone who has never touched a cigarette. But this might not be a good principle for organising society. We are not prisoners of history. The recovering addict, the reformed criminal and the repentant sinner are all ideals that reflect the belief that people can change their lives for the better. Our myths and stories are filled with the euphoria of unforeseen events – the sacrifice of the 300 at Thermopylae, the escape at Dunkirk – that teach us to take charge of our own destiny, even against the odds.[6] Tennyson wrote that it is not too late 'to seek a newer world'. But if data is used to shackle us to the past, a new world will always remain out of reach.

Finally, unlike humans, profiling systems are utterly uninterested in the why of their predictions. Patterns are not always a useful guide to action. The fact – and it is a fact – that there is a 93 per cent correlation between tea consumption and the number of people killed by lawnmowers tells us little of how we ought to regulate tea or lawnmowers.[7] Intriguingly, an algorithmic analysis of senior politicians from fifteen post-Soviet states found that their 'median estimated body-mass index' correlated strongly with 'conventional measures of corruption'.[8] That is to say, the chubbier the politician, the more likely he was to be corrupt. I'd still rather have a vote.

Massive datasets are not wise. They can be good at helping us find out what is, but cannot tell us what ought to be. Suppose that an algorithm discovers that, for some reason, the sum of the digits in a person's date of birth is a good indicator of their creditworthiness. Those with a total above x are more likely to pay back their loans than those below x. Would it be right to reject the mortgage application of someone whose number was below x? Not if you believe each candidate should be considered on their own merits. Nor if you think it's wrong to make judgments on the basis of patterns we can't explain.

Shouldn't things of social value – jobs, credit, homes – be allocated on the basis of morally acceptable factors, not (or not only) statistically relevant ones? It's understandable that a person's credit score might be affected by their income and debt payment history. But should their loan application be determined by their social media posts or web-browsing activity?[9] This question is increasingly being answered for us. One system for determining creditworthiness already records 'how quickly a loan applicant scrolls through an online terms-and-conditions disclosure', and uses it as an indicator of how 'responsible' she is.[10] US credit card companies have cut customers' credit limits when payments appear for marriage counselling because divorce is highly correlated with loan defaults.[11] Facebook has filed a patent for a system that would enable lenders to ascertain the 'average credit rating' in a person's social media network – that is, whether their friends are creditworthy – to help decide whether they should get a loan.[12] Car insurers have increased the premiums of people who use a Hotmail email account because Hotmail users are apparently prone to more accidents than others[13] (as well as more than twice as likely to miss debt payments[14]). Customers who enter their names on online forms using all lower case letters (*jamie susskind* rather than *Jamie Susskind*) are more than twice as likely to default on their loans.[15] But should such matters really feed into whether I get the mortgage I need to purchase a house?

Systems like these are fascinating, but they seem to be arbitrary in at least three ways: they are unpredictable, they do not necessarily tally with accepted norms and morals, and they are unanswerable to the people they affect. Should we really entrust significant decisions to such systems without some form of oversight or appeal if things go wrong?

***

The case against the computational ideology is not a slam dunk. The truth is that humans also judge whole classes of people on the basis of characteristics that only some of their members possess. A sensible teenage boy will usually pay more in car insurance than a reckless middle-aged woman, even though he is personally a lower risk.[16] We make generalisations all the time. We generalise that bagels are delicious (though some are not), that rugby is dangerous (though not all players get hurt) and that Japanese cars are reliable (though some break down). Almost all medical diagnoses are derived from a process of generalisation (cancerous freckles tend to have uneven edges; older men are more at risk of prostate cancer).[17] And even our laws are based on general assumptions about the average person. That's why speed limits apply to good drivers as well as bad.[18] In short, rules of thumb are a necessary part of simplifying the world. Computer systems that generate their own generalisations are not, therefore, as different from us as we might suspect.

Moreover, there may be no such thing as a truly personalised assessment. In a job interview, all the interviewer can do is make predictions about the candidate's aptitude based on the available data. If they turn up on time, they are likely to be well organised; if they look smart, they are likely to be professional; if they answer questions logically, they are likely to be a good problem-solver; if they have good grades, they're likely to be intelligent. We make these generalisations even though we know that the general pattern doesn't always hold.[19] In reality, what we call an individualised analysis will often be 'simply an aggregate of stereotypes'.[20] Again, that's not so different from what profiling systems do – and in many ways they are more accurate and precise than we could ever be.

***

What emerges from this chapter is that the computational ideology poses a number of challenges to the way that we use digital technology. There are new questions to answer. When should predictive systems be used and when should they be prohibited? (Would you accept a trial verdict based on a prediction of what a jury would have found? Or a college grade based on a prediction of your performance?) Is generalisation improper in all contexts or just when it worsens the treatment of historically disadvantaged groups? What should happen when an automated determination appears to be unjust? Should there be ways to challenge it?

We will seek to answer some of these questions later in this book. To give a flavour of the republican approach, however, it is worth noting that political communities generally try to build social institutions that reflect their shared moral principles. The appropriate principle might depend on the context – sometimes people should get what they *deserve*; sometimes people should get what they *need*; sometimes people should get what they have the *right* to; sometimes people should have an equal *opportunity*; and so forth. Those who subscribe to the republican philosophy do not always agree on what the good life looks like. But they do agree that decisions about the good life, and who gets to live it, should not be left to unaccountable bodies over whom citizens have no control.

When the internet first arrived, some commentators expected that its decentralised, networked structure would cause society itself to become more decentralised and networked, with power dispersing to the peripheries and away from traditional elites. In fact, although the internet did give power to some at the margins, it also made already-powerful institutions like states and large corporations even more mighty. Technologies are stamped by the character of the societies they are born into, just as they in turn shape those societies. Their values and priorities are not inevitable or immutable. They can be chosen and changed. The question is whether the rest of us should play a part in determining them.

TEN

Technology and Domination

Those who design and control digital technologies have become permanent residents of what Max Weber called the 'house of power'. As we saw in part two, those who write computer code can shape the way we live. They increasingly determine what can and cannot be done; what may and may not be said. They create and enforce hard-edged rules. They churn oceans of data to reveal the hidden dynamics of individual and social behaviour. Softly, they cajole and nudge. They filter and refract our perceptions of ourselves and each other, shaping our preferences and desires and setting the agenda for public discourse. And as we have seen in this part of the book, the power is *political* in nature, even if it doesn't resemble ordinary politics. It is permeated with values, biases and ideologies. 'The doing of our deeds, and the thinking of our thoughts,' as Derek Parfit put it – none is exempt from the touch of technology.[1] And tomorrow's technologies will be much more powerful than today's. While we experience the passage of time in a linear way, technological progress is speeding up. Any new system of governance will have to reckon with the fact that technology's social power, along with its capability, is growing at an accelerating rate.

Of course, not every occupant of the house of power is a global player. There are countless tech minnows alongside the tech giants. But from the perspective of ordinary people, it is the *combined* effect that matters. We will still be unfree if we are the passive subjects of rules written by others, or subject to moral codes that are alien to us – regardless of who writes the rules or moral codes. We will be unfree if made to live under constant scrutiny or dependent on others for the quality of our public deliberation. We will be unfree if we live at the mercy of hidden algorithms for our access to social goods. One product or platform may have little effect on our freedom, but the sociotechnical system as a whole may be quietly oppressive.

As time goes on, to borrow a metaphor from Roger Brownsword, society could become a kind of massive airport; a permanently controlled zone in which we are constantly surveilled and nudged with no point of escape.[2] Ironically, no one needs to *want* such a future for it to become a reality. Without public oversight, a thousand technologies might combine to form an atmosphere of oppression, even if, individually, each technology seems perfectly reasonable.

Some of the risk flows from the fact that digital technology is such an efficient medium for imposing order. A rule that is unobtrusive when enforced 20 per cent of the time may become oppressive if enforced 95 per cent of the time. Motorists expect to encounter speed cameras on occasion, but the act of driving would be very different if cameras were every thirty yards on every street in every town. Now imagine an autonomous vehicle that refuses to go over the limit at all, and you have a glimpse of what the future might look like. Writing in the seventeenth century, Thomas Hobbes wrote that regulating 'all the actions, and words of men' was 'a thing impossible'.[3] But it seems much more possible now – and the rules are coming from private authorities, not just public ones. The new industrial revolution is a political revolution too, and we are all caught up in it.

***

Why does it matter that politics is being subsumed by private corporations, big and small? To begin with, there are three important points of principle.

The first concerns liberty. People ought to be able to live under rules of their own choosing, according to norms and values that they have selected for themselves, or at least had the chance to influence. This was the principle that motivated colonial Americans to declare: *No taxation without representation.* To which our generation may respond: *No automation without legislation.*

Then there are concerns about legitimacy. Economic prowess is not a good justification for the wielding of power by a private person or company. That's why we don't allow the rich to buy votes or bribe public officials. '[There] is no more degenerate kind of state,' wrote Cicero, 'than that in which the richest are supposed to be the best.'[4] Money has been a source of political influence for a long time, but this fact has rightly been treated as unsatisfactory. In the US, where commercial and political interests are arguably more closely aligned than elsewhere,[5] there is a long tradition of opposing the political activities of what Theodore Roosevelt called 'mighty commercial forces'.[6] Why should command of technical resources translate into control of human beings?

Finally, there are concerns about quality and safety. Rules made in the private sector are rarely subject to the procedural safeguards that are the foundation of legitimate authority. In a free society, rules of public importance should not simply be asserted and enforced. As Jeremy Waldron notes, they ought to come from proposals subjected to public debate – in homes, workplaces, universities, pubs, on the media and elsewhere – that are then translated into principles and norms, and communicated to the people in ways they can understand. Where disputes of interpretation arise, there should be independent procedures of resolution and appeal. And rules should never be eternal. It should be possible to change them through open and peaceful channels.[7]

These safeguards are not generally present when it comes to private rulemaking, which takes place behind the closed doors of corporate boardrooms.

***

What would it be like to live in a world where there are more rules than ever before *and* increasingly perfect methods of enforcement? Brownsword and Goodwin ask whether it will be possible 'to do the right thing for the right reason' rather than simply because the technology restricts our choice.[8] It will be harder to teach ethics to our children when the world makes them ask not what is *right* but what is *possible*.[9]

We risk becoming a society in which serendipity and spontaneity are replaced by the mechanistic ranking, sorting and indexing of human life. Friedrich Hayek, the father of neoliberalism, wrote that the advance of civilisation depends 'on a maximum of opportunity for accidents to happen'.[10] What would he make of predictive systems that seek to iron out serendipity altogether, reducing human affairs to a bundle of probabilities?

These are not easy issues. What should be beyond argument, however, is that a new political settlement is needed: one that can only be achieved at a systemic or structural level. While each device, app or algorithm taken alone might have a negligible impact on freedom or democracy, their cumulative effect could be crushing. If that's right, it makes no sense for digital technologies to be governed at the level of individual corporations, or according to the choices of individuals clicking 'I agree' to terms and conditions they haven't read. *The political community as a whole* must choose its priorities, and a public system of laws and norms must be developed to meet those priorities. That is the republican way.

As matters stand, however, we do not have the kind of political and legal systems that we need. Worse, for decades we have been in thrall to a 'solution' that is the source of many of the problems.

The market.

PART IV

THE MARKETPLACE OF IDEALS

A public is not every kind of human gathering, congregating
in any manner, but a numerous gathering brought together
by legal consent and community of interest.
Cicero

ELEVEN

The Market's Place

This part of the book is concerned with the market economy and its role in the evolution of digital technology. There is no question that a well-functioning market economy, in a stable legal environment, can be explosively productive. This is a good thing. But a more difficult question is whether economic forces alone can properly regulate the power of what they unleash. Some argue that market forces encourage good behaviour. Others say that industry can be trusted to regulate itself *against* market forces. Still others say that consumer choice, and consumer empowerment, are the best bulwark against corporate power. This part seeks to explain why these arguments are wrong, or at least overly simplistic. The problems caused by the market cannot be solved by the market alone. Something more is needed – good governance.

***

In the *Communist Manifesto* (1848), Karl Marx and Friedrich Engels marvelled that a hundred years of capitalism had 'created more massive and more colossal productive forces than have all preceding generations together'. Since then, with a few interludes, market economies have continued to generate economic growth.

How?

In a pure market economy, there is no central authority directing the allocation of resources. Instead, each person owns property that they can exchange with others for money, goods or labour. If everyone pursues their own self-interest, trading and bargaining with others, then the market will put a price on each of society's resources and everything will find its way to where it is wanted most. That, at least, is the theory. The Physiocrats of the eighteenth century (like many others after them) saw markets as self-correcting ecosystems that needed little external intervention.[1] They believed that, left to themselves, markets would enable the accumulation of wealth where it was most deserved *and* regulate the conduct of those who worked within them. More than two centuries later, this remains the orthodox view in Silicon Valley.[2]

This book takes a different perspective.

There is no doubt that markets can be an efficient way of allocating resources, and for that they are a vital form of social organisation. But they are less reliable when it comes to pursuing society's collective aims, like freedom, justice and democracy. This was, in fact, forcefully pointed out by Adam Smith in *The Wealth of Nations* (1776), perhaps the most quoted celebration of the market system in history. Contrary to Smith's mythologised status as a free-market doctrinaire, he did not argue that the market alone could sustain a civilised society. On the contrary, he believed that 'some attention of government' would always be necessary 'in order to prevent the almost entire corruption and degeneracy of the great body of the people'.[3]

***

The traditional defence of the market, and of the market's capacity to encourage useful behaviour, begins with the argument that even the largest companies will always be incentivised to provide goods and services that are consistent with people's values and desires. If they don't, consumers will reject their offerings and move to competitors. Capitalism will flush out products that do not make the grade.

This perspective must be taken seriously for several reasons. The first is that it reflects the status quo, and any alternative prospectus must explain why change would be worth the cost. Moreover, it is an immensely popular point of view, not only in Silicon Valley but in Washington DC and wider Anglo-American culture. Faith in the market belongs to a highly respectable intellectual tradition that includes republican thinkers. Because the market is a reliable generator of competitors, the argument runs, it prevents consumers from being captured by any single corporation. Indeed, even the threat of consumers switching encourages businesses to do what it takes to retain their share of the market.[4]

There is evidence that markets sometimes incentivise better behaviour on the part of businesses. When Apple declares that 'data privacy is a fundamental right' and 'personal data belongs to you, not others', it is making a point of principle but also trying to differentiate itself from the other tech giants in a bid to usurp their share of consumers.[5] In a similar vein, Microsoft has made the protections of the EU General Data Protection Regulation available to all its customers, not just those strictly covered by the regulation.[6]

Let us begin our critique of this orthodoxy, however, with the fact that many of our interactions with technology take place without us having chosen them at all. Workers in Hitachi and Amazon warehouses can scarcely opt out of digital monitoring unless they wish to lose their jobs. Pedestrians cannot opt out of surveillance as they walk down the street. Consumers cannot opt out of digital payment platforms when they make purchases online. In each case there might be a theoretical choice – not to get a job, not to walk down the street, not to buy things online – but these choices are illusory. In reality, living a full and meaningful life increasingly means interacting with digital technologies designed and controlled by others. There is no opt-out.

Next, it will often be the case that where there is a choice between competing products, that choice is not meaningful. If a consumer has a choice between five providers, all of them repugnant, then that's not much choice at all.

Supposing for the sake of argument that people *were* largely free to choose between technologies, and that a real range of choices *did* always exist, the next difficulty is that consumers do not always have sufficient information to evaluate competing systems. This is not a slight on consumers; it is a well-known form of market failure. For competition to flourish, people must have adequate information about the available alternatives (what's out there?) and the key differences between them (what difference would switching make?). As to the first, it can be hard for ordinary consumers to know the full extent of what the market has to offer. But even if they did, most would not be in a good position to compare complex technical systems, even if they weren't kept secret: the myriad weightings and algorithmic models; the processes for data scraping, cleaning, sorting and analysis; the vagaries of the code. Engineers themselves often struggle to understand the operation of their own systems, so the rest of us don't have much chance. It is well known, for instance, that people consistently overestimate how much privacy they are afforded on social media platforms. Sixty-two per cent are unaware that social media companies make money by selling data to third parties (although, more accurately, the big ones sell advertising capability).[7] If we expect the market to offer users 'as much privacy as they actually value', then consumers will undersell themselves every time.[8]

Supposing, again for the sake of argument, that consumers *were* largely free to choose between technologies *and* there was a meaningful range of options available *and* consumers had sufficient information to make an informed choice, what would they choose? Unfortunately, the ethical choice is not always the obvious one. In fact, the market often requires people to weigh their needs against their principles in an invidious way. As a citizen, I may deplore algorithms that use personal data in an unethical way. But as a consumer on a tight budget, I may have no choice but to go with the cheapest mortgage provider, despite its shady data practices. There is nothing irrational about placing consumer priorities over civic preferences – that's the market in action – but it's another reason why markets alone cannot be expected to prioritise the common good.

Furthermore, economists now accept what the rest of us already know: that people do not always make decisions in a rational way. In 2020, privacy activists were aghast when crowds of people downloaded FaceApp, a free application that alters photographs to show users what they would look like as an elderly person or a different gender. Not coincidentally, the app also furnishes its Russian owners with an 'irrevocable, nonexclusive, royalty-free, worldwide, fully-paid, transferrable sub-licensable license' to use images of people's faces for whatever purposes they see fit.[9] Opinions will differ about the value of this trade-off, but it is naïve to suppose that consumers will always make the civic or even the sensible choice, often because they haven't thought through the wider ethical and political implications of what appears to be an amusing but inconsequential decision.

So far, we have assumed that with the requisite will and knowledge, consumers will switch freely between products according to their preferences. But even that is not so simple. There are often steep costs involved in switching. If I have spent ten years on a social network – building connections, establishing a following, curating groups and albums – then leaving that network to start over elsewhere may mean losing everything. If I run a business that depends on social media traffic, it could mean financial ruin. The costs of exit are particularly pronounced where the technology derives its value from the size of its network. For consumers, the question changes from *is that system superior?* to *how many other people are also prepared to make the change?* If the answer is *not enough*, then self-interest will usually mean staying put.

***

It is not merely that the market cannot safely be relied on to improve the ethical or political functioning of digital technologies. The pressures of the market often make matters *worse*.

Take the problem of algorithmic injustice, touched on in the previous part of the book. We know that companies have begun to use profiling systems to make decisions about jobs, credit, insurance and the like. Let's assume for a moment that the market does its job and kills off algorithms that yield economically useless results (because the algorithm is flawed, or the data is bad, or for any other reason). The market will be left with sophisticated systems that produce reliable predictions. If one of these systems suggests that someone's post code is a good indicator of whether they will be a productive employee, then employers will understandably want to take that data into account. If they don't, their competitors will. The fact that this might lead to discriminatory outcomes – perhaps because a higher concentration of one racial group lives in a particular area – is, in market terms, irrelevant. Here the market *actively encourages* discriminatory behaviour rather than reducing it. Again, the market rewards efficiency, not justice.

An analogous problem arose in the pre-civil rights era in the US. For 1950s law firms whose clients didn't want a black lawyer, or restaurants whose clientele didn't want black waiters, the *economically rational* choice for businesses was the *immoral* one: to hire only white staff. And many did just that. The only way to escape the cycle of discrimination was through collective action. The result was the Civil Rights Act 1965. By outlawing discriminatory conduct across the board, the Act insulated businesses from market pressure to discriminate in their hiring decisions.[10] Today, there are too few laws in place to protect tech firms that try to act more ethically than their rivals.

***

In the early days of the Roman Republic, Italy was invaded by the Greek general Pyrrhus. After some skirmishing, the two sides entered into negotiation. Rome sent an emissary to the Greeks called Fabricius. When Fabricius arrived at the Greek camp, Pyrrhus offered him 'so much silver and gold that he would be able to surpass all the Romans who are said to be most wealthy'. But Fabricius rejected his offer. He explained (one can only assume smugly) that in a republic it mattered more to hold high office and the esteem of the citizenry than to amass great wealth. Why accept riches at the expense of 'honour and reputation?'[11] Fabricius's devotion to public duty became republican legend. He was a citizen first, a consumer second.

The Enlightenment era saw the emergence of a more moderate strain of civic republicanism than that favoured by Fabricius. The rise of 'commercial society' had unleashed unprecedented economic growth. But with prosperity came anxiety. The Scottish thinker Adam Ferguson, whose writings were popular among the Founding Fathers in the US, worried that commerce might come to supplant politics as the dominant mode of social ordering. He did not think this was a good idea. 'Nations of tradesmen,' he wrote, 'come to consist of members who, beyond their own particular trade, are ignorant of all human affairs.'[12]

Ignorant of all human affairs may be too harsh for Silicon Valley's more thoughtful denizens, but Ferguson's central point was structural anyway: was the market really the best way to hold the powerful merchant class to account? Mark Zuckerberg unwittingly summarised the danger in a three-word mantra: 'Company over country'.[13] At some point the pursuit of private profit will always collide with the public interest.[14]

The next chapter examines a different defence of the market deployed by Silicon Valley market enthusiasts. They say that technology companies can be trusted to behave *despite* the market rather than *because* of it. Or put another way: they argue for self-regulation.

TWELVE

Selfie

Reid Hoffman is the billionaire behind PayPal and LinkedIn. He is a Silicon Valley legend. His book *Blitzscaling* (2018) argues that companies should be grown at such a 'furious pace' that they knock the competition 'out of the water'.[1] The German armies of the Third Reich, he writes:[2]

> abandoned the traditional approach of moving at the slow pace at which they could establish secure lines of supply and retreat. Instead they fully committed to an offensive strategy that accepted the possibility of running out of fuel, provisions and ammunition, risking potentially disastrous defeat in order to maximise speed and surprise.

According to this philosophy, Hoffman argues, the way to build a business is 'speed over efficiency'.[3] That's why venture capital investors demand 'hockey-stick' growth from the fledgling companies in which they invest – so that their competitors never stand a chance.[4] Move fast and break things.

This might make sense in commercial terms, but the political implications of what Hoffman says should give us pause. If technology entrepreneurs are really 'building the future', as he claims, should their guiding philosophy really be borrowed from military strategy, still less the Wehrmacht?[5]

***

Every business wants to make money. Those dealing in technology are no different. Milton Friedman said the general duty of the corporate executive was 'to make as much money as possible while conforming to the basic rules of the society'.[6] Not everyone agrees with the Miltonator, but he evidently has some fans in Silicon Valley, where success is largely measured by the aggressive, steroidal pursuit of growth.

Of course, for a long time Silicon Valley sought to convince itself and the wider world that the business of digital innovation was different from ordinary market activity. 'Make tools that advance humankind' was Apple's mantra. 'Don't be evil' was Google's. A cynic would dismiss these as hollow marketing statements, but they do reflect some important truths. Technologists really are building the future. Many of them are well meaning. And many of the mighty oaks of Silicon Valley look nothing like the acorns from which they grew.

The question, however, is whether entrepreneurs can be automatically expected to build a future that serves the interests of humankind generally or one that chiefly serves their own commercial purposes. There is no criticism here. It is simply the type of challenge that any powerful economic entity should face.

So long as digital technology is left to develop according to market forces alone, decisions that we might regard as political in nature – how to treat personal data, whether to ban certain types of social media content, what kinds of algorithms to deploy – will never be entirely free from the need to make money.[7] That doesn't mean, of course, that every decision will be seen in crude terms of profit and loss. It might be that the best long-term strategy for profitability is cleaning up a platform. But it might also mean ingratiating the company with influential state actors or turning a blind eye to the posts of powerful figures who breach the community guidelines.[8] The general lesson is we should not be surprised when technology companies make choices that benefit themselves but not the wider public; there is no inherent reason why the two should always align.

The last chapter examined the argument that market forces can be trusted to rein in the power of those who design and control digital technologies. We saw that the opposite is often true: the market *empowers* technology companies at the expense of consumers, and can *incentivise* unjust behaviour rather than constraining it. So, when defenders of the free market argue that platforms need no regulation because they are subject to the discipline of market forces, they are both right and wrong. Right because market forces do shape the conduct of businesses. Wrong because the effect of such forces is often to make matters worse.

Some defenders of the free market take a different tack. They argue that the technology industry can be trusted not because of markets but *despite* them. They say that, where necessary, corporations will *resist* market forces and do the right thing, even at the expense of profit. This is the ideology of self-regulation.

***

There is a germ of wisdom in the idea of self-regulation. Many industries – medicine, law, journalism – are given latitude to set their own rules and norms. And some non-profit bodies have historically done a good job of handling technical matters like internet protocols and the distribution of domain names.

But self-regulation has its limits.

The practice of law provides a helpful analogy. Every day, lawyers face choices between right and wrong. Should I bring this unhelpful document to the attention of the court or bury it in a stack of irrelevant material? Should I advise settlement (good for the client) or push on to trial (good for my fees)? Unsurprisingly, society does not just leave lawyers to identify the 'right thing to do' and blindly trust that they will do it. It places strict requirements on them. To become a lawyer, a person must undertake a specified programme of study and become duly qualified and certified. To practice, she must be under the auspices of a regulator and fully insured. Every lawyer is taught that expertise cannot be separated from social duty – and that like doctors, lawyers owe their clients special duties of trust. Part of practising law is accepting that profit must come second to professional obligations like honesty, integrity and a commitment to the rule of law. It's for these reasons that law, like medicine, is said to be a profession rather than just a job.[9] Putting aside any cynicism about whether the legal profession always lives up to these lofty ideals (obviously it does not), there is undoubtedly a functioning system of self-regulation with appropriate norms and rules. And when those norms and rules are broken, there are serious consequences: fines, public censure and even the loss of livelihood.

The technology industry, at least in the private sector, works according to a different model. There are no mandatory qualifications for most roles. There is no widely accepted code of conduct. There is no obligatory certification, even for sensitive functions. There is no duty to put the public ahead of profit. There is no requirement to behave more ethically than an ordinary businessperson. With some exceptions, there are no regulatory authorities to set standards or conduct investigations. The crucial difference is that there is no accountability – no sanctions for gross failures of ethics, no fear of being suspended or struck off. A doctor or a lawyer can expect to face serious consequences for egregious moral failures. In the technology world, this is not the case.[10]

So, when people in the tech industry speak of 'self-regulation', they are not really describing self-regulation as it is understood by lawyers or doctors. They are exploiting an ambiguity in the term 'self-regulation' itself. Self-regulating professions usually involve government oversight in drafting statutory rules, constituting oversight bodies or involving the public in consultation.[11] But in

Silicon Valley, 'self-regulation' means powerful technologies being left almost entirely to the wisdom of those who design and control them. In truth, this is not regulation at all.

Experience suggests that, without enforcement mechanisms, technology companies will not reliably put public welfare ahead of profit. Yaël Eisenstat, a CIA officer for thirteen years, was hired by Facebook as their global head of elections integrity operations. She left after six months. 'Not only was I not empowered to do the job I was hired to do,' she says, 'but I was intentionally sidelined and never able to effect any sort of change or even participate in most of the conversations that were directly tied to the job I was hired to do.' The reason? '[The] things I tried to push, all actually were things that went against their core idea of how it could make money.'[12]

Ross LaJeunesse, formerly Google's head of international relations, writes of his unsuccessful struggle to set up a human rights programme within the company. He found, in his words, that Google's priority was 'bigger profits and an even higher stock price'.[13] Another gesture to social responsibility, Google's ethics advisory council, was abolished nine days after it was announced, following uproar at the inclusion of the head of a controversial think tank and the chief executive of a drone company.[14] Google has since caused controversy by allegedly forcing senior ethicists out of the company.[15]

Should we be surprised at these accounts? No. Corporations owe duties to their shareholders – not to the citizenry at large and not to philosophical principles. Self-regulation in the tech industry is a non-starter, both in the literal sense that it has not started, and in the broader sense that it is unlikely to work in the short or medium term. It basically boils down to hoping tech companies will do the right thing, even when powerful incentives are yanking them in the other direction.

THIRTEEN

Ethics Washing

It takes a very special type of business to be able to operate in a competitive marketplace while maintaining a meaningful commitment to the public good. Silicon Valley is not a hospitable environment for this kind of business, partly because of the ferocity of the competition, partly because of the demands of investors, and partly because of its crippling lack of diversity.

The problem begins at university. Three in four students in artificial intelligence courses, and four in five professors, are men.[1] Since 1985, the number of women receiving undergraduate computer science degrees has halved. It now stands at 18 per cent.[2] Only a quarter or so of PhDs in computer science, mathematics and statistics are awarded to women[3] and more than 80 per cent of the papers published at machine learning conferences are written by men.[4] There is no general tradition of teaching young software engineers about the moral challenges of their work. Some colleges offer an 'AI ethics' class[5] but this is usually an elective.[6] STEM students are not generally required to take courses in philosophy, literature, music, history or art.[7]

In the industry itself, more than half of the women who work in science and engineering end up leaving their jobs, usually before the age of forty.[8] In the start-up world, more than 90 per cent of venture capitalists are men.[9] Of founders who have raised funding, just one per cent are black.[10] At large Silicon Valley firms, black and Latino coders are found in numbers as low as 1 or 2 per cent.[11]

Silicon Valley is not exactly a monoculture, but there is a recognisable 'engineering mindset' guided by the principles of optimisation, scale and efficiency.[12] It is hard to define the 'in' culture, but those who are 'in' certainly seem to know who is 'out'. Google allegedly has its own word for the people it likes to hire: *Googley*. In Silicon Valley more generally, 'culture fit' matters. It means 'hiring people like everyone else in the company'.[13]

In a recent experiment, 5,000 CVs were sent to technology employers. When candidate names were removed, 54 per cent of women were selected for interview. When names were shown, that figure fell to 5 per cent.[14] No wonder that the Valley is 'overwhelmingly white, overwhelmingly male, overwhelmingly educated, overwhelmingly liberal or libertarian and overwhelmingly technological in skill and worldview'.[15] Black applicants to Facebook have complained of being turned away because of a lack of 'culture fit'.[16]

The technology world has come a long way from the California counterculture of the 1960s and 1970s. Back then, there was a 'fiercely anti-commercial' milieu that sought to challenge powerful interests while protecting the privacy of ordinary people.[17] Now the tech industry is host to many brilliant people who might previously have sought their fortune in law or banking. It has maintained its hostility to *government* power but quietly shed its mistrust of *corporate* power.[18]

Of course, every industry has its quirks and foibles. Some companies are more civic-minded than others, and – usually in response to public outrage, government pressure, employee protests or leaks – some have shown a degree of willingness to change. But in the final analysis, the problems with digital technology cannot be cured by the market. And the problems with the market cannot be cured by tech culture.

***

Following public outcry, the technology industry has begun to develop ethical principles and practices for the development and use of advanced computing systems. (The modern field of medical ethics, too, was the product of scandals, which often drive progress.[19]) Microsoft, Google, IBM, Amazon and a host of other tech giants have all started or funded initiatives concerning AI ethics.[20] In a weird irony, ethics has been hailed as 'arguably the hottest product in Silicon Valley's hype cycle today'.[21] The most common ethical principles concern privacy, accountability, safety and security, transparency and explainability, non-discrimination, human control, professional responsibility and 'human values'.[22]

At a superficial level, the ethical turn in Silicon Valley is to be cautiously welcomed. The more energy spent contemplating the moral dimensions of digital technology, the better. But the 'ethics' paradigm is also problematic for at least six reasons.

First, the majority of ethical principles coming out of Silicon Valley (and even academia) are so vague and contestable that no one agrees what they mean and they cannot easily be put into practice. (Privacy is a good example. Even privacy scholars argue ferociously about what it means.)

Second, principles are all very well, but the ethical exercise only really begins when principles come into conflict with the need to make money. On this, tech firms' ethics charters tend to be suspiciously silent. Unsurprisingly, research shows that staff tend to discount the ethical considerations of certain behaviours when the consequences are 'favorable to the organization'.[23] As Groucho Marx put it: 'These are my principles, and if you don't like them – well, I have others.'

Third, ethics charters rarely make clear what should happen when principles come into conflict with *each other*. This often happens when well-meaning generalities are listed in no particular order.[24] (What should be given priority, privacy or transparency?)

Fourth, ethics principles are often couched in unsuitably dry and technical terms. It is all very well to speak of reducing 'bias' in algorithms but that is not enough. Sometimes bias might be in the interests of justice (in affirmative action, for instance), while other times it might be unacceptable. Deciding between these options is not a technical activity. It is political.

Fifth, most of the ethics documents proliferating in the industry are themselves the creation of a closeted elite, whether in Silicon Valley or the academy. This doesn't make them wrong, necessarily, but it does dent their legitimacy. You don't need to be a hardcore republican to believe that citizens themselves ought to have some say in the rules that govern the technologies that, in turn, govern them.

Finally, almost all of the ethical frameworks assume that nonhuman systems are appropriate in any context, as long as they are used ethically. This dodges the deeper question that ought to be asked in almost every situation: is this something a machine should be doing at all?[25]

A cynic might say that these defects reveal a deeper truth: that technology companies are going to resist putting ethical principles into practice if it might seriously affect their profitability.[26] Tellingly, one study found that just one in sixteen AI ethics documents contains any 'practical enforcement mechanisms' at all.[27] The rest simply contain guidance.

What's left is the nagging concern that voluntary codes of ethics merely give the *appearance* of ethical behaviour while masking the same old harms. This is problematic. It gives the comforting impression of ethical correction without the substance.

These concerns form part of a broader critique of 'corporate social responsibility'. According to this critique, private codes of conduct, councils of ethics and declarations of principle are often used as smokescreens to give the impression that all is well while hiding a more sinister reality.[28] They are, at heart, a public relations exercise, fending off government regulation with gestures towards meaningful change.[29] This practice even has its own name: 'ethics washing'.[30]

This is not a cynical book. But you don't need to be cynical to believe that self-regulation in the technology industry is unlikely to work. Even if you believe, as I do, that many who work in tech are brilliant and well meaning, it is sensible to consider that given (a) the immense power they hold, (b) the nature of Silicon Valley culture, and (c) the pervasive need to grow and make money, wider society should have more say in how technologies are developed and deployed.

To be a digital republican is to understand that human beings – even the best of us – mess up sometimes. There are limits to our altruism. We are subject to fierce social, institutional and economic pressures. A terrific programmer or entrepreneur might not also be a great arbiter of truth or justice. Some of the issues raised by technology are so large, so inherently *political*, that they should be the subject of democratic debate, not corporate policy. Even a perfectly formed charter of ethics might be out of kilter with public morality. If you want to hold the powerful to account, you shouldn't let them mark their own homework. It is time to move from ethics to law.

***

What about consumers? Defenders of the market argue that individuals are best placed to make choices about their own lives – whether to use a particular technology, agree to a platform's terms of service, allow personal data to be gathered, and sold on. Each one of us, it is said, is able to consent (or not) to digital technology in its various forms. The idea is attractive in theory, but in practice consent can be a kind of trap: an all-too-easy way to sign away our freedoms, and those of others, without even realising. That's the subject of the next chapter.

FOURTEEN

The Consent Trap

Close your eyes and try to recall every time you have clicked 'I agree' in the last ten years to online terms of service, privacy notices and the like. If you're like me, you can probably recall a handful but know there have been thousands more. Each click seemed insignificant at the time – a momentary inconvenience, quickly forgotten. Believe it or not, many defenders of the current system say that those clicks were your primary line of defence against the power of technology corporations. This chapter seeks to show that, instead of protecting us, the idea of consent has been twisted to leave us more exposed.[1]

***

For the last two decades, technology policy (particularly in America) has been characterised by a failure to recognise that some of the biggest challenges thrown up by digital technology cannot be resolved, and in fact may be worsened, by everyone separately pursuing their own interests. Among the most significant of these failures has been the spread of what I call the consent trap:[2]

> Instead of regulating technology at a social level, individuals are left to negotiate the terms of their engagement with digital services, leaving them at the mercy of the more dominant player.

Consent, however, is a concept with distinguished pedigree. In liberal thought, it has been regarded as the foundation of political legitimacy since at least the time of John Locke (1632–1704). The idea of the state as a 'social contract' has been popular since the Enlightenment. More recently, but still more than 130 years ago, Samuel Warren and Louis Brandeis argued that every individual possesses a 'right to be let alone',[3] and over time, this metastasised into a general assumption that individuals should be left to determine for themselves, 'when, how, and to what extent information about them is communicated to others'.[4] This assumption lies at the heart of many contemporary theories of privacy, in academia and law. It is also the governing logic of the existing regulatory regime in the US. The paradigm is known as *notice and choice*: individuals are given *notice* of what is going to be done with their data and a *choice* about whether to accept those terms.

Where possible, individuals should indeed be able to decide important matters about their lives for themselves. This respects their dignity and autonomy. But in the context of technology, consent is often a trap masquerading as a safeguard. It does nothing to rebalance the relationship between consumers and powerful technology corporations. Instead, it entrenches the domination.

***

Let us open the case against the consent trap by stating the obvious: most consumers do not read the terms before giving their 'consent'.[5] Can you remember the last time you did? If so, you are in a tiny minority. Only one in a thousand people click to see an end-licence agreement before purchasing software online.[6] On average, those who *do* look spend just fourteen seconds on documents that would need at least forty-five minutes for adequate comprehension.[7] This is not because people are fools. It is because many privacy policies are thousands of words long – several times the length of this chapter. The vast majority are difficult to understand, even for the most educated. According to research, Facebook's privacy policy – often said to be among the clearest – is almost as difficult to understand as Immanuel Kant's 1781 treatise *Critique of Pure Reason*, a book so impenetrably dense that philosophy students tremble at the very thought of it.[8] And *thousands* of digital contracts are thrust under our noses. A 2007 study concluded that it would take seventy-six days to read all the privacy policies a person encounters each year.[9] That period would be longer today. In fact, it is fair to say that it is impossible for an ordinary person to read even a minority of the terms and conditions laid down by tech companies while still leading an ordinary life.

There is another reason why no one reads the fine print. Those who bother only tend to discover terms that are so imprecise – 'we share your information with third parties' – that they offer no meaningful indication of what they are agreeing to.[10]

The idea of 'consent' is also unsuited to a world of ambient technology that works without screens or other interfaces that would allow a person to provide meaningful input.[11] Often we have no idea that data is being gathered about us. As Shoshana Zuboff has noted, a 'single home thermostat' would give rise to nearly a thousand contracts for the user to review if all the devices and apps associated with it came with proper terms of service.[12]

It is not merely 'notice' that is illusory. So is 'choice'. A choice between equally unpalatable options is no real choice at all. Nor is a choice to leave a platform when doing so would cause irreparable damage to one's interests. Technology companies usually hold (and deal) the cards. The idea of a negotiation between consumers and technology providers is farcical. Ordinary people are presented with pre-determined terms that are drafted to protect the companies themselves. Every major platform, for instance, has a clause stipulating that an individual's access may be terminated 'at any time for any or no reason'.[13] What are you going to do – instruct lawyers to open negotiations with the platform? Solemnly inform them that you won't sign up until the clause is removed?

When we first engage with a new digital product or service, the temptation to click 'I agree' is usually irresistible. Normally this is because we have already decided to use it. But it's also because businesses are incentivised to design interfaces that encourage us to agree to everything. Saying yes is made easy with big juicy buttons. Saying no is made unattractive with dull hidden options and privacy 'settings' of baffling complexity.[14] One employee of Google says that its Settings function feels like 'it is designed to make things possible, yet difficult enough that people won't figure it out'.[15] One medical app asks whether users would like to receive notifications – the alternative being 'no, I prefer to bleed to death'.[16]

In truth, informed consent is not only a myth. It is an impossibility. To have any real idea of what is to be done with our data we would need large amounts of granular information. But most of us can only make meaningful choices based on terms that are reasonably simple and easy to grasp. This is the 'transparency paradox'.[17] Too little information means no meaningful notice; too much information means no meaningful choice. The problem is not the technology. Nor is it the individual clicking 'yes' or 'I agree'. It is the legal mechanism – consent – used to mediate between the two.

***

Say you visit the dentist in the daytime and a bar in the evening. At the dental surgery, you pay in cash. At the bar, you pay by credit card. You probably don't know that people who spend money at bars are statistically more likely to default on their loans than those who spend money at the dentist (it's true).[18] Thus if you later 'consent' to your credit card data being used to calculate your credit score, you will be consenting to negative consequences that could never have occurred to you at the time.

The lesson is that meaningful consent would still be impossible even if every consumer were a highly trained lawyer with an insatiable lust for reading boring legal documents. It is simply not possible for people to know what their data will reveal once it has been aggregated with everyone else's.[19] Data that appears perfectly innocuous in one context might yield valuable insights in another.[20] That's often the point of using machine learning systems – they find patterns that humans cannot.

The consent mechanism assumes that consumers can hold in their heads the *cumulative* meaning of all the data they have surrendered in the past. Can you recall even a fraction of the data you have consented to give away in the last week, let alone in your lifetime? Of course not. So why pretend that you can consent to the consequences of each new parcel of data being analysed alongside the rest of the personal data that is already out there?[21]

Finally, it is worth remembering that the consent trap does not affect everyone equally. Certain groups can be less well equipped (in terms of time, education or other factors) to make certain legal decisions than others. Instead of protecting them, the consent paradigm exploits them.

What personal data about you is out there? Who holds it? What are they doing with it? Can you answer any of these questions? Of course not. No one can. The idea that 'consent' empowers the individual is nonsensical. It is hard to conceive of a system that would give ordinary people less power than they currently have.

The fundamental flaw with the consent model is that it grew out of a set of assumptions that rarely apply in the context of consumer technology, where there is no bargaining, no give-and-take, no barter or exchange. There is a hopeless imbalance in power between the parties. The knowledge, understanding and information available to consumers is incomplete. Yet as law students learn in their first semester, what matters is the objective appearance of agreement. Once you click 'I agree', the law does not care for your misgivings or misapprehensions. The deal is done.[22] This uncompromising doctrine makes sense in a hard-nosed negotiation between commercial counterparties, but not between individuals and the tech firms that increasingly dominate their lives.

Contracts, agreement and consent are not the best way of regulating the relationship between the strong and the weak. They take an existing relationship of domination, and instead of correcting it, they fortify it with the force of law.[23]

***

There is yet another, still more fundamental, reason to object to the consent paradigm. As we know, data is gathered at an individual level, but it is processed with the data of multitudes. One person's data is of almost no economic or political worth, but when combined with the data of millions of others, it becomes susceptible to useful analysis.

It follows that the data choices we make for ourselves do not just affect us. They affect other people too. John may not wish to have his location tracked and so disables the location-sharing apps on his smartphone. But if his wife Jane keeps her location-sharing apps on, it's rather easy to predict John's location based on Jane's – just as the New York Times was able to track the whereabouts of Donald Trump based on the location of his bodyguard.[24] Likewise, John may click 'no' every time he is asked to share his data, but that will count for little if Jim, Jack and James – people with similar attributes to John – hand over their personal data which can be used to build a highly effective model of people like John, and which can therefore be applied to John himself.[25] Data really is the new oil. Using it has spillover consequences that affect other people.

***

'A society in which men recognise no check upon their freedom,' wrote the renowned American judge Learned Hand, 'soon becomes a society where freedom is the possession only of a savage few.'[26] We live according to a myth that individual liberty is best achieved by people acting individually. But if you really believe in freedom of the individual, you don't leave individuals to fight for it alone.

Digital republicanism, as we will see, holds that individuals can only flourish in the right social conditions. In particular, ordinary people must be free from the predations of powerful others who can bend them to their will.[27] It's why the great republican thinker John Milton spoke of 'common liberty'.[28] But if freedom is social *and* individual, why treat our relationships with technology companies as something that should be managed by individuals on their own, rather than according to a shared set of values and safeguards? The irony is that even if every consumer wanted less overall intrusion from technology, the aggregation of their individual choices might still lead to collective serfdom. Being autonomous should not mean standing apart from everyone else. We lose little and gain much by working together. 'We must all hang together,' said Benjamin Franklin, 'or assuredly we shall all hang separately.'[29]

***

We have reached the end of this part of the book. It can be summarised briefly: the market alone will never rein in the power of digital technology, and often it will encourage undesirable uses of that power. So what is the correct republican posture towards free-market capitalism? Not outright rejection, for sure, or anything close to it. That would be to sacrifice the immense forces of innovation and creativity that arise from free enterprise, and to forego the freedom that lies in the availability of genuine choice between alternatives. It would also be to forget James Madison's lesson that the ownership and protection of private property is itself an important bulwark against the power of an overbearing state.[30]

Nevertheless, when it comes to technology, the balance between capitalism and democracy has swung too far to one side, partly because we have persisted in seeing technology as a matter of economic rather than political concern. In the long run, a system shaped principally by market forces will always prioritise consumer impulses over civic values, efficiency over justice, comfort over liberty, and entertainment over democracy. If the digital is political, then to leave technology to market forces is to invite the commercialisation of politics itself.

Silicon Valley's unqualified faith in the market is misplaced. Digital technology is not immune from the ordinary failures that would trouble any part of the economy. There is, in truth, no reason to assume that technology companies – or their advertisers or shareholders – will do the right thing with the power at their disposal (even if we could agree on what the 'right thing' was).

So, what's the alternative? It is to *govern* technology with suitable laws and institutions, drawing on the sophisticated legal and regulatory apparatus that is used in other industries. In part five, we look at tech governance as it currently stands and ask: is it fit for purpose?

PART V

THE GHOST OF GOVERNANCE

In questions of power, then, let no more be heard of
confidence in man, but bind him down from mischief by the
chains of the Constitution.
Thomas Jefferson

FIFTEEN

Making Our Own Laws

In the distant past, people believed that laws were discovered, not made. They saw themselves as playthings of unseen forces, governed by rules that existed independently of their own lives and that could only be found in prayer, contemplation or nature.[1] The remarkable idea that people could make their own laws came late in human history.[2]

Today, in democracies, it is taken for granted that laws are written by people rather than found in nature. Laws are just one part of *governance* – a term derived from the Greek *kybernan*, meaning 'to pilot, steer, or direct'.[3] In this book, the word *governance* is used to describe any systematic method of structuring social behaviour, from treaties and constitutions to legislation. Nowadays, governance often takes the form of *regulation*: setting standards, monitoring compliance and punishing misconduct.[4]

Before the modern era, many spheres of human activity were governed only loosely.[5] Most daily activities were effectively unregulated; much of the rest fell under the purview of morals, norms, byways, customs and traditions. That changed in the last few centuries. Governance now pervades society. There are governance regimes for the air we breathe, the water we drink, the food we eat, the roads we drive on, our systems of agriculture, industry, education and healthcare, our places of work and almost every other imaginable field of human activity. Technology is no exception.

To begin, it is worth recalling how governance has evolved over time. As the essayist Walter Bagehot put it in the nineteenth century, 'you must take the trouble to understand the plan of an old house before you can make a scheme for mending it.'[6]

***

The history of governance is not a story of smooth progress. It zigs and zags. That's because there has rarely been agreement on the role the state should play in the economic and social spheres. In the US before the Civil War (1861–65), the federal government was a remote presence in most Americans' lives, felt mostly through the postal system. By the end of the war, it was the nation's largest employer, with more than 50,000 employees.[7] The modern American regulatory state was born in 1887, when Congress established the Interstate Commerce Commission to oversee the railways.[8] Three years later, prompted by fear of the great industrial robber barons, the Sherman Act introduced federal supervision of antitrust abuses.[9] The Federal Reserve Board and Federal Trade Commission followed soon after.

This was an era of startling expansion for a small number of capitalist infrastructure providers: gas and electricity companies, telegraph, banking and railroad interests. One political grouping, known as the Progressives, questioned the legitimacy of these great concentrations of wealth and power.[10] If corporations wanted to take responsibility for important human needs, the Progressives argued, then they ought to be subject to proper checks and balances. A public utility is still a public utility, even if privately owned.[11]

A similar movement evolved in the UK in the late nineteenth century. Successive governments rolled out measures to alleviate the squalid conditions that had accompanied the industrial revolution.[12] By the time of World War I, Britain had an army of inspectorates – of 'factories, mines, explosives, prisons, police, reformatory and industrial schools, aliens, anatomy, animal welfare and inebriate retreats' – all staffed by highly trained officials.[13] The 1930s saw regulation of corners of the economy that had never previously felt the touch of the law.[14]

Anti-business sentiment intensified in the Depression era and was accompanied by calls for more state intervention. In America, the New Deal summoned forth a constellation of new federal agencies, from the Securities and Exchange Commission to the Federal Deposit Insurance Corporation and the National Labor Relations Board.[15] The years 1933–34 alone saw the creation of more than sixty new agencies endowed with the power to make and enforce rules.[16] This was a revolution in governance.

After World War II, a degree of consensus emerged among Western European intellectuals that the state should continue to play an energetic role in the peacetime economy, as it had done during the war. Governments assumed more and more responsibility for welfare, education and healthcare, most of which had previously been entrusted to private and voluntary institutions.[17] Key industries like coal, steel, water, telecommunications and the railways were brought into public ownership. Even in non-socialist countries, governments were happy to orchestrate economic investment, production and consumption.[18]

The European 'postwar consensus' never made it to the US, but Franklin Roosevelt's 1944 call for a Second Bill of Rights set in motion the 'Rights Revolution' that followed in the 1960s and 1970s. In this period, the locus of regulation expanded from economic to social concerns. Americans came to enjoy rights that far surpassed those named in the Constitution: rights to 'clean air and water; safe consumer products and workplaces; a social safety net including adequate food, medical care, and shelter; and freedom from public and private discrimination'.[19] In this era, national governance came to mean more than the mitigation of market failures. Laws and public schemes were seen as the way to pursue collective goals based on a shared conception of the good life.

The early 1970s marked the zenith of state activism on both sides of the Atlantic. Thereafter the state's involvement in the economy began to wane. As the 1970s progressed, European governments struggled to control inflation and unemployment despite the powers at their disposal. Policymakers lost confidence in the Keynesian orthodoxies that had dominated postwar economic thought.[20] The 1980s and 1990s saw public assets sold to business corporations to be run privately. Ministerial direction was replaced with light-touch oversight.[21] Social goals were reassigned to charities and voluntary institutions. Ronald Reagan gave this countermovement a credo: 'Government is not the solution to our problem, government is the problem.'

The last few decades have seen the emergence of another relatively stable consensus, this time over the 'regulatory state'. Nowadays, governments tend to plump for legal standards enforced by independent agencies rather than state ownership or central command. In normal times, the state is reluctant to involve itself in active economic management. It 'governs at a distance'.[22] In America there remains a deep suspicion of government intrusion and a presumption that the market should be the default mode of social ordering.

The most significant recent innovation in global governance has been the European Union. Unique in modern history, it is a legal order unto itself, distinct from the member states that comprise it. And when it comes to tech regulation, the EU is ahead of the pack. But as we will see in the next chapter, even the EU does not yet have the necessary tools to hold the tech industry to account.

SIXTEEN

The Mild West

It is often said that digital technology is unregulated and that the internet is a Wild West beyond the reach of government. This is a myth. Digital technology is, in fact, governed by a baroque entanglement of overlapping laws, regulations and norms. The question is not *whether* technology is governed at all, but *who* benefits from the governance regimes in place. Too often, the answer is the tech industry itself.[1]

Much of the technical side of internet governance – the nuts and bolts of protocols and standards – has historically been left to specialised bodies that most people have never heard of. ICANN, for instance (the Internet Corporation for Assigned Names and Numbers) oversees the distribution of internet domain names (.com, .net and others).[2] The IETF (Internet Engineering Task Force), W3C (World Wide Web Consortium) and IEEE (the Institute of Electrical and Electronics Engineers) are not-for-profit standard-setting agencies that have maintained the technical functioning of the internet for years. These bodies are remnants of an early ideal of internet governance: 'self-regulation by engineers through standards'.[3] The work they do is often quite esoteric. Their power is usually constrained by bureaucratic processes. By and large, they work well, although consensus may be harder to come by in the future, as China seeks a growing influence over global standard-setting.

On the commercial side, technology corporations are subject to the usual laws that apply from sector to sector – advertising and labour regulations, national security, intellectual property, anti-discrimination laws and so forth. For tech in particular, the law offers an alphabet soup of acronyms. In America, the FDA oversees medical devices, the SEC regulates 'robo-advisers', and the NHTSA governs autonomous vehicles. For credit reports there's the FCRA. For financial information there's the GLBA. Patient health data is governed by HIPAA and the use of information in schools is regulated by FERPA. For children's online safety there's COPPA, for data monitoring there's ECPA, and for hacking there's CFAA.[4] Then there's the Federal Trade Commission (FTC), which has the power to raise complaints against companies for 'unfair methods of competition' and 'unfair' or 'deceptive' practices in or affecting commerce.[5]

American technology companies are also subject to antitrust laws. There is, however, no single definitive antitrust authority in the US federal government. That responsibility is divided between the US Justice Department and the FTC, which have been known to take different sides in technology antitrust cases.[6]

As well as federal laws, tech firms are subject to state laws. Some add little to the federal menu but others are proactive. California, for instance, has introduced its own data protection scheme – a watered-down version of the European model – and laws requiring chatbots to disclose that they are not human.[7]

For the more discerning legal nerd, American law also offers a medley of venerable causes of action – 'intrusion upon seclusion', 'public disclosure of private facts', 'false light' and 'appropriation' – which have sometimes been held applicable in the technology context. These are known as torts. A tort is a legal wrong for which a victim can seek recompense. Trespass and defamation are examples. It is not to be confused with a torte, a type of German cake.

The sheer volume of laws and regulatory bodies shows that the tech industry is not ungoverned, even in the US. But it would be wrong to assume that tech firms are robustly held to account. The present regime arguably combines the worst of excessive complexity and overall inefficacy.

**

It is no accident that power has flowed into the hands of those who design and control digital technologies. This is partly because of the technologies themselves. But it is also because of laws which, instead of restraining the power of corporations, only give them more. Looking first at the US, they do this in three ways.

First, American law in this area is principally organised around the consent trap. We saw in chapter fourteen that consent is a largely useless and often harmful way of governing the relationship between individuals and technology companies. Yet consent is said to be 'the dominant approach to regulating the Internet',[8] the 'dominant regime of privacy regulation'[9] and the 'linchpin' of enforcement.[10]

Secondly, a few sectoral regimes aside, American companies are subject to few restraints on what they can do with personal data that has been lawfully gathered. This makes America an anomaly. It is one of the only advanced democracies in the world without omnibus data protection legislation.[11] American corporations have little to fear from the FTC, the main agency expected to enforce data rules. It has just fifty staff tasked with 'privacy' issues. (By contrast, the UK's data protection authority has more than 700 for a population around a fifth of the size.[12]) On average, the FTC has lodged just ten privacy complaints each year since 1997. Its largest ever penalty was a $5 billion fine against Facebook in 2019 – a significant sum, but still just a third of Facebook's revenue that quarter. When it was made public, Facebook's share price *rose*.[13] Although the FTC reached a settlement with Facebook under which it would improve some of its privacy practices, no leaders were sanctioned.

While elderly torts like 'intrusion upon seclusion' might provide occasional excitement for legal connoisseurs, the reality is that these causes of action are not widely known, seldom invoked, and largely ineffective at protecting individuals against machine learning methods.[14]

Finally, the US legal system shields many technology companies from legal risks to which they would otherwise be exposed. That's right: instead of *enhanced* levels of liability, powerful technology companies are protected from laws that apply to everyone else. If you have seen the film *The Wolf of Wall Street* (2013), you might recall the cocaine-snorting, Quaalude-popping, prostitute-hiring, alcohol-bingeing portrayal of the investment firm in the film. In real life in 1995, the firm's president (played by Jonah Hill in the film) sued an online message board for hosting derogatory messages, including one saying that the firm was a 'cult of brokers who either lie for a living or they get fired'. At trial, the Court rejected the host website's defence that it did not write the posts in question. Under the law of defamation, it was still a 'publisher' of the alleged libel and therefore legally responsible.

This judgment, along with a couple of others, led to concern in Washington DC. Holding websites liable for the unlawful posts of their users seemed short-sighted. It could encourage websites to remove *any* content that appeared legally problematic, which would stifle free speech and hold back growth in the digital economy. In response, Congress enacted Section 230 of the Communications Decency Act of 1996. Little noticed at the time, this law offered two bespoke protections for the tech industry that turned out to be revolutionary. The first made sure that intermediaries would almost never be liable for the speech of their users. The second ensured that they would almost never be liable for 'good faith' efforts to restrict access to 'obscene, lewd, lascivious, filthy, excessively violent, harassing, or otherwise objectionable' material – even if that material was 'constitutionally protected'. In short, the law protected anyone offering 'interactive computer services' both when they were *moderating* platforms and when they *chose not to moderate* platforms. The Courts have interpreted the legislation broadly.[15]

The impact of Section 230 in the last couple of decades is hard to exaggerate. It let platforms grow without the constant risk of being sued. It freed them from the toil of pre-screening content. It removed any legal incentive to censor excessively. It created space to experiment with moderation techniques. And it fostered an explosion in internet activity. In Silicon Valley, Section 230 became 'a kind of sacred cow – an untouchable protection of near-constitutional status'.[16] Certainly the internet, and the world, would have been very different without it.

Times have changed since 1996, however, and Section 230 no longer enjoys universal approbation. Indeed, it is in the firing line.[17] When Congress passed the law, the websites in its sights were personal homepages, community message boards, small commercial and retail websites and the like. Back then, only around 40 million people had access to the internet.[18] Now there are around 25–30 billion devices connected to the internet – four for every person on the planet.[19] Facebook alone counts nearly 3 billion people among its members, roughly two-fifths of the world's population.[20] The issue now is not that platforms are too vulnerable, but that they are too powerful. Laws granting them extra protection seem archaic. Congress is waking up to this fact.

In truth, there have always been principled reasons for discomfort about Section 230. When platforms profit from amplifying the posts of bad actors, it is hard to see them as blameless.[21] And it seems strange that they can profit from terrorist groups using their sites,[22] that victims of 'revenge porn' cannot take legal action against them to take down footage,[23] or that they can facilitate the sale of illegal firearms with impunity.[24] But those have been some of the consequences of Section 230.

The most common criticism is that instead of incentivising platforms to protect their users, Section 230 does the opposite. It prevents victims of revenge porn, libel, trolling, harassment and the like from taking action against them.[25] It is, as Rebecca Tushnet puts it, 'power without responsibility'.[26] Or as the republican would put it: unaccountable power.

***

In general, the American inclination has been to seek to minimise the liability of digital technology companies, whether for data use and abuse, anti-competitive behaviour or algorithmic injustice. This blind-eye approach has been part of a broader strategy to spread US soft power abroad through the ubiquity of American technology. American companies dominating the international markets is good for government as well as industry.

The position in Europe is different. While America places almost no limits on how lawfully acquired data can be used, in the EU the processing of *any* personal data is subject to harmonised rules.[27] In 2016, the EU introduced the General Data Protection Regulation (GDPR), a many-headed beast of a law that took a decade to assemble. The GDPR regime is serious about enforcement too, at least in theory. It allows supervisory authorities to impose fines of up to 4 per cent of a company's annual worldwide turnover.[28]

The democracies of Europe have also taken a different approach to platform immunity, as have Canada and Australia. In most of Europe, platforms enjoy some immunity conferred by law, but this immunity is conditional rather than absolute. If someone posts something unlawful and a platform has knowledge or notice of it, they must expeditiously respond to requests by users or governments to take it down.[29] The German *Netzwerkdurchsetzungsgesetz* law (try saying that after six flagons of German pilsner) requires platforms with more than 2 million users to remove 'obviously illegal' hate speech posts within twenty-four hours or face a fine of 50 million euros.

The EU has also been faster to stand up to the power of large technology corporations, including with antitrust actions that routinely involve billion-euro fines.[30] But while there is much to be praised in the European approach, there are real shortcomings too. The GDPR, for instance, is only properly understood by a small cadre of nerds. It is expensive to operationalise and enforce.[31] Authorities in Ireland and Luxembourg, where many of the largest technology companies are based, have been painfully slow to act.[32] Ninety-eight per cent of major GDPR cases referred to Irish authorities remain unresolved.[33] Like in the US, cash-strapped regulators often lack the means to enforce the rules properly.[34] The evidence suggests that most bad deeds go unpunished.[35]

Most seriously of all, even the European regime adopts a version of the consent trap that suffers from the same defects as its American counterpart.[36] Europeans sign away their rights every day with countless 'I agrees'. Consent operates as an ejector seat that deprives people of protections they would otherwise enjoy. For instance, the GDPR forbids companies from using data in a manner that is incompatible with the purposes for which it was collected.[37] But this safeguard dissolves if the individual gives consent.[38] Likewise, the GDPR gives people the right not to be subject to decisions that significantly affect them 'based solely on automated processing'[39] – again, unless the individual clicks 'I agree'.[40] No matter how tough the European definition of consent[41] – and it is tougher than the American one – the truth is that individuals still click away too many of their protections.

SEVENTEEN

Private Order

The last chapter showed that the tech industry, particularly in America, has benefitted from a legal environment that entrenches its power. Sometimes that is through an absence of regulation, as in the lack of federal data protection laws; other times through regulation that benefits the powerful, like the consent trap; and occasionally through shields that give the industry special protection, like Section 230.

The tech industry's umbilical reliance on the law runs even deeper, however. It would never have reached its current size or sophistication without markets to propel its growth. But commercial markets are not spontaneous products of nature. They do not occur in the wild. They thrive only in countries that provide intricate legal rules to sustain them. Take something as basic as the law of property. 'A law can't be right if it's 50 years old, like it's before the internet,' said Larry Page, the co-founder of Google.[1] But where would Google or any tech company be without intellectual property laws that safeguard their inventions and allow them to hide their algorithms from public view?[2] It may seem natural to Mr Page that people can own and exchange items, including intangible things like ideas, interests and debts. But there was a time when no one could say 'this is mine' and expect anyone else to respect their claim. As Thomas Jefferson wrote in 1813, 'stable ownership' is 'the gift of social law' – and one that is given 'late in the progress of society'.[3]

Or take the business corporation, the mighty building block of the tech economy. This is also a legal construct. Corporations are fictional persons created by law. The idea that a nonhuman body could count as a 'person' first arose in Roman times. In the medieval era, Pope Innocent IV accepted that a corporate body could be a *persona ficta* – a fabricated person.[4] In the early modern era, corporations were permitted only for public purposes like building roads, maintaining waterways or trading with foreign lands. As the jurist William Blackstone put it in the eighteenth century, they were 'created and devised by human laws for the purposes of society and government'.[5] Before they were colonies, Massachusetts, Virginia and other states were companies chartered by the English state. They began not as private enterprises but as 'bodies politic'.[6] It was only in the nineteenth century that governments in America and Western Europe allowed corporations to be started without government approval. The law granted them an indefinite lifespan and let their officers and agents be replaced over generations. Then the law gave them more powers – the power to own assets, the power to sign contracts, the power to sue others.[7] The law allowed them to skip between legal systems without having to move.[8] The law made them answerable only to their shareholders, and not the general public.[9] The law shielded shareholders from the liabilities of the companies they owned (whether for wages, debts or taxes) and protected bankrupt companies from the claims of shareholders and creditors.[10] Today, the law continues to endow corporations with rights that had previously been seen as the preserve of flesh-and-blood persons.[11] Recently, the US Supreme Court held that the 'speech' of corporations enjoyed the protection of the First Amendment.[12]

There is nothing natural about the power of modern business corporations. They accrue power because the law allows them to. They *exist* because the law allows them to. To say that corporations are, can or should be free from state intervention is an absurdity: they are themselves creations of the state and they benefit handsomely from the present legal arrangements.

The same goes for the humble contract, the prime instrument of modern commerce. Contracts took us from a world in which transactions were governed by strength and honour alone. Technology corporations create contracts in their thousands, and when they do so they are utterly reliant upon the protection of the courts.[13] Without detailed legal rules about what a contract is, how it should be interpreted and what happens when it is breached – and a legal system competent to enforce those rules – it would be impossible to structure business relationships with the complexity that modern commerce demands.

***

The main way we currently govern digital technology, particularly in America but also elsewhere, is to create social conditions in which it can be ordered and organised through private arrangements.[14] It's not that there is *no* law, it's just that the law tends to let businesses decide for themselves what the rules should be. It is a 'private law-making system'.[15] This can be an effective way to organise some aspects of social life, but where there is a significant imbalance between the power of the regulated group and those whom the law seeks to protect, it lets the powerful ride roughshod over the rest. We see this all around us. Our lives are increasingly governed not only by the rules embedded in technologies themselves (part two), but by the boilerplate in the contracts we signed to use them.[16] Millions, even billions, of people live under legal regimes devised entirely by technology companies, in which they have pretty much no say at all. We are visitors in the 'legal universe' they have created.[17]

This makes republicans uncomfortable because it allows private corporations to lay down their own law and enforce it in many important areas of life. We have seen some of the threats to democracy, liberty and social justice. Where are the procedural safeguards? Where is the democratic oversight? Where is the public supervision? What options are there for independent appeal and review?

American lawmakers can't exactly be faulted for failing to circumscribe the power of technology companies. Until recently at least, that was never their aim. The current governance regime was designed to encourage technology companies to develop their own norms of conduct. But the cost has been an almost total surrender of control by ordinary people.

***

Digital technology is not 'unregulated'. It is governed by laws, including the laws that constitute the market. This simple point matters more than you might think.

Even those who profess to be sympathetic to the need for better governance portray the current arrangements as the natural order of things, unsullied by the hand of government. They reduce the debate to a choice between 'government intervention' and 'no government intervention', or 'regulation' versus 'deregulation'. But the real question has always been different. *What kind of government intervention is best?*[18] The current system is not the moral or intellectual default. Nor is it the product of chance. It is a human-made creation, and it can be changed however we choose. Why not a system that limits the unaccountable power of the tech industry, rather than enabling it?

'If I could wave a magic wand,' wrote Ira Magaziner, a senior aide to President Clinton in an influential 1999 article, 'I would say we should go through a complete deregulation' of the internet, and 'let the market go'.[19] By now you should see the problem with this kind of talk. It falsely equates 'deregulation' with 'letting the market go'. But the author was *not* arguing that technology should be freed from governance altogether. Property law, corporate law, contract law – he didn't want to deregulate *those*. No, by 'deregulation' he meant keeping laws that marketised and empowered the technology industry, and stripping away those that did the opposite. His vision has so far prevailed. Why? Because the entire system we have been discussing – combining powerful technologies, market norms and laws geared towards private ordering – rests on a bedrock of outmoded ideas; a bedrock of market individualism that must be eased aside if we are to progress.

PART VI

FOUNDATIONS OF THE DIGITAL REPUBLIC

Thus, it is an easy matter for anyone who examines past
events carefully to foresee future events in every republic
and to apply the remedies that the ancients employed, or if
old remedies cannot be found, to think of new ones based
upon the similarity of circumstances.
Niccolò Machiavelli

EIGHTEEN

Four Principles

We have reached the end of the diagnostic stage of this book. Let's pause to recap the argument so far. There are five main points.

The first is that digital technologies exert power. They contain rules that the rest of us have to follow. They condition our behaviour, often without us realising. They frame our perception of the world, determining which information reaches us and in what form. They subject us to near-constant scrutiny. They set the rules of public deliberation, deciding what the rules are, when they are enforced and when they don't apply.

Secondly, technology is not neutral, objective or apolitical. Digital systems are soaked through with biases and prejudices.

Third, under the current system, digital technology is ordered primarily according to the logic of the market economy. This brings benefits, like economic efficiency. But it also has drawbacks. Instead of reining in the might of corporations, the market empowers them. Instead of curtailing the worst instincts of those in the technology industry, the market encourages them. Instead of fostering innovation in the public interest, the market rewards the pursuit of private gain. Instead of empowering citizens, the market strips them of individual and collective agency.

Fourth, there is nothing natural or inevitable about the current system. It is, in large part, the product of a legal regime that prioritises private ordering over public safeguarding. The law gives companies, particularly in America, too much freedom to do whatever they want with our personal data. It cements the consent trap as the dominant mode of regulation. It makes technology companies immune from legal action without asking for anything in return.

Fifth, both the development and regulation of digital technology have been conditioned by the ideology of market individualism, which holds that social progress is primarily the result of individuals pursuing their own interests.

To these five propositions we may now add a sixth: that the current system is unsustainable because it allows for too much unaccountable power. As citizens, we ought to live under rules of our own making, according to norms and values that we have chosen for ourselves or had the opportunity to influence. There is no good reason why private corporations should be able to wield great political power simply because they design and control digital technologies. It is unsatisfactory that there are no procedural safeguards of the kind that are the cornerstone of legitimate authority in other aspects of life.

In short – and this is the essence of the republican critique – our shared future is increasingly at the mercy of those who design and control digital technologies. And this means we are unfree.

The rest of this book asks: *what can we do about it?* The answer I propose is *digital republicanism*, a system of ideas for governing technology that would mark a departure from the last few decades. We need new ideas, new democratic processes, new rights and standards, new ways to contest digital power, new certification systems and regulatory powers, new duties of openness and new laws for governing social media.

These policies are not randomly generated, however. They are derived from a set of principles that form the backbone of digital republicanism. I call them the *preservation principle*, the *domination principle*, the *democracy principle* and the *parsimony principle*. They are shorthand reminders of *why* we should govern technology and *how* we should do it. They are the subject of this chapter.

***

The most basic goal of digital republicanism is the survival of the democratic state itself. In short, the law must enable people to live together peacefully in a free and stable political system. It must enshrine the rights, duties and institutions needed to govern ourselves in a democratic fashion. There must be freedom to ponder and mull, to object and criticise, to gather and protest and to choose the people, principles and policies by which we are ruled. This is the *preservation principle*.

While the preservation principle is obviously important, it is also the least likely to crop up in day-to-day governance of technology. Despite the hype, most technologies will never pose an existential threat to the survival of democracy or the legal order on which civilization rests. But it is not fanciful to think that some future technology, not yet invented, could erode the integrity of the democratic process. The first priority of digital republicanism is to make sure this doesn't happen.

***

The second principle of digital republicanism is the *domination principle*. This holds that we should reduce the unaccountable power of digital technology and keep it to a minimum. No one should need to rely on the benevolence or wisdom of private corporations in matters that affect their basic liberties or the health of the political system.

The aim of the domination principle is not to eliminate the power of technology. That would be impossible. It is, instead, to *expose* that power, to make it *answerable*, to *disperse* it, to *restrain* it and to make sure it cannot be used by one group to dominate others.[1]

***

The third principle is that the design and deployment of powerful technologies should, as far as possible, reflect the moral and civic values of the people who live under their power. This is the *democracy principle*. It takes some explanation.

While most technologies will never threaten the existence of the republic, some will give rise to new harms that the existing law is unable to address. Take a simple example. Most countries introduced laws long ago to protect people from harassment. These laws were geared towards physical stalking or threatening. But as the internet age unfolded, these laws were increasingly unable to tackle online harassment, which differs from offline harassment. No one had decided *harassment is OK now*; on the contrary, the law still sought to prohibit it. But technology made new forms of harassment possible, and the old law became somewhat obsolete.[2] And the democratic will of the voters who had sought to outlaw harassment was undermined. The law needed to be updated.

Consider a second example. In many jurisdictions, the law prohibits insurers from asking applicants certain questions about their private lives. This policy reflects a democratic choice that disadvantaged people should not be forced to pay higher premiums because of certain issues in their private lives. But in the future, prying insurers won't need to *ask* for that kind of information. They may be able to buy or gather data which is far richer in detail, if not about *us*, then about people *like* us. Our online purchase histories may reveal our underlying health conditions. Our search queries may disclose our lifestyle habits. Unless the law changes, its purposes will be undermined. Perhaps it will be necessary to prevent insurers from buying or using certain types of data at all.[3]

For a third example, consider the trade-off that most legal systems strike between the right of workers to unionise and the desire of businesses to operate without encumbrance. Most countries have found a balance between the two, with rules about what constitutes a lawful union, what kinds of balloting must precede industrial action and so forth. This balance is often controversial, but it also usually bears the legitimacy of the democratic process. Technology now threatens to overturn the careful balances that have been struck. Large companies can use dozens of data metrics to identify union 'hot spots', keeping tabs on efforts to organise.[4] Computer systems rank employees by their likely attitude to unionisation, using data about them gathered in a different context.[5] Facebook has even considered a chat platform in which the very word 'unionise' is blacklisted, such that it cannot even be uttered by staff.[6] These developments are indirectly quite undemocratic. If a community has assembled a delicate legal regime to govern a social activity – in this case, industrial relations – then it should not be shattered by digital technology without careful democratic scrutiny.

The democracy principle recognises that social change is inevitable and often desirable, but holds that it should be driven by democratic politics, as well as by technology.

The democracy principle also requires careful attention to how laws are *enforced*, not just what they say. Professional sport provides a useful analogy. The rules of football were devised long ago. They include simple prohibitions against players handling the ball, and more complex rules like the notorious 'offside' rule. For generations, referees enforced these rules as best they could, but with a natural degree of error and laxity. Occasionally they missed or overlooked small transgressions. Nowadays the rules are the same but technology is increasingly used to enforce them. Match officials use precise recording systems that can examine play in microscopic detail. Suddenly, minor technical breaches that would never previously have been noticed are penalised with ruthless efficiency. To critics, the game feels unnatural, disjointed and sometimes unjust. The freedom of the sport is lost.[7]

The lesson applies to society more generally. In 2018, I wrote about the 'hinterland of naughtiness' enjoyed by people in a truly free society:

Have you ever grabbed a shopping bag at the supermarket without paying for it? Or paid someone cash-in-hand knowing they won't declare it for tax? Perhaps you've streamed an episode of *Game of Thrones* without paying, dodged a bus fare, nabbed an illicit refill from the soda dispenser, 'tasted' one too many grapes at the fruit stand, lied about your child's age to get a better deal or paid less tax on a takeaway meal then eaten it in the restaurant. These are all illegal acts. But according to a poll, as many as 74 per cent of British people confess to having done them. That's hardly a surprise, and not because the British are scoundrels. People aren't perfect. It would be absurd to suggest that the world would be better if every one of these indiscretions were routinely detected and punished.

All civilized legal systems offer a slender stretch of leeway in which people are sometimes able to get away with breaking the law without being punished. A surprising amount of liberty lives in this space. Its existence is a pragmatic concession to the fact that most of us can be trusted to obey the law most of the time.

My concern when I wrote these words was that technology was beginning to shrink this precious hinterland of naughtiness. These days, digital rights management (DRM) technology makes it very hard for ordinary punters to stream content online without paying for it. Soon the bus fare will be automatically deducted from your smart wallet when you board. The soda dispenser will recognise your face and refuse additional service. The 'smart' fruit stand will screech when its load is lightened by theft. Your child's age will be verified by biometric scans. In many spheres of life, enforcement regimes are being quietly tightened. But any system of law, no matter how enlightened, will become oppressive if it is enforced much more stringently than was anticipated at the time it was created.

Confusingly, some technologies make enforcement harder rather than easier. Anti-discrimination law offers a good example. Most democracies have laws to prevent discrimination in the hiring and firing of workers. Those laws can be enforced in courts and tribunals. But the procedures – rules of discovery and evidence, the burden of proof and so forth – were designed with human discriminators in mind. But where *nonhuman* systems take or inform decisions, then it might be very difficult for people to show they have been the victims of discrimination. It is not merely that machines discriminate differently from humans (chapter eight), but also that their internal processes are often quite opaque.[8] Again, it is unsatisfactory for a democracy to have settled on a principle – that workers should not be disadvantaged because of certain characteristics – only to see that principle undermined by a technology that makes it harder to enforce. The answer is to reform the enforcement mechanisms.[9]

The democracy principle envisages a world in which the most powerful digital systems are shaped by the norms and values of the communities into which they are born. This might mean specifying what technologies may not do or be used for. But it could also mean engineering social values and goals into the technologies themselves. Imagine platforms engineered to improve democratic deliberation rather than eroding it, or algorithms that reduce injustice rather than amplifying it.

***

The final principle is a restraint on the other three principles. It is that a republican system of governance should place firm limits on the power of the state, and give no more power to the state than is absolutely necessary to perform its regulatory functions. This is the *parsimony principle*.[10]

Friedrich Nietzsche called the state 'the coldest of all cold monsters'.[11] In all the furore about the tech industry, it is easy to lose sight of the fact that states have been supercharged by the power of digital technology. They badger intermediaries to block content and remove websites.[12] They wheedle and hack their way into private communication systems and databases.[13] They use algorithms to sort, order and surveil the populace. The Russian government recently demanded that Apple and Google remove a tactical voting app, made by supporters of the opposition leader Alexei Navalny, from their online stores. The companies did what they were told.[14]

The iron fist of government is often hidden inside the velvet glove of 'enforcer-firms' doing the state's work.[15] The doorbell camera Ring, owned by Amazon, has video-sharing partnerships with more than 400 law-enforcement agencies in the US. Clearview AI provides images of billions of faces to 'train' police facial-recognition systems.[16] States get tech companies to enforce the law online.[17] The Five Eyes surveillance regime (the US, the UK, Canada, Australia and New Zealand) gathers information from corporations and 'launders' it through other governments, to circumvent laws prohibiting them from spying on their own citizens.[18] The Snowden documents revealed not only that the US National Security Agency had infiltrated 'every conceivable form of computer and communications system around the globe', but that they had also been successfully demanding information from tech giants that collected data about individuals abroad.[19] And while the US constitution might prohibit mass government surveillance, American law has little to say about what the government can do with data it has acquired from corporations.[20]

A system of governance that merely replaced the dominium of technology corporations with the imperium of the supercharged state would be actively harmful. As Madison put it, 'You must first enable the government to controul the governed; and in the next place, oblige it to controul itself.'[21] In practice this means, firstly, that any action by the state should be strictly proportionate to its purpose. Then, where possible, the work of governance should be done at arm's length from the central state. And finally, the law must place firm restrictions on what the state itself may do with digital technology.

***

The essence of digital republicanism is simple. Digital technology should not be able to threaten the foundations of democratic order, or serve as an instrument of domination, or reconfigure society in ways that fail to respect the interests or values of the citizenry, or undermine the conditions needed for democratic self-rule. Some old laws will need to be tweaked and others will need to be minted, both in a way that keeps the state's power to a minimum.

With the theory under our belt, we can finally turn our minds to practical measures. We begin with ideas for making technology more *democratic*.

NINETEEN

Technology and Democracy

In a 2009 blog post, Mark Zuckerberg announced that Facebook was to become more democratic. He wanted the platform's terms of service to reflect 'the principles and values' of the people who used it. A new voting process would allow users to veto policies they didn't like.[1] It was to be a grand experiment in digital democracy.

In 2012, the experiment was put to the test. Facebook unveiled new terms and proposed them to its users. Eighty-eight per cent of voters opposed them. But their votes counted for nothing.[2] Facebook had set the minimum threshold for a veto at 30 per cent of *all users*, which, in 2012, would have required 300 million votes.[3] That's just slightly less than the entire population of the US, and nearly four times as many people as voted for Joe Biden in 2020 (who received more votes than any other presidential candidate in history). In the event, less than a million took part. Facebook later jettisoned the new voting procedure. And in a whitewash that would have made Stalin proud, the byline of the 2009 blog post was quietly altered from Mark Zuckerberg's name to that of another employee.[4]

It is easy to laugh at this abortive attempt at DIY democracy, but, as Tarleton Gillespie has observed, there may be a lesson lurking in it.[5] To be a digital republican is to hold that digital technologies should be accountable to the people over whom they exert power, and shaped according to those people's values. Democratic processes of one kind or another will be vital in making this happen. Forget Facebook's ham-fisted effort. Drawing on thousands of years of political theory and the best empirical research, we can find better ways to democratise the digital. That's the subject of this chapter and the next.

***

What is democracy's role in the digital republic – conceptually and practically?

Well, democracy has at least three basic things in its favour. First, *liberty*: we are only truly free when we live under rules of our own making. That applies (at least in principle) to the rules embedded in powerful technologies as it does to the law of the land. Second, *equality*: everyone subject to intrusive power should have an equal opportunity to make their interests and preferences heard. Again, that principle applies to digital power as it does to legal and political authority. Third, *efficacy*: democratic deliberation is a good way to reach certain types of moral and political decisions. Superior, in any event, to relying on a narrow elite.

More democracy is not always better, however. There is no need to bring the tech industry into public ownership or hold elections for the CEOs of Facebook and Google. Innovators must be free to come up with ideas without public officials watching their every move. Commercial enterprises operating in benign political conditions are still our best vehicle for producing awesome things that people don't know they want until they see them. (That said, as Mariana Mazzucato has showed, capitalist innovation does not itself take place in a vacuum. It is often the product of heavy state investment and involvement.[6])

There is always a balance to be struck between the moral primacy of democracy and the productive power of capitalism. This balancing exercise is not new. Since the nineteenth century, the term 'social democracy' has been used to describe a system of redistributive taxes, workplace rights and anti-discrimination laws – each symbolising the imposition of democratic values over commercial ones.[7] But while the balancing exercise may be old, the context is new. Other products and services in a capitalist economy may occasionally pose risk of harm, but few have the capacity to shape and guide the course of human affairs.

So, when considering the role of the democratic state in governing capitalist innovation, there are two schools of thought. One is that the state should step in only to curtail the worst excesses of the market. As we have seen, this is the approach currently adopted by many governments in the developed world. It has not produced satisfactory results, and will only become more inadequate as time goes on. A bolder approach holds that we should sometimes go further than making second-order corrections to the market. Certain forms of innovation should be guided not only by profit, but by shared social objectives and collective agreement about the good life.[8] This approach would require that those affected by powerful technologies – the people – should have their voices heard in the boardrooms and workspaces of Silicon Valley, either directly or through the voice of the law.[9]

Take the now-clichéd 'trolley problem' as applied to self-driving cars:[10]

> You are motoring down the freeway in a self-driving car, and a small child steps into the path of your vehicle. If you had control of the car, you would swerve to avoid the child. You know that this would cause a collision with the truck in the adjacent lane, probably killing both you and the trucker – but to preserve the life of the child, that's a sacrifice you are willing to make. Your car, however, has different ideas. Whether by intentional design or through 'learning' from others, it considers that saving two

lives is better than saving one. And so you have no choice: you are forced to plough into the child, killing her instantly.

If automated vehicles become commonplace, this kind of moral dilemma will present itself countless times a day. What should the cars do? They might be programmed to do what other humans have done in similar situations in the past. Or to maximise the number of human lives saved in any incident. Or to treat the life of a child as being worth twice that of an adult. Or never to change course, even to save more lives. Reasonable people will disagree on the right approach, but one thing seems clear: this choice (and choices like it) probably shouldn't be left to engineers and business executives at Tesla or Ford. Under the right conditions, big moral decisions with no obvious answer should be decided by the people, directly or indirectly. We, after all, are the ones who live with the consequences.[11] That's not to say, however, that individuals should necessarily be able to choose settings for themselves. In a 'tragic choice' situation, should a feminist be able to choose a setting that prioritises the life of a woman over that of a man? Or a racist setting that prioritises white over black?[12] No – as Antje von Ungern-Sternberg points out, the equal value of every human life is a 'central tenet of modern societies'. Personal choice cannot trump that tenet.[13]

As at least one author has observed, the task is not as simple as identifying a 'true or correct moral theory' and implementing it in machines. No such moral theory exists – at least not one that would command the agreement of everyone. Rather, people have different beliefs about moral issues, and what we need is a mechanism of balancing, reconciling and choosing between those beliefs.[14] That is the business of politics. And the best political mechanism we have is democracy.

***

That's the theory. Now for the practice. How can citizens be involved, or at least represented, in the shaping of powerful technologies?

The elected legislature is the obvious starting point. People elect politicians to make laws on their behalf. The laws should, at some level, reflect the popular will. Therefore legislation should plainly be used to govern powerful digital technologies. But it also has limits. The law-making process is cumbersome and opaque; plagued by grandstanding, filibustering, logrolling and pork-barrelling. Time is usually short, with different causes clamouring for priority on the agenda. Trust in politicians is low. While the ordinary sausage-making process of legislation will play its part in the digital republic, it will need to be updated or supplemented with other forms of decision-making.

One modest way to improve legislative performance would be to introduce, revive or expand the scope of bodies like the US Office of Technology Assessment (OTA), which was inexplicably shuttered in 1995. As Bruce Bimber writes, the OTA was 'a public think tank, joined institutionally to the policy process but vested with no political authority of its own'.[15] Created by Congress in 1972, it was staffed by scientists and experts, not political hacks. It acted as a nonpartisan advisor to members of the US Congress.[16] Instead of urging one course or another, it laid out options for elective representatives to consider.[17] At least one presidential primary candidate in 2020 argued for the reintroduction of the OTA, and rightly so: the legislature will never rival the technical firepower of the tech industry, but it should at least have its own dedicated experts to inform policy.[18]

An alternative is for the legislature to palm responsibility off to government ministries, agencies, bureaus and administrative bodies, which offer qualities that are often lacking in legislatures – like expertise, adaptability and a capacity to plan for the long term.[19] Officials are selected on aptitude and expertise, not popularity or patronage. Policy is developed away from the public gaze.[20]

In each democracy, it would be useful to have at least one body tasked with thinking systematically about the future of tech. Such a body – informed by expert opinion, the best available research and public opinion – would try to anticipate problems coming round the corner and what might be done to forestall them.[21]

Of course, trying to predict the future is a fool's game, but not nearly so foolish as sitting around waiting for it to happen and hoping for the best. Bruce Schneier imagines a National Cyber Office modelled on the Director of National Intelligence, with other organelles like a National Artificial Intelligence Centre, a National Robotics Centre and a National Algorithms Centre under its umbrella.[22] The charity Doteveryone argues for an Office for Responsible Technology with a remit for anticipatory governance – looking ahead and trying to 'future-proof' the regulatory system.[23] These are sensible proposals.

The amount of independence granted to unelected officials is a matter of national taste. In the UK, they are generally expected to put 'flesh on the bones' of decisions made by Parliament, not to pursue designs of their own.[24] In the US, by contrast, many agencies are operationally independent from both Congress and the rest of the federal government. They are given broad powers to do as they please.[25]

Either way, while they have many qualities, unelected organs of the state don't offer much democratic legitimacy.[26] Indeed, their value lies in the very fact that they are insulated from the hurly-burly of politics. So, while a digital republic will certainly have need for sophisticated civil servants, their role must ultimately be subordinated to other more democratic approaches, particularly where there are value judgments to be made. Civic involvement cannot be hived off to experts.

What about referendums? Direct democracy has its place (just ask the Swiss) and plebiscites have already been used to decide matters affecting the tech industry. In California in 2020, voters chose to exempt certain gig economy firms from stricter laws governing the use of flexible workers – a big win for Uber and Lyft. In the future, technology could make direct voting much easier. We could do it using handsets from the comfort of our homes, ten times a day.[27] In reality, however, referendums on their own are a narrow form of the democratic art. A binary vote does little to facilitate fruitful deliberation, encourage consensus or develop civic virtue.

***

At first glance, then, the machinery of modern democracy looks necessary but not sufficient for the task of democratising the digital. We need new ways for citizens to have a say. The next chapter introduces the concept of *deliberative mini-publics*, new micro-institutions that might supplement our existing ones.

TWENTY

Deliberative Mini-Publics

'The single hardest thing for a practising politician to understand,' writes Tony Blair, 'is that most people, most of the time, don't give politics a first thought all day long.' He's right. A lot of folks don't find politics very stimulating. Even political nerds like me (and you?) rarely have time to think deeply about political issues. We spend our days worrying about other things: 'the kids, the parents, the mortgage, the boss . . . friends . . . weight . . . health'.[1] But do you ever wonder whether we could be more useful citizens if we were given the chance? Whether, in the right conditions, we could make better decisions for ourselves and each other?[2] In a *deliberative democracy*, people have a proper chance to weigh competing arguments on a given political issue before reaching a decision. They are able to find reliable information. They have time and space to ask questions, to ponder and mull, to listen and speak in small and large groups. They have access to experts who can inform without imposing their own views.

And their decisions are taken seriously. Crucially, they sometimes change their minds and those of others.[3]

***

If all our opinions were fixed and immutable, there would be no point in reading the news or participating in public debate. It would be impossible to resolve disagreements or find common ground. Citizens would simply cast their votes according to pre-formed views and politics would mean no more than counting the ballots.

The premise of deliberative democracy is that people are, in fact, capable of changing their minds,[4] and that in the right conditions they can listen to others, weigh competing arguments, examine evidence and reach conclusions in a spirit of mutual respect. This is very republican. It is more than a procedure for voting. It is a system of public reason in which citizens test their arguments in open debate. The constitutional design of the American republic reflects the belief that a political system should be able to accommodate even the fiercest differences between people. The Framers did not fear disagreement; they welcomed it as a 'creative and productive force' that would 'encourage the emergence of general truths'.[5]

In a well-ordered republic, how political decisions are taken matters as much as which decisions are taken.[6] If citizens know that they have a meaningful say in the rules that govern their lives and a fair chance to convince others of their point of view, then when they lose the argument they can grit their teeth and respect the outcome. Freedom means being able to challenge power – not just at the ballot box, but in the realm of ideas too.

Even in ancient polities, there was never one single site for public deliberation. Citizens would argue in homes, taverns, markets and forums. In the last few centuries, deliberation typically took place in parliaments and coffee-shops, at conventions, rallies and debates, in books, pamphlets, journals and newspapers and on radio and television.

Of course, it's important not to romanticise the history of deliberation. Political argument has always been riddled with falsehoods, irrationality, myopia and prejudice. Raucous disagreement is a fact of life and incivility is sometimes the price of change. But we are currently in a period of transition. Deliberation is migrating from the analogue world to the digital. This presents us with the chance to mould a system of public deliberation that, if not perfect, at least encourages our better instincts rather than our worst. That has not happened so far, in part because so few of us have any meaningful say in the rules that govern online discourse. That power is reserved to the business corporations that design and control the platforms on which the rest of us rely.

***

Deliberative democracy differs sharply from polls and surveys in which participants offer their own opinions without discussing them with anyone else. It also differs from public consultations in which respondents file independent submissions that pass like ships in the night. It is unlike election campaigns and referendums, in which the whole polity clamours to debate at once. And it is nothing like viral internet discourse, in which popularity alone determines prominence.

Real deliberation is, in truth, only possible among relatively small groups of people. In ancient Athens, laws were decided by an Assembly of all the citizens. But before an idea could even be put to the Assembly, it had to be approved by a council of 500 randomly selected delegates. Then the Assembly's own decisions were submitted to a special court, again made up of 500 citizens selected by lot, with the power to prosecute 'illegal or unwise' proposals. And decrees of the Assembly would only become law if they received the approval of legislative commissions, again comprised of ordinary citizens chosen at random.[7] Instead of mobilising the entire citizenry at every turn, the Athenians relied on sortition – 'microcosms' or 'mini-publics' – to make decisions on behalf of the people at large. Their decisions were binding.[8]

Mini-publics are an underrated device in the democratic toolbox. They will have an important part to play in the digital republic. They are not meant to replace elections and referendums or usurp the primacy of the legislature. Their purpose would be to supplement existing institutions and offer a rich new layer of legitimacy to tech policymaking.

***

Thousands of deliberative mini-publics have now been conducted across the world, and a body of research is emerging about their different forms and functions.[9]

There are *citizens' assemblies*, in which a hundred or so people meet regularly over the course of a year in order to generate detailed recommendations on big constitutional questions. Citizens' assemblies have been used in Canada for more than a decade now. The Irish Citizens' Assembly (2016 to 2018) was empanelled to address several intractable issues, including same-sex marriage and abortion. Based on its recommendations, the Irish government called a referendum that led to an overhaul of Ireland's abortion laws.[10]

Citizens' assemblies tend to follow a particular procedure. Once assembly members have been chosen at random (or for representativeness), they enter a 'learning phase' in which they study the policy issues and are presented with different perspectives by experts and affected groups. They then have the chance to pose questions and request further information. Next there is a 'consultation phase' in which input is sought from the public at large. Thereafter there is a 'deliberation phase' in which assembly members discuss the evidence, weigh the arguments and develop recommendations. These sessions, led by trained facilitators, are conducted according to rules that allow everyone to be heard. Only then is there is a vote – and while the majority generally gets its way, there are also minority reports for the views that do not attain consensus.[11]

Citizen juries or citizen panels comprised of thirty or so people offer a compact alternative to citizens assemblies. They meet every day for a few weeks to come up with recommendations on specific problems. Again, they go through phases of learning, consultation, deliberation and decision-making. Citizen juries have been used to decide the location of a UK hospital and a ten-year budget for the city of Melbourne.[12]

So-called *consensus conferences* involve an even smaller panel (usually sixteen or so) meeting over the course of two weekends. The first is devoted to learning, and the second to an expert panel and deliberation. The conference's recommendations are then discussed by experts and politicians. This model has been used more than fifty times around the world.[13] *Citizens' dialogues* (also known as *citizens' summits* and *deliberative workshops*) are two-day processes in which citizens are asked for feedback on particular proposals, rather than asked to make recommendations themselves.

There is no definitive form of deliberative mini-public. They can be designed according to the problem they are asked to address. They can be big, like the Dutch G1000 summits, with nearly a thousand participants, or small, like the *citizens' councils* of fifteen used in Austria.[14] They can be permanent standing bodies, like the Ostbelgien Citizens' Council, or ad hoc panels convened to address a specific issue.[15] They can last for days or months. Their decisions can be binding, trigger a parliamentary debate or referendum or be treated as recommendations. The key point is that they give people the chance to have a say in matters that affect their lives.

The available evidence suggests that, in the right conditions, ordinary citizens will deliberate competently and civilly. In the carefully constructed conditions of a mini-public, it is hard to form echo chambers or huddle into groups of shared ethnic or ideological identity.[16] With balanced information, expert testimony and careful facilitation, people grow less extreme, less wedded to the dictates of identity, less reliant on partisan framing and more inclined to change their minds.[17] Research suggests that deliberative processes also reduce susceptibility to fake news, because participants cannot rely on intuition alone.[18] What's more, people place a high degree of trust in their own capacity to make decisions.[19]

***

Deliberative mini-publics are not just for improving the quality of big decisions or breaking political deadlocks. They have the weight of legitimacy. It is easier to live with difficult choices when we know they have come from our fellow citizens, thinking hard under ideal conditions, not from a political or industrial elite.

This is well known to the citizens of Taiwan, whose vTaiwan process allows them to pose and discuss a variety of difficult policy questions. It starts with a crowdsourcing stage for gathering facts and evidence.[20] There is then a 'reflective' stage in which participants use a special social media platform called Polis, developed in Seattle, to create a visual 'attitudes map' of views and feelings. This map clusters participants together according to areas of agreement. It promotes content that wins broad support. This means that to get attention, contributors must appeal to opponents, not just supporters. 'People compete to bring up the most nuanced statements that can win most people across,' Taiwan's Digital Minister has said. 'Invariably, within three weeks or four, we always find a shape where most people agree on most of the statements . . . People spend far more time discovering their commonalities rather than going down a rabbit hole on a particular issue.'[21] The purpose is to identify 'consensus statements' – broad statements that all sides are largely comfortable with. These form the primary basis for policy.[22] The vTaiwan process has led to many new laws and regulations. Famously, it was used to resolve a bitter dispute between those who wanted to ban Uber and those who welcomed it. The process ended with compromise: Uber could operate, but only with licensed drivers.[23]

Taiwan offers a glimpse of how deliberative mini-publics might be used in the digital republic. But each republic should choose its own path. Research suggests that deliberative mini-publics are particularly effective in addressing three types of question: 'value-driven dilemmas' (moral questions to which there is no self-evident answer), 'complex problems that require trade-offs' and 'long-term issues that go beyond the short-term incentives of electoral cycles'.[24] Fortuitously, these are precisely the kinds of problems thrown up by the power of digital technology. In the UK, the Ada Lovelace Institute has already successfully empanelled a fifty-person Citizens' Biometrics Council to deliberate on the use of biometric technology. After sixty hours of deliberation, the Council came up with a number of impressive conclusions.[25] In Canada, too, the Canadian Commission on Democratic Expression is harnessing deliberative methods to come up with recommendations for reducing digital harms.[26]

***

The digital republic would be well served by a range of new deliberative bodies. A permanent citizens' assembly, reconstituted by sortition every year, could be tasked with answering large principled questions. *What principles should guide social media moderation? Should data processing be taxed?* In each case, the assembly could draft parameters, guidelines and best practices.

More specific questions could be delegated to citizens' juries or consensus committees. What data should be forbidden for use by employers, or insurers? What liability should attach to the manufacturers of self-driving vehicles which make fatal errors? During the 'learning' phase, participants would be given the results of audits, impact statements, useful data and the like, accompanied by expert guidance. If policy in a particular area is being led by government, citizens' dialogues could be used to scrutinise any plans.

Sometimes, like the Greek *dikastēria*, deliberative mini-publics could be deployed as actual juries – to deliver verdicts on particular systems, features or functions. Should this acquisition be permitted? Did this search engine fail to meet the legal standards applicable to it? Does this face-recognition system discriminate against people of colour?

Instead of constituting mini-publics at the political centre, the law might require large firms to incorporate deliberative mini-publics into their own design processes, overseen by an independent third party. Where possible, mini-publics should be arranged and facilitated by accredited private-sector bodies, themselves subject to oversight.[27]

***

There is no need to pretend that the design and implementation of deliberative mini-publics is straightforward. They cannot be treated as carbuncles on the ship of state. They must be vested with appropriate authority and dignity. They must be properly resourced and appropriately housed. They must be truly representative of the public at large, and their workings must be suitably transparent.[28] As the OECD recommends, their tasks must be clearly defined, and their decisions implemented in a timely manner.[29] Deliberative mini-publics will not always function perfectly. But that is not the test. The test is whether they might be superior to what we have now: corporate elites taking important political decisions behind closed doors.

Some in the tech industry might bridle at the idea of a deliberative process that introduces a degree of friction to the process of innovation. Friction, after all, is often seen as something to be smoothed over and engineered away. But friction is partly the point of deliberation. The challenges of technology are too difficult and subtle to be treated as design kinks to be ignored or plastered over. Introducing deliberation doesn't add to the challenges; it merely recognises the complexity of the issues that inevitably arise and offers a way to resolve them.[30]

It's possible a deliberative mini-public might seem a little burdensome for those called up to serve. But in time it could be recognised as just another civic duty, like jury service. We should not be ashamed to ask a little more of each other. Aristotle regarded a citizen as one 'who shares in judging and in office'.[31] Cicero wrote that 'when it comes to preserving the people's freedom, no one is just a private citizen'.[32] Madison called it 'a chimerical idea' that 'any form of government will secure liberty or happiness without any virtue in the people'. By comparison to the republics of antiquity, the digital republic will not ask much of us. The price of liberty is not eternal vigilance, but occasional involvement. It's not too much to ask.

***

Once the people are given a say in the development of digital technology, what should they do with that power? That's really what the rest of the book is about – new regimes for governing data, algorithms, social media and more. But the next two chapters are less concerned with the *substance* of new laws than with describing the available legal *mechanisms*. The two explored here are *rights* and *standards*, both of which can reduce the unaccountable power of technology.

TWENTY-ONE

Republican Rights

Imagine that Grace is unhappy with something that has been done by a digital system. Perhaps her mortgage application was rejected by an algorithm. Or a facial-recognition security system keeps locking her out of her building because it was trained on white faces and can't 'see' her darker skin tone.[1] In an ideal world, Grace would be able to do something about her problems. From the mortgage lender, she might want an explanation or a fresh decision. From the security company, she might want an apology or compensation.

To pursue these grievances in a court of law, however, Grace would have to show that a legal rule had been broken. Put another way, the law must have recognised some kind of duty on the system providers, and given Grace the right to sue them if they failed to comply with that duty. The word *duty* matters. It demarcates between wrongs that are legally actionable and those that are merely upsetting. If Grace wants to be able to challenge the power of digital technology, she needs *enforceable legal rights*.

For as long as there has been civil law, there have been rights. There is no magic to them. They are the product of the general law. A political community can choose to introduce whatever rights it likes, from the right to free speech to a right to a free burger every Wednesday. The community can also set the rules by which new rights may be generated, and old ones terminated. The law of contract, for instance, allows parties to create new rights between them by mutual agreement.

Advocates of tech reform sometimes declare the need for new 'bills of rights' without enough thought about what that actually means in legal terms. Rights come in a variety of forms, some more useful than others, and it's important to think about them with as much precision as possible. This chapter sketches out some of the rights currently known to the law, and the functions they might serve in the digital republic.

***

Let's start with *human rights*. Human rights are often invoked as a generically good thing, like apple pie or David Attenborough. Although such laws have featured in Western discourse for centuries, the modern practice of human rights law can be traced back to the end of World War II, when the United Nations adopted the Universal Declaration of Human Rights.[2] This Declaration listed the rights that we are said to enjoy simply by virtue of being human, as opposed to being the citizen of a particular state, or a particular class, caste, colour or gender. Scholars disagree about the ultimate source of this entitlement — whether it comes from God, or from nature, or whether human rights are simply artifices of human civilisation, like other legal constructs.

Many human rights — privacy, freedom of expression, non-discrimination, the right to a fair trial — can be threatened by digital technology.[3] In 2020, for instance, the District Court of the Hague ordered the Dutch government to halt a benefit fraud-detection algorithm because it undermined poorer people's human rights to privacy and social security.[4]

But human rights have limits. The main one is that they are usually enforceable only against public authorities, rather than corporations or people.[5] This makes them a clunky tool: if your human rights are violated by a company, your only recourse may be to sue the government for failing to stop the violation, not the company itself.

***

Contractual rights are different. They are born out of agreements.[6] As we saw in chapter fourteen, contracts (or the cousin concept of consent) are often the main source of rights between individuals and tech firms. However, we've also seen that consenting to digital power is problematic. Corporations dictate the terms. Few of us read them. Even fewer understand them. When we click 'I agree' we are usually surrendering, not agreeing.

There are, however, some ways that contractual rights, and the idea of consent, could be strengthened for the future. In European law, for instance, 'consent' already requires a 'freely given, specific, informed and unambiguous indication' of a person's wishes by a 'clear affirmative action'.[7] Clearer policies, digestible summaries, standardised language, graphics and explainers could all make consent more meaningful. The US Congress has looked at banning 'dark patterns' that seek to manipulate users into consenting away their rights.[8] And scholars have suggested lurid privacy warnings like those on cigarette packets that show images of diseased lungs.[9] (Although it is hard to imagine what these would look like. The subtle and gradual compression of human freedom is not easy to express in a single image. Indeed, it can be hard to do it justice in a full-length book.)

The law can also specify the terms of certain contracts, rather than leaving them to be dictated by corporations. Thus, certain terms could be banned (like catch-all data permissions), and others might be made mandatory (like clauses that commit platforms to a certified complaints procedure, or enable class actions). So-called 'smart contracts' could also make it easier for citizens to vindicate their rights. Say, for instance, the law gave social media users the right to have complaints of abuse or harassment dealt with within twelve hours. Smart contracts would make platforms automatically pay one-dollar micropenalties to users every five minutes after the twelve-hour period elapsed.[10]

Contractual rights, then, are not useless. But they do have limitations that will never be fully overcome. Other rights will be needed too.

***

Doctors owe their patients a duty to perform their work with reasonable care and skill. Motorists owe other drivers a duty not to drive recklessly. Neighbours owe each other a duty not to unleash noxious fumes or play loud music at 3 a.m. These duties do not come from contractual agreements, or human rights. They come from the law of tort, which allows people to seek compensation for wrongs done to them by others.[11] Over time, society has developed a range of torts. Negligence, nuisance, trespass and defamation are the most famous.

In the digital republic, the law of tort could be used to place a higher burden on technology companies than they assume under contracts. One scholar proposes that individuals harmed by the repeated or wilful use of faulty data should be able to claim for extra damages.[12] Another argues that automated vehicles should be required to be as safe as a reasonable human driver – and that manufacturers should be insulated from liability if they meet that standard, but liable for damage if they fall below.[13] Tort law could impose a duty of reasonable care on those who use machine learning systems for important social purposes, like mortgage or recruitment decisions. With such a duty in place, individuals would be able to claim compensation for harm caused by the negligent use of sketchy data, faulty algorithms or dodgy hardware.

***

Even sterner duties arise out of fiduciary law. The term may be foreign to the non-lawyer's ear, but fiduciary relationships are all around us. Parents owe fiduciary duties to their children. Directors owe fiduciary duties to their corporations. Bankers owe fiduciary duties to their clients.[14] The term 'fiduciary' comes from the Latin fiducia, meaning trust or confidence.[15] Fiduciary relationships tend to arise where one party has significantly more power than the other, or where one party has assumed a special duty towards another by virtue of their relationship. [16]

Fiduciary law is strict. It requires those in a position of trust to put others' interests ahead of their own. Some of the sharpest minds in the field of digital governance see fiduciary law as a promising way to rein in the tech industry.[17] Making tech companies into 'information fiduciaries' would allow individuals to sue them for collecting, using, selling or disclosing personal data in ways that promoted the corporation's own interests over theirs. Harvard professor Jonathan Zittrain gives the example of the traffic navigation app Waze. Imagine that Waze plots a route to your destination that takes you past a Burger King. It does this not because it is the best route for you, but because Waze gets commission from Burger King. Waze would be using your data to put its interests ahead of yours. That would be a breach of its fiduciary obligations.[18]

The fiduciary principle need not be limited to data. Firms whose machines could physically harm consumers – household robots, for instance – could be deemed fiduciaries of those with whom their devices interact, increasing their liability under the general law.[19]

***

The types of right sketched in this chapter – human, contractual, tortious, fiduciary – are the ones we have inherited from the past. They have proven to be durable and adaptable, even if we have still not made the most of them in the context of digital technology. But there is nothing to stop us from generating entirely new categories of right, as well as adapting old ones. We can create just about any right through legislation and regulation. In parts nine and ten we will look at some more. Why not, for instance, hold large tech firms directly responsible for human rights abuses? Or give people a statutory right to algorithmic decisions that are technically sound and morally coherent? In the next chapter, we look at an alternative to rights – *standards* enforced not just by individuals, but by the state itself.

TWENTY-TWO

Republic of Standards

Think of some of the problems commonly associated with social media, like the polarisation of political culture or the coarsening of public discourse. Now imagine that a judge asked you to put a monetary figure on the harm you have personally suffered because of them. I would struggle to do so – not because the harms are trivial, but because they principally affect the community as a whole, rather than the individuals within it. Sometimes technology can damage the social fabric without violating anyone's rights, at least not seriously enough to bring a legal claim.[1]

As well as being *felt* collectively, some digital harms are *caused* collectively. In the future, a thousand technologies could combine to create an atmosphere of stifling unfreedom. But no single company would be culpable.[2] Individuals might complain of being deprived of their right to freedom, but there would be no one to sue.

Now consider that some harms can only be *countered* collectively. If a factory dumps toxic waste into a river, local residents should be compensated. But there should also be a public inquiry, official guidance to stop it happening again, and a new oversight regime. None of these remedies would usually be granted by a court to an individual litigant seeking to vindicate her rights.[3] They need government action.

We saw in the last chapter that different kinds of rights could give individuals the chance to challenge wrongful exercises of digital power. But there is an inherent limit to what rights can do – because some harms are *caused* collectively, *felt* collectively and can only be *countered* collectively. After a point, introducing more rights makes little difference. Mature legal regimes recognise this by providing for *collective* enforcement of certain rules and standards. We aren't expected to check the bus engine before we get on board, inspect the café kitchen before we order or test the chemical properties of tap water before we drink. We expect these matters to be taken care of by public authorities. Where digital harms resemble 'public health' risks,[4] spread among the community, they demand a social response as well as an individual one. We need *standards*.

***

According to lore, the term 'standards' first entered common usage at a battle between English and Scottish armies nearly a thousand years ago. As the fight raged, blood and facial hair everywhere, the English rallied around 'the mast of a ship' to which they gave the name 'the Standard'. The clash became known as the Battle of the Standard.[5]

Nowadays, standards are rather more mundane, used to 'prescribe the behaviour or characteristics of people or inanimate objects'.[6] Many standards are voluntary and non-binding – more like guidelines than laws. But some are binding and there are consequences for breaching them. I refer to these as *legal standards*.

In the UK, the British Standards Institute (BSI) is appointed by the government under a Royal Charter. It publishes thousands of standards every year: environmental standards, construction standards, health and safety standards, security standards and the like. The BSI does not work alone. It develops its standards in tandem with industry experts, trade associations and businesses.

There are international bodies too, working to harmonise standards across borders. The International Organization for Standardization (ISO) is a sort of intergalactic federation of 165 national standards bodies. It has more than 200 technical committees,[7] including one with the catchy name ISO/IEC JTC 1/SC 42, working on 'Standardization in the area of Artificial Intelligence'. Industry associations like the Institute of Electrical and Electronics Engineers (IEEE) are also working on standards for data privacy, transparency and algorithmic bias.

In theory, legal standards should combine democratic priorities and technocratic knowledge. The people set the broad parameters; experts translate them into workable standards. As we have seen, the legal standards governing important and sensitive areas of digital technology are not currently fit for purpose, to the extent that they exist at all. That will have to change.

***

Choosing how to structure a standards regime is a surprisingly delicate art. Imagine a scheme to reduce the visibility of dangerous anti-vaccination content on social media platforms. One form of standard could concern only the desired outcome, not how it is achieved:

> 99 per cent of false or misleading content regarding the efficacy or safety of vaccinations must be removed within one hour of posting.

Or it could be phrased more generally:

> Platforms should have reasonable systems in place to reduce the visibility of false or misleading content regarding the efficacy or safety of vaccinations.

In both cases, the platforms themselves would be left to work out how to comply.[8] An alternative would be to specify the processes to be used, using strict rules that leave no room for discretion:

> Platforms must employ at least 1 certified moderator for every 5,000 users.

> Moderators must be certified according to Standard X-150 concerning the competencies of anti-vaccination moderation.

> When moderating anti-vaccination content, moderators must follow the steps in Checklist Y-620.

There is vigorous (and, you will be shocked to learn, not always scintillating) scholarly debate about the optimal format for standards regimes. One study compared the regulation of nursing homes in Australia and the US. The Australian regime used broad performance standards, including a requirement to provide a 'home-like environment'. American providers, by contrast, had to follow extremely detailed rules for every aspect of the care they provided. The American regime ultimately performed worse because it deprived carers of creativity and empathy, leading to a box-ticking culture.[9] The main lesson from the literature, however, is that there is no general rule. Sometimes a box-ticking approach will be called for, particularly for products that could hurt humans. Other times, it will be best to let firms themselves find ways of reaching desired outcomes. These are pragmatic issues, not ideological ones.

Contrary to what is often assumed, legal standards benefit businesses. They provide a basic way to show they are trustworthy. They create a level and durable playing field, so responsible enterprises aren't undermined by cheaper but more dangerous rivals. And they create what economists call 'network externalities': products and services increase in value when they are similar or complementary to an accepted standard.[10]

***

A well-ordered republic will offer an appropriate menu of new legal rights and standards. But these would be useless unless they can be properly enforced. We need new institutions to hold digital power to account, to challenge and expose it in the name of freedom. That's the subject of the next part.

PART VII

COUNTERPOWER

As a general rule, every time we see everyone quiet and
peaceful in a state which calls itself a republic, we can be
sure that liberty is not present.
**Charles Louis Secondat, Baron de la Brède
et de Montesquieu**

TWENTY-THREE

Tech Tribunals

This part of the book is about *counterpower*.[1] Similar terms have been used by other authors to describe different things. I use the term here to describe the ability to challenge the exercise of power, usually through legal channels.[2] Investigations, inquiries, reviews, audits, appeals, complaints and litigation are all forms of counterpower.[3]

Why is counterpower so important? Well, in the future, tech firms are going to make decisions that significantly affect our lives. Sometimes those decisions will be problematic. People will be inexplicably booted off social media platforms. Products for sale on online marketplaces will disappear for no reason. Applicants will be denied credit because of erroneous data. Errors will creep in from faulty algorithms, unrepresentative datasets and human incompetence. Mistakes are inevitable; the question is what we do about them. One option is to give people legally enforceable rights to challenge those mistakes. Another is to impose legal standards on the tech industry. That was the last two chapters. The next step is to make sure that rights are actually vindicated, and legal standards enforced. That's what we turn to now. That's counterpower, and it requires new institutions, beginning with a new way to enforce legal rights.

***

The stereotypical image of a court of law is a grand building in the neoclassical style. Pinstriped lawyers march up and down marbled floors. The walls are lined with rows of leathery tomes. Becloaked judges thrash their gavels on mahogany plinths.[4]

This is not a million miles from the reality.

In my work as a courtroom advocate, I wear a robe and horsehair wig when I rise to address the judge. I pass my days in gothic courtrooms, surrounded by litigants on splintering wooden benches which look and feel – and sometimes smell – like nineteenth-century torture devices. But the justice system is not always this grand. One much-loved tribunal in southeast England, where I have represented many clients, is situated inside a shopping mall, overlooking a car park. The meaty fumes from a nearby burger outlet do not distract from the justice that is served there. What matters is the fair and consistent application of the law, not the architecture (or ventilation system) of the courthouse.

At present, however, there are serious problems with the way we resolve legal disputes. Litigation can take months or even years, with backlogs of tens or even hundreds of millions of cases in some countries.[5] Bringing a claim is expensive. Court fees, lawyers' fees and hours of preparatory work can make the whole enterprise a waste of time and money. More of the world's people 'have access to the internet than access to justice'.[6] Even in countries that offer legal aid, the courts are more hospitable for the wealthy and leisured than the rest.

There must be a better way to resolve small disputes between individuals and tech firms, with manageable costs and navigable procedures. Ask yourself: what do we really want when challenging a digital decision?[7] We want to know why something has been done to us. We want a chance to say why we think it was wrong. We want our plea to be considered by someone independent and trustworthy. And if that person agreed with us, we would want them to do something about it.[8] If they disagreed, we would want to know why, so we can understand and move on. This isn't much to ask. It should be achievable at scale without being too intimidating, expensive or time-consuming. And it doesn't require neoclassical architecture or gothic furnishing.

A natural first step would be to require technology companies to put in place internal review processes that are fit for purpose. Many of the complaints processes presently offered by tech firms are Byzantine in design and Kafkaesque in operation. Complaints take ages to be resolved, and the results are often baffling, inconsistent and illogical. Encouragingly, the EU's draft Digital Services Act would require platforms that take down content, or suspend or terminate an account, to provide 'access to an effective internal complaint-handling system' that enables 'complaints to be lodged electronically and free of charge'. These systems would have to be 'easy to access' and 'user-friendly'.[9]

Internal appeal processes, however, are not enough for certain kinds of decisions. There are now many algorithms – in employment, education, credit, healthcare and criminal justice, to name a few – that are of serious social importance. They share some traits. First, they are likely to have a significant impact on the lives they touch. Second, they make decisions that are moral or political in character, in that they necessarily prioritise some values over others.[10] Third, they tend to be used in scenarios where the decision-maker is in a position of relative power to the person who is the subject of the decision. I call them *high-stakes algorithms*. (The EU's draft Artificial Intelligence Act adopts a slightly different definition of 'high-risk' algorithms, but along similar lines to these.) For high-stakes algorithms, there must be a channel for appeal that is independent of the original decision-maker. 'All would be lost', wrote Montesquieu, if the same people made the laws and judged 'the disputes' arising out of them.[11] The same goes for digital power. Those who hear appeals against decisions should ideally be independent from those who made them.

Facebook has offered an intriguing answer to the problem of independence by introducing an Oversight Board to review some of its content decisions. The Board has attractive features. It is composed of distinguished public figures. It has the power to instruct Facebook to allow or remove content, or uphold or reverse an enforcement action.[12] It appears to be operationally independent of Facebook, even if the company retains some say in its composition.

The Board is set up to function as a private judiciary. Its remit is not to make rules or apply the general law, but to judge whether decisions made by Facebook are consistent with its own 'content policy and values', albeit with 'particular attention' to 'human rights norms protecting free expression'.[13] The Board's Charter establishes a doctrine of precedent: 'For each decision, any prior board decisions will have precedential value and should be viewed as highly persuasive when the facts, applicable policies, or other factors are substantially similar.'[14]

Many commentators are understandably cynical about the Board, particularly as Facebook has not earned a reputation for acting in good faith when it comes to its own content moderation. Critics say that the Board is just a PR safety valve, allowing Facebook to offload its most controversial decisions without taking the heat. There may be some truth in this. And it doesn't help that the Board only hears a small number of cases and can take months to decide each one. As for how the Board will function and the quality of its decisions, it is still early days, though it is noteworthy that in four of its first five cases the Board overturned Facebook's decisions.[15]

To fixate on the minutiae of the Oversight Board, however, may be to miss the bigger picture. We do need new arbitral bodies that are independent of both the general public *and* technology corporations. They need to be highly specialised and able to earn the trust of all parties. In this respect, the Oversight Board looks like progress. The idea that people have the right to challenge Facebook's decisions before an independent tribunal is very republican. But what is very unrepublican is the fact that people only have that right because Facebook granted it to them. In reality, the company could just as well have chosen not to introduce the Board. And no other tech giant has followed suit.

A digital republic must offer its citizens a fair and speedy forum in which to resolve disputes concerning high-stakes algorithmic systems. Such a system would need to be cheap and easy to use. It would need to be committed to rigorous standards of integrity and probity. And it would have to be regarded as part of the apparatus of the state, an essential public service for a free citizenry, not the gift of any corporation or CEO. This is the vision for *tech tribunals*: specialist independent tribunals which would enable citizens to contest high-stakes algorithmic decisions.

***

Expert independent tribunals are not a new idea. Nearly a century ago, industrialised countries began to introduce them for disputes relating to employment, welfare benefits, immigration and tax. These tribunals were specialised and less formal than normal courts. They offered a sensible way to govern new and complex areas of modern life. So why not tech tribunals, staffed by independent public servants, to resolve disputes arising out of relationships between citizens and tech firms?

Tech tribunals should be able to operate entirely online. As Richard Susskind has argued for years, court is a service, not a place. If a dispute can be resolved without everyone having to congregate in the same building at the same time, then we ought to make it happen. An unglamorous but instructive example is the UK's Traffic Penalty Appeal Tribunal (TPT), a tribunal that hears appeals against traffic enforcement penalties. Every year, just thirty adjudicators decide around 25,000 cases.[16] The TPT has recently introduced a 'Fast Online Appeals Management' procedure that brings the parties together online:[17]

Accessible via any internet-enabled device, including smartphones and tablets, appeals can be submitted and processed entirely online, while evidence – photos, videos, PDFs – can be uploaded and revised as the case progresses. The parties can communicate in various ways, including messaging and live chat. The adjudicators can communicate with the parties at any time, seeking clarification and providing updates. Once evidence and arguments have been submitted, the appellants can choose between having an 'e-decision', which involves a decision by the adjudicator without any kind of hearing; or they can ask for a telephone conference hearing with the adjudicator and a representative from the other side.

Some 95 per cent of those who have used this procedure would be 'very likely' or 'highly likely' to recommend it to others.[18] (Though a better strategy might be to stop driving through red lights.)

Online dispute resolution, once a novelty, is now commonplace. British Columbia boasts a Civil Resolution Tribunal that resolves small claims (under CAN$5,000) and car accident and injury claims up to CAN$50,000. It involves a series of simple stages, starting with a tool that helps users understand their legal position. Then there are stages to encourage negotiation and settlement, followed finally by an online adjudication process.[19]

Purists may balk at the idea of legal or quasi-legal disputes being resolved outside bricks-and-mortar courtrooms, but such objections are behind the times. If the Covid-19 crisis taught us anything, it is that much more can be done online than we expected.[20] Most disputes are not big or complicated. And online resolution systems can deal with them, freeing courts to focus on issues that can't be handled online.

We should consider the development of private-sector adjudication bodies, strictly regulated but competing to offer the best platform for hearing small disputes. The online marketplace eBay already resolves tens of millions of internal disputes online every year – more than the US justice system – and should serve as inspiration.[21] Again, the EU is ahead of the curve. Its draft

Digital Services Act anticipates the emergence of out-of-court dispute settlement bodies that are 'impartial and independent', possess 'the necessary [legal and technical] expertise', are 'easily accessible' using technology, and capable of settling disputes in 'a swift, efficient and cost-effective manner' and 'in accordance with clear and fair rules of procedure'.[22]

Just as courts already prescribe different procedures for different kinds of disputes, tech tribunals could offer different streams for different types of case. Some would prioritise quick-and-dirty resolution, like a simple complaint that a platform has been too slow to respond to a notice of illegal content, while others would require disclosure of evidence and careful deliberation, like a complaint that a credit algorithm appears to have made an unjustifiable decision. Some contexts would require a light standard of review (*was this a decision which no reasonable firm could have reached?*) while more serious complaints would prompt heavier scrutiny.

Tech tribunals would not replace the ordinary judicial system. Complex cases, and appeals against decisions of tech tribunals themselves, would still be held by conventional courts. But most cases are not hard cases. Most could be disposed of justly without too much cost or fanfare.

***

Those who have suffered an adverse decision from a high-stakes algorithm should have a forum in which to vindicate their rights. That forum need not be a traditional courtroom. Indeed, it *cannot* be a traditional courtroom – not with all the cost, time and worry. If citizens choose not to avail themselves of tech tribunals when they are wronged, that's fine. What matters is that they could if they wanted to. For the republican, freedom lies in the ability to challenge the decisions of even the mightiest powers.

TWENTY-FOUR

Collective Enforcement

In 1935 and 1970, Congress passed laws giving safety protections to American workers.[1] But these laws did not give workers the right to sue their employers when the safeguards were breached. Instead, they gave public officials the power to investigate and punish wrongdoing on behalf of the workforce.[2] By contrast, the laws that protect Americans against workplace discrimination *can* be enforced by individual workers. And they almost always are. Some 98 per cent of discrimination actions are brought by individuals.[3]

A scheme of governance for digital technology will rely partly on individual or class-action, and partly on collective enforcement by the state. Imagine, for instance, a law that made platforms give their users reasonable warning before expelling them for misconduct. On an individual enforcement approach, users would be given a right (probably contractual) not to be deplatformed without reasonable warning. If that right was breached, they could complain to a tech tribunal. On a collective enforcement approach, a regulator would investigate whether platforms gave their users reasonable warnings, either in individual cases or taken in the round, depending on what the standard required. If platforms fell below the standard, the regulator would issue sanctions.

Of course, it is perfectly possible to have individual and collective enforcement at the same time. In the EU, the GDPR provides for both. A digital republic should offer mechanisms for personal redress *and* collective enforcement, depending on the context. On the one hand, individual rights take pressure off cash-strapped regulators. Letting people stand up for themselves limits the risk of state overreach. And it can be a real deterrent. A thousand citizen lawsuits can be just as effective as one enforcement action by a central regulator.[4] (The Office of the Attorney General of California has stated that it will prosecute no more than three cases each year under the California Consumer Privacy Act, calling for an expansion of private rights of action that it termed a 'critical adjunct to governmental enforcement'.[5])

On the other hand, regulators can penalise wrongdoing without the expense of going to court. In Australia, for instance, the eSafety Commissioner can demand that social media platforms take down abusive content and revenge porn, and issue fines for noncompliance.[6] Public regulators can also make strategic choices about their priorities. Unlike individual litigants, who can only argue the facts of their particular case, regulators can act with the public interest in mind. Regulators also have flexibility when it comes to how they enforce legal standards. There are a number of approaches, ranging from deterrence (using penalties, prosecutions and prohibitions to discourage violations[7]) to compliance (working patiently and collaboratively with businesses before cracking the whip).[8]

Even if tech tribunals lowered the cost of individual enforcement, people shouldn't always have to fork out their own money to vindicate their rights. And that's assuming their rights have been violated to begin with. As we have seen, where harms are felt collectively, it may be that no individual has sufficient standing to bring a claim.[9] And where harms are caused collectively, individuals may have no one to sue. In these cases, only a regulator will be able to enforce the law.

In recent years, American lawmakers have favoured individual over collective enforcement. This is reflected in the fact that more than 97 per cent of lawsuits to enforce federal statutes in the decade to 2010 were brought by private claimants.[10] This makes the US unusual among advanced democracies, which generally seek more of a balance between the two.[11] Some put the American approach down to the litigiousness of the American people, the assertiveness of the American judiciary or the aggression of the American bar.[12] These factors play their part. But the American fetish for individual enforcement also reflects a darker truth about American politics. One reason why lawmakers would rather leave citizens to enforce the law is their fear that regulatory bodies might fall into the hands of their political opponents. This fear has grown over time. Between the 1960s and 2000, the rate of private lawsuits enforcing federal statutes increased by around 800 per cent.[13]

I'm calling for a mix of individual and collective enforcement. A digital republic needs citizens ready to stand on their rights *and* public authorities ready to protect the republic as a whole. Though there is no general rule, the former will tend to be best for contesting particular *decisions* that have affected individuals *personally* in a substantial way,[14] and the latter for governing *systems* which harm *society*.

***

Most breaches of the law can be resolved using ordinary civil law procedures. But some wrongs are so serious that society calls them 'criminal' and punishes them with sanctions that can include the loss of liberty.[15] Because of its severity, the criminal law tends to have more safeguards, including a higher standard of proof – *beyond reasonable doubt* rather than *on the balance of probabilities* – and, sometimes, the right to a jury trial.

What should count as a crime? There is no definitive list of criteria. A crime is an act that the community identifies as a public harm so serious that it requires particular censure. Societies differ in where they draw the line. To the republican, an unpunished crime is offensive because it represents a pure form of unaccountable power. It suggests that the strong can predate on the weak at their pleasure.[16] Part of the public outrage after the financial crash of 2008-09 lay in the lack of criminal sanction for the bankers who had wrecked the lives of untold millions. The tech industry should be different. Extreme, dangerous, morally odious and systemic failures should give rise to criminal responsibility. In Australia, for instance, there are criminal penalties for content and hosting services that fail expeditiously to remove 'abhorrent violent material', with jail sentences for executives of up to three years.[17]

Those who assume senior positions of responsibility in tech firms – chief executives, senior counsel, compliance officers and the like – should have to be much warier of criminal sanction than they are now. In extreme cases, the community should be able to require the removal of certain leaders as a consequence of criminal misconduct. Perhaps we will see proceedings to 'impeach' the CEOs of large tech firms.

These days, criminal prosecutions are mostly brought by public officials acting on behalf of everyone else (the People, the State, the Commonwealth, the Queen). That is to say, the criminal law is a form of collective enforcement. Private prosecutions, in which an ordinary citizen asks a criminal court to issue a criminal sanction, do occur in some jurisdictions but they are rarer.

What is the republican take on criminalising wrongful conduct? With the *parsimony principle* in mind, we should be cautious about criminalising bad behaviour unless it is quite necessary to do so. After all, criminal punishment is the state at its coldest and most domineering. Only the most serious wrongs should be treated as criminal.[18] Singapore's Protection from Online Falsehoods and Manipulation Bill, for instance, makes it a crime to spread 'false statements of fact' in circumstances in which that information is deemed 'prejudicial' to Singapore's security, public safety, 'public tranquillity' and various other ends. That goes too far. It is possible to govern social media without slinging individuals into jail for what they say.

***

Counterpower means being able to challenge unlawful exercises of digital power. But is there a form of counterpower that can prevent things going wrong in the first place? That's the subject of the next chapter.

TWENTY-FIVE

Certified Republic

Certification does not look like a revolutionary word. It suggests humourless bureaucrats in grey suits, shuffling paperwork in dreary office cubicles. It is hard to imagine it as part of a political slogan (*Give Me Certification or Give Me Death!*)

In truth, however, certification is an important, even radical, idea. To say that a thing or person is certified is to say that society has agreed it can be trusted.[1] In fact, certification is so ubiquitous that we take it for granted. The food we eat, the cars we drive, the products we use, the professionals we hire – they are all certified in one way or another. But as recently as the nineteenth century, certification was almost unheard of. There was no formal way of knowing whether the person removing your tooth had any medical training, whether the egg you were eating had been laid six days or six months earlier or whether the machinery you were operating was even remotely safe.

Certification is the process by which a thing or a person is deemed to have met an agreed standard. As we have seen, a standard might relate to any kind of quality, and so can certification: competence, reliability, safety, trustworthiness, integrity, probity. Certification can concern detailed specifications (*a car engine must have at least the following components*) or general qualities (*a drug must be safe and effective*) or procedures (*a lawyer must have a law degree or diploma*). Once something or someone has been certified, it will usually have something to show for it: a certificate, logo, kite-mark, seal or the like.[2] Importantly, certification is usually done by an independent third party, preferably one with no skin in the game.[3]

Although certification is sometimes associated with red tape, more often it is a social and commercial lubricant. It allows us to trust the work of others without getting bogged down in investigations and negotiations. Tamsin Allen offers a useful analogy. When you enter a building, she says:

> you do not sign away your rights to enter it safely. You do not sign a form with 14,000 pages that tells you how the building was built and that says you have to accept the risk. You rely on the fact that the architect, the engineer and the builder will be subject to regulation, and that there will be insurance and public liability requirements on the building because it is open to the public, and you will feel that you can then walk into that building safely.

You expect the building and those who built it to be certified.[4]

***

In the digital republic, certification would enable people to place their trust in digital systems, knowing that they conform to the legal standards set by the community.

To begin with, certification can be used to determine who or what may compete in the market at all. Quality control, assurance tests, data checks and algorithmic impact assessments would all play a part in the certification of high-stakes algorithms.[5] In Australia and New Zealand, developers of slot machine software must already submit their algorithms for pre-market approval. The applicable standards are specific and exacting. The 'Nominal Standard Deviation' of a game must be 'no greater than 15'. The 'hashing algorithm' for the verification of gaming software must be 'the HMAC-SHA1 algorithm'.[6] When the stakes are high – when it is vital that a product, system or person consistently meets a certain standard – certification is a priceless tool of governance. Take facial-recognition systems. If they are to be used at all, these systems should plainly have to be certified as fit for use, and those who operate them should need some kind of licence.

Pre-market certification could also be used to certify that certain values or principles have been engineered into products and systems. This would be in the tradition of privacy by design, which holds that designers should 'hard-wire' privacy into technologies from the outset rather than merely compensating people when their privacy is breached, or punishing those who do the breaching.[7] There's no reason why this tradition should be limited to privacy: justice by design for high-stakes algorithms would be a perfectly realistic goal. If the community has determined that certain values should govern certain systems – recall the example of the self-driving car from chapter nineteen – then such systems should have to be certified as compliant before being rolled out. Of course, certification is not a one-time affair. It is usually subject to renewal, and so represents an ongoing form of protection.

In food and drug regulation, very high-risk products are already subject to the precautionary principle.[8] As its name suggests, this principle emphasises caution in the face of risks that are unknown but potentially significant. It asks producers to prove the safety of their wares before unleashing them on the public.[9] Technologies that pose a heightened threat to individuals or the republic should be subject to extra pre-market scrutiny. Likewise technologies that could pose grave moral or political risks.[10] In these cases, pre-market certification could be subject to approval by a citizens' jury convened for that purpose.[11]

Certification can also be used to score things (this algorithm has an 8/10 anti-bias rating. This social media platform has a grade 'C' anti-misinformation *system*). And it can be used to rank things (*of 500 automated vehicles on the market, this model ranks 254 in road safety*).[12] To the extent that individuals have a choice between digital products, it should be an informed choice. Certification offers a standardised form of information.

Almost anything can be certified, not just technologies and people. Platform terms of service, for instance, could be certified as conforming to a relevant standard. Bodies already exist that offer (voluntary) certification for privacy terms contained in boilerplate contracts.[13]

In Europe, the law already expects companies to perform 'data protection impact assessments' every time that data processing 'is likely to result in a high risk to the rights and freedoms of natural persons'.[14] It also anticipates new forms of certification, voluntary rather than compulsory.[15] In the digital republic, certification would not be optional.[16] And assessments would not be left to corporations themselves. Nor, in most cases, would they be kept secret. Republican liberty requires that citizens be told whether a digital system has met the appropriate standard – as a matter of law, not choice.[17]

***

Certification has a whiff of the nanny state. It sounds stultifying and bureaucratic. For republicans, who fear the power of government as much as that of corporations, this is a concern. But as Gillian Hadfield has pointed out, governance need not be performed by the central state directly. She proposes the dispersal of regulatory powers to a system of private operators overseen by an overarching public regulator (a 'superregulator'), in which the operators compete to offer the most efficient and elegant forms of oversight.

In Hadfield's system, private regulators would be strictly accountable to government, bound by targets and constraints. They would be required to apply the letter of the law strictly and without fear or favour. Their staff would be expected to uphold professional standards. And as secondary regulators, these businesses would themselves be subject to oversight by the primary regulator, the state. Within these constraints, however, businesses could compete to find the best ways to certify digital technologies. 'To succeed in this competition,' Hadfield writes, 'a private regulator would have to offer a system that simultaneously meets government criteria for approval and is attractive, relative to the market alternatives, to regulated businesses.'[18] Businesses would be incentivised to find ingenious systems of certification, just as professional auditors look for new ways to audit financial accounts. Private regulators could attract first-rate talent, respond with more agility to new technical developments, and build institutional expertise across disciplines and jurisdictions.[19]

As Hadfield concedes, however, this model is not foolproof. For one thing, private auditors might try to take a soft touch within the rules – even to turn a blind eye – to attract business. Once profit becomes a factor, the wrong incentives begin to creep in: cost-saving, corner-cutting, even outright corruption. To sceptics, the whole scheme will sound a lot like government subcontracting more generally, which is not always a great success. Hadfield points to the failure of the US state to regulate private prison operators as a case in point.[20] These pitfalls are real and even obvious, but they are also preventable. Without proper resourcing, no system of governance will succeed for long, be it direct or indirect as Hadfield proposes. Either society takes governance seriously or it doesn't. It can't be done half-heartedly or on a shoestring. With care and investment, there is no reason why the governance of technology cannot be done in a manner that befits its importance, whether by the state or by non-state entities exercising its functions on a delegated basis.

Of course, those who certify must themselves be subject to legal oversight. The process by which 'certifiers are themselves certified' has its own name, *accreditation*.[21] Again, it's hard to imagine this forming a rousing political slogan (*No Certification Without Accreditation!*) but it's important. The framework for a better system of tech governance already exists. We just need to try it properly.

TWENTY-SIX

Responsible Adults

One way to govern technology more effectively would be to improve how we govern *people* in the tech industry. We can't hope to review or litigate every important decision, and we wouldn't want to. Total enforcement would be intrusive and stifling. But we can use the law to foster a culture in which those who take important decisions are obliged to think carefully about their social responsibilities before they act. As Herbert Hart put it in *Concept of Law* (first published 1961):

> The principal functions of the law as a means of social control are not to be seen in private litigation or prosecutions . . . It is still to be seen in the diverse ways in which the law is used to control, to guide, and to plan life out of court.

We want tech executives to fret about potential injustices in their systems. We want corporations to treat questions of ethics as questions of compliance. We want organisations to have institutional mechanisms in place to prevent ethical failures.[1] And rather than hoping they will do these things despite all the forces pulling against them, we should encourage them to do so by making it their duty under the law.

**

Mary Wollstonecraft wrote that 'the character of every man is, in some degree, formed by his profession.'[2] But for those who work in tech, there is generally no formal profession to speak of: no mandatory qualifications, no shared norms governing conduct, no duty to serve the public good, no unitary industry body, no personal sanctions for wrongdoing. In other areas of life, those who hold positions of social responsibility – doctors and lawyers, pharmacists and pilots – work under binding codes of conduct enforceable by law. Tech is the exception.

The modern idea of 'professionalism' recognises that expertise is a kind of power. We require professionals to work for the good of others, not just themselves. We expect them to hold themselves to a higher standard. We require them to be rigorously credentialed before they conduct their work. We set standards relating to ethics and competence – and if they fall below those standards, they are punished.[3] And we ask them to prove, throughout their careers, that they remain fit to hold the title of lawyer, or doctor, or whatever profession they belong to. At the same time, professionals enjoy the benefit of exclusivity, in that no one else can do their work without the necessary certification. They generally enjoy high social esteem. At their best, they are seen as representing a higher calling or vocation, 'not so much a job as a way of life'.[4]

Many in the tech industry share at least one thing with traditional professionals: their work requires specialised skill and knowledge. And for the last couple of decades, many (but not all) tech workers have enjoyed the social esteem and remuneration that comes with performing a valued social function. But that power has not come with corresponding responsibility. They are unburdened by legal expectation. This has to change.

***

Imagine two people. One is a software engineer, developing recruitment algorithms that will accept or reject thousands of applications in hundreds of companies. Another is a senior executive at a social media company, with responsibility for the platform's moderation systems. At present, both do their work without appropriate regulatory supervision of their personal conduct. Though their decisions affect the public quite profoundly, they are accountable only to their bosses. How might we hold these people a little more responsible for the important work they do?[5]

A first step would be mandatory education or training before being licensed to perform their functions. The software engineer might need a diploma in algorithmic injustice. The social media executive might need to attend a certification course in which participants study the latest legal and ethical issues arising out of platform moderation. Not too burdensome, but at scale these steps could make a difference.

Next, they might be expected to work under codes that place them under quasi-professional duties. These could be minimal, like a basic requirement to become re-certified every two years. Or they might be more onerous, like the fiduciary responsibilities placed on doctors, lawyers and financial advisers.

Then there might be disciplinary mechanisms by which to hold these individuals to account if they fall below the legal standards placed on them: public complaints procedures, ombudsmen, review boards and the like. A fair disciplinary process could conclude with written warnings, fines, temporary suspensions or disqualification from further activity in the field.[6]

Finally, there could be mechanisms for wronged citizens to seek redress from regulated persons, either individually or collectively, perhaps in a tech tribunal. There might be formal compensation schemes for the damage done by faulty algorithms. To fund compensation, regulated persons could be required to pay into an industry-wide insurance fund or compensation pool.[7]

The system described here would not amount to full professionalisation of the tech industry. But it would make it more accountable. And it would bring tech into alignment with comparable industries like financial services. The UK financial regulator identifies certain roles of particular significance, called Controlled Functions. There's a 'director' function, a 'compliance oversight' function, a 'money laundering reporting' function, a 'system and controls' function, an 'actuarial' function, a 'significant management function' and others. To perform a Controlled Function, an individual must become an Approved Person by satisfying the regulator that they are 'fit and proper' to do so. Then they must behave consistently with a set of legal standards.[8] Why should tech be any different? There is no reason not to introduce controlled functions and 'fit and proper' tests in the tech industry – at least for applications and functions that have achieved social significance.

***

The tech world has taken some fledgling steps towards professionalisation, albeit mostly in academia. University researchers form a 'broad, cooperative international community, glued together by shared interests, conferences, cooperative agreements and professional societies'.[9] There is a growing body of scholarship on what it means to be an ethical software engineer, computer scientist or data analyst. The trouble is that there is no hard core of agreement – which is unsurprising given the essentially political nature of digital technology – and no way of enforcing the norms that are agreed.

There is also evidence to suggest that many academic researchers still see their work through a rose-tinted lens. 'There clearly is a massive gap,' writes one group of scholars,[10]

between the real-world impacts of computing research and the positivity with which we in the computing community tend to view our work. We believe that this gap represents a serious and embarrassing intellectual lapse . . . it is analogous to the medical community only writing about the benefits of a given treatment and completely ignoring the side effects, no matter how serious they are.

As we saw in chapter twelve, the self-regulatory initiatives that have sprouted within the tech world tend to have no teeth. Codes of ethics published by two of the largest professional associations for software engineers, the Institute of Electrical and Electronics Engineers (IEEE) and the Association of Computing Machinery (ACM), for instance, are 'comparatively short, theoretical and lacking in grounded advice and specific behavioural norms'.[11]

The long-term aim should be to generate a coherent body of rules, standards and expectations for people working in the tech industry, backed up by robust institutions of the kinds found in other socially important industries.[12]

TWENTY-SEVEN

Republican Internationalism

It has become a cliché to speak of the global reach of digital technology. Data moves seamlessly across borders. Tech firms span continents. The architecture of the internet lies between, or outside the control of, any single state. Although the internet has a physical existence within borders – cables, servers and computers are all located *somewhere* – the actual flow of information between humans, other humans and computers is largely decentralised and 'border agnostic'.[1] This makes it hard to enforce domestic laws, for a number of reasons.

To begin with, when something goes wrong, it can be tricky to point to a single entity or company and say *they are responsible*, or point to a country and say *that is where it happened*.[2] In the absence of comprehensive global frameworks for cooperation, it is often difficult to hold companies responsible for wrongdoing that may technically have taken place elsewhere. Global companies can escape local laws by moving around, legally or physically. In 2018, for instance, Facebook announced that the legal 'data controllership' of more than one billion users from outside the EU would move from Ireland to the US. This transfer took place without physically moving anything or anyone. The legal intent appears to have been to strip GDPR protections from those users overnight.[3]

Then there is the risk of regulatory arbitrage. It makes sense for businesses to move to places where they are subject to less oversight and taxation. If one place introduces new laws, businesses may simply move elsewhere, depriving their former home of the economic benefits of their presence or even the service itself. That's why governments are often more concerned with their countries being more hospitable to tech businesses rather than curbing their power. Businesses are always liable to make commercial calculations about whether the cost of relocating would outweigh the benefits – and legal regimes play a part in that analysis. The result is a natural tendency towards competition between blocs, states and regions, sometimes called the 'Delaware effect' or a 'race to the bottom'.[4]

On top of these challenges, many governments are concerned with the geopolitical implications of technological advance. Vladimir Putin made the point with Bond-villain theatricality: 'The one who becomes the 'leader' in artificial intelligence will be the 'ruler of the world.'[5] In America it is common to read headlines like 'Washington Must Bet Big on AI or Lose Its Global Clout'.[6] There is talk of a new cold war, based around the race to superintelligent AI, with influential voices calling for democracies to align in an anti-China alliance.[7]

China and the US, and to a certain extent the EU, take the AI race seriously. The rise of 'digital nationalism' could see a range of government measures which have little to do with increasing the freedom of the citizenry. Some, like funding domestic research and blocking foreign acquisitions, do not necessarily impede the governance measures sought by digital republicans. Others, like efforts to strip away regulation altogether (save for market-making laws), or regulating for geopolitical rather than domestic goals, could act as retardants on the republican agenda. There is an

obvious tension between reining in the power of digital technology
and wanting to use it to make one's country more powerful.

***

What, then, is the republican take on the international question?

Before considering whether global governance is workable
in practice, let's first ask whether it's desirable in theory. Recall
that the *democracy principle* of digital republicanism holds that the
design and deployment of powerful technologies should, as far
as possible, reflect the moral and civic values of the people who
live under their power. Those values change from place to place.
As we will see in part ten, for instance, America and France have
very different traditions when it comes to free expression. Things
that may permissibly be said in America – shouting 'Heil Hitler'
outside a synagogue, for instance – would be prohibited in France.
So, when it comes to social media, why should French people
be governed by American speech norms, or vice versa? For that
matter, why should a Canadian court be able to order that citizens
of the UK can't see a particular Google search?[8]

To say that citizens should set the rules that govern their lives is
to prompt another question: *citizens of where?* Some people find their
principal identity at a local or regional level, others at a national
level, while others see themselves either as citizens of nowhere or
citizens of everywhere. In many countries there are fierce battles
over who gets to call themselves a citizen and who gets to benefit
from citizenhood at all. There is no universal or definitive unit of
political organisation. In the past, we organised ourselves in tribes
and clans, now we mainly use states, with local and transnational
governance too. Some imagine a borderless future in which the
walls that divide us – real and invisible – are torn down forever.

It would be obtuse, however, to ignore the fact that we still live in a
Westphalian world in which most of the world's regulation is done
at a national or subnational level. This is not an accident of history.
While a banker in the City of London may have much in common
with a banker in New York, the reality is that national identities,

traditions, customs and folkways are real and meaningful. The risk with global governance is that, by treating everyone the same, we paper over the real differences in national identity that still exist between countries. Different cultural norms, different traditions, different structures of government, different histories – these can't just be ignored. This isn't an argument for cultural relativism, i.e. reserving judgement on other systems in the name of cultural sensitivity. It is, rather, a point of pragmatism. Like donor organ tissue that does not match the recipient's biology, it is unwise to transplant laws and institutions from one place to another without checking they are compatible.

Nor am I making the case for nationalism. Even in transnational blocs like the EU, significant areas of law and regulation (like tax and financial regulation) are left to the constituent countries, for good reason. And in the rest of the world, though there is cooperation and a growing body of international laws and customs, people are quite content to have their laws made at a national or local level. As Brexit showed beyond any doubt, in the UK at least, the urge to retain national control can trump even the most obvious economic and legal efficiencies of international cooperation.

This rather conventional way of looking at politics is at odds with 1990s internet utopianism. Tim Wu identifies two 'duelling visions of the internet':[9]

> The first is an older vision: the idea that the internet should, in a neutral fashion, connect everyone, and that blocking and censorship of sites by nation-states should be rare and justified by more than the will of the ruler. The second and newer vision, of which China has been the leading exponent, is 'net nationalism,' which views the country's internet primarily as a tool of state power. Economic growth, surveillance and thought control, from this perspective, are the internet's most important functions.

Digital republicanism offers a third vision: a system of national governance aimed neither at global interconnection nor national dominance but at the self-determination of the people who live within each republic, or collection of republics.

Of course, some regulatory regimes will always have more global clout than others.[10] The rules set in America, the EU and China will matter more than those elsewhere simply because they are the largest consumers, suppliers and regulators of technology. What happens in these places will have an outsized impact on the rest of the world – and so there will always be a natural limit on the extent to which other republics are truly self-governing. This can be a good thing: if American and European standards are high, companies from all over the world will raise their standards in order to access these markets. This is what happened when California adopted stricter emissions standards for cars. Other states adopted them too, in order to sell into the large Californian market.[11] The GDPR has had some similar successes.

When it comes to republican self-governance, therefore, there is a balance to be struck. We've seen that we can't do it at the level of individual consent. But it wouldn't be desirable to make all our laws globally either. The law would be too remote, too distant, even too foreign. The state lies somewhere between these two extremes: small enough to reflect a shared sense of identity, large enough to resolve problems of collective action, yet not so large that it becomes entirely alien to those who live within it. Unfashionable though it is, therefore, the digital republic is most likely to be realised at the national level, first and foremost – while working (as ever) to develop the international organisations, treaties and protocols that make international cooperation possible. Being realistic, a digital republic should also avoid being overly reliant on foreign powers, particularly those whose interests are not aligned with its own, for the provision of its essential digital infrastructure.

This may seem heretical to those who lived through the heady years of the 1990s, when the internet was seen as a force that would inexorably erode national boundaries. But technologies are shaped by the societies into which they are born, as well as shaping those societies in turn. The internet emerged into a world with hard borders, large disparities in national power and different political cultures. No wonder that it is possible to speak of at least four 'internets': a 'bourgeois' European internet subject to norms of civility; an 'authoritarian internet' in places like China where technology is principally used as a means for social cohesion and control; a 'commercial' internet favoured by American elites; and a 'vulnerable' internet susceptible to hacking by rogue states and agents.[12]

***

If governing technology primarily at the national or bloc level is acceptable in theory, what about in practice? No doubt it creates an extra burden for transnational technology companies to have to adjust their products and services for each place they do business. Their lives would be much easier if the law was the same in London as in Los Angeles. But abiding by local laws is an unavoidable part of trading in a globalised economy, not just for technology companies but for all companies. There is no inherent reason why the tech industry should be treated differently.

Moreover, tech firms are already quite used to adjusting their offerings to align with different legal regimes. Social media platforms, for instance, are perfectly capable of blocking or filtering content in one country while leaving it visible in another. They do it all the time.[13] Likewise, Google screens out pro-Nazi content in Germany (where it is illegal) but not in the US. eBay prohibits the sale of cigarettes to consumers in America, but not elsewhere.[14]

The risk of arbitrage, too, can be overstated. Government regulation and taxation are not the only factors in determining whether there is a hospitable business environment. A country with generous research subsidies, a well-educated population, stable government, an efficient legal system and solid infrastructure will still be an attractive business location even if there are regulations and taxes that do not apply elsewhere.[15]

Finally, even though it is decentralised, the internet does have 'regulatory access points', located in real places, that can be made subject to the law.[16] There are often hooks on which to hang legal liability for transnational providers, such as assets or people within the jurisdiction.[17] And – subject to the usual concerns about accountability – big companies can be used to help enforce liability for smaller ones that use their platforms.[18] The law evolves. As Nicolas Suzor points out in his fine book *Lawless*, when the International League Against Racism and Anti-Semitism sued Yahoo in France for allowing users to auction Nazi memorabilia, the company retorted that since its servers were based in the US, French law should not apply. The French high court dismissed this argument and ordered Yahoo to block French access to auctions of Nazi memorabilia.[19]

One of the advantages of predominantly national regulation is that nation-states can be 'natural laboratories' for different forms of institutional design.[20] This is a time for experimentation and innovation, and the digital republics of the world can learn from each other's successes and failures, reacting more nimbly than a transnational supertanker on its own.

National enforcement will never be perfect. But it isn't perfect when it comes to tax or environmental regulation either, and that doesn't stop countries from introducing taxes, or laws to protect the environment. In practice, enforcing the law is a pragmatic exercise that evolves and hopefully improves over time. If we wait for international consensus, we will be waiting a long time (which might be why, a cynic would say, the tech firms are so keen on the idea).

***

We have reached the end of part seven. By now the legal infrastructure of the republican system should be visible in outline: new democratic processes, new rights and tech tribunals, new legal standards and public enforcement, new certification schemes and new duties on tech workers who perform sensitive functions. Democracy and counterpower would combine to bring technology under the control of the people. For these new laws and institutions to be effective, however, citizens and regulators need to know what is actually happening inside the tech industry. The next part of the book confronts the issue of transparency. It imagines a new way of inspecting powerful technologies and a new duty of openness on the tech industry itself.

PART VIII

OPENNESS

Liberty lies in the hearts of men and women; when it dies
there, no constitution, no law, no court can save it.
Billings Learned Hand

TWENTY-EIGHT

A New Inspectorate

Scrutiny is the soul of governance. Chefs keep their kitchens clean when they know there might be a surprise inspection. Bankers take more care with client money when they know the financial regulator is monitoring their trades.[1] To govern technology – to govern anything – it is necessary to gather information about the systems, corporations and people that are the target of the law. This is an ongoing process. It never stops.

In the digital republic, legal standards need to be monitored to be enforced. But in a world of code that can be altered at any moment, machine learning systems that evolve over time and algorithms that moderate content a thousand times a second, oversight is not straightforward. We will need a new corps of professionals for the task. Their function? To assess, investigate and audit those who design and control digital technologies – and make sure they are compliant with the law.[2]

***

In the French revolution, inspired by Jean-Jacques Rousseau, philosophers dreamed of a society in which the rich and powerful had no secret places to hide. Foucault described it as 'a transparent society, visible and legible in each of its parts, the dream of there no longer existing any zones of darkness, zones established by the privileges of royal power or the prerogatives of some corporation'.[3] Digital technology has distorted this dream. While ordinary citizens are more exposed than ever – their inner and outer lives surveilled, logged and processed – powerful technologies operate behind fortresses of encryption, intellectual property and corporate secrecy.

In recent years, there has been growing interest in the idea of algorithmic audit: inspecting algorithms to check their compliance with a particular value or principle. Some firms already offer auditing services on a voluntary basis, but in the future it should not be optional. In fact, we need something broader – a body of professionals competent to examine not only algorithms and data practices, but social media moderation, robotic safety, cybersecurity and whatever else the law seeks to govern. For every right or legal standard, there must be some trusted person or body capable of discerning whether it is being complied with or not.

Methods of inspection will vary according to the technology and the purpose. To test whether an algorithm is compliant with anti-discrimination laws, an inspector might take a 'black box' approach, examining the inputs and outputs (e.g. the data used to 'train' a recruitment algorithm, and data about who was recruited),[4] or a 'white box' approach, examining the workings of the software itself.[5] This could be done on a 'static' basis (examining the code without running the programme) or a 'dynamic' basis (examining the programme while it is running).[6] Or to test cybersecurity, an inspector might 'red-team' the system, assuming the role of an attacker to find chinks in its armour.[7]

None of this work is straightforward or routine. It requires specialised technical and legal knowledge. In general, however, it need not be adversarial. Inspectors and innovators should be able to work together to ensure that the latter understand and meet the relevant legal standards. The purpose is not to catch people out. Instead of waiting to be burned, tech firms should have access to frank guidance from regulators – just as American taxpayers can seek 'private letter rulings' from the IRS.[8]

For the citizenry, there are obvious benefits to an effective system of oversight. For one thing, it gives comfort that the law will be enforced. What's more, oversight can prevent harm before it happens, rather than leaving it to individuals to claim compensation after the damage is done. The point is to ensure that digital systems are maintained in good working order. As Richard Susskind likes to say: better a fence at the top of the cliff than an ambulance at the bottom.[9]

From an industry perspective, oversight may sound like a drag. But it's no more than we ask of other critical sectors. We expect nuclear power plants, dams and hospitals to accede to external regulatory oversight. Why not large tech firms? Professional oversight has advantages for the industry too. One is that it can be done discreetly. There is no need for everyone to see proprietary algorithms, trade secrets, personal data or business plans and records. What matters is that the outcome of the process can be trusted by the wider public.

Purists will ask, however, whether a system of professional oversight amounts to replacing one unaccountable technocratic elite (tech firms) with another (the inspectorate). Are we being asked, yet again, to place our trust in the wisdom of powerful figures whose work we can scarcely understand, let alone control? What if inspectors act in an arbitrary or capricious manner? What if they become too cosy with the industries they are supposed to regulate, like the credit rating agencies before the 2008 crash?

There is force in these concerns. They have troubled republican thinkers for centuries. *Quis custodiet ipsos custodes?* Who will guard the guardians? No neat answer is available. An endlessly recursive scheme of oversight – inspectors inspecting other inspectors, in turn inspecting other inspectors – would be a political Ponzi scheme in which the buck ultimately stops nowhere.

There are, nevertheless, reasons for optimism that a new inspectorate would not itself become a concentration of unaccountable power. For one thing, the work of inspection would be professionalised. Inspectors would be subject to legal duties of their own: strict rules, controls and disciplinary mechanisms.[10] We cannot demand perfection, but we can expect inspectors to embody the same standards of independence, expertise and probity as other professionals like doctors or lawyers.[11] And if they fall short, we should expect them to be sanctioned, suspended or struck off. Moreover, a professional system of public inspection, with all its challenges, would still be preferable to no oversight at all.

A new inspectorate is all very well, but what about the duties on technology companies themselves? What should they have to do to enable citizens and regulators to enforce the law? That's the subject of the next three chapters.

TWENTY-NINE

Zones of Darkness

In the beginning, the laws of Athens were passed down by word of mouth alone, stored in the minds of a few elders who jealously guarded their knowledge. Then, nearly three millennia ago, a lawgiver did something extraordinary. He decided to record the laws in writing for the first time. At his instruction, they were carved into wooden tablets for all to see. Athens had its first written constitution.

Unfortunately for the lawgiver, whose name was Draco, his crowning achievement was overshadowed by his rather uncompromising attitude to criminal justice. The laws he introduced were severe. For minor infractions like theft, people were sentenced to death. Thus, when we describe something as *draconian* today, we remember the harshness of Draco's laws, not the revolutionary step of writing them down in the first place.[1]

Draco's innovation did not trigger universal approval for the idea of written law. (Moses had met with some resistance too.)

As recently as the introduction of the printing press, philosophers sincerely doubted that laws should be shown to everyone. What if people read them without really comprehending them?[2] Wouldn't it be better to let the experts study the rules and tell others what they are? Nowadays, legal scholars tend to agree that laws should be published, clear, prospective and stable – even if that is an ideal to aim for, rather than something that is always achieved in practice.[3]

But what about digital technologies? They too contain rules that the rest of us have to follow. They shape and mould our lives. They exert power. Should the rules embedded in digital technology not also be, to some extent, published, clear, prospective and stable?

***

Digital transparency has been going in and out of fashion among computer scientists for decades. Most of us will never fully grasp the technical intricacies of the technologies that shape our lives, but we ought to be told about the values they encode, the data they rely on, who designed them and the purposes for which they were made. There are four reasons why.

First, our dignity as persons is undermined when we cannot understand what is being done to us. When important decisions about our lives are taken by systems we cannot see, using methods we cannot fathom, on the basis of criteria we are not shown, we cease to be free agents charting the course of our own destinies. We become the playthings of unseen powers, buffeted around by mysterious forces outside our comprehension. We grow helpless and childlike.

Next, so long as digital technologies are obscured from the public gaze, they cannot be properly challenged. There would be little point in tech tribunals, inspectors, certifiers and public enforcers if no one had any idea what was going on inside the tech industry.

Then there is the fact that we need a base level of knowledge to discharge our duties as citizens. If we do not know the facts, we do not know the risks. And if we do not know the risks, we cannot make informed choices about what is acceptable, both for ourselves and society.[4] We need to be able to tell between good and bad products. As is often the case, the best ones may cost more – but if we don't know why, they just look like bad value for money.[5]

Finally, sometimes the power of technology derives from the very fact that its inner workings are hidden. Revealing its workings can diminish its spell. Thus, research suggests that chatbots can outsell inexperienced salespersons when pitching to consumers – but only until they reveal their true identity. Then sales drop by 80 per cent.[6] Online 'dark patterns' are used to nudge and cajole users. Exposing them would be part of reducing their potency, like showing how the magician performs his tricks.

***

A world in which technologies were operated openly as a matter of course would be very different from the one we know. Today, the inner workings of the digital world are mostly invisible. On social media, for instance, users are shown general 'community standards' but not how those standards are made or applied.[7] It is often unclear why some conduct is penalised while other, more egregious, conduct is left unpunished. Decisions by human and machine moderators are made in almost total secrecy.[8] Algorithms that promote or suppress certain content are understood only in general terms, if at all.

Tech firms, like most private enterprises, prefer not to explain themselves unless required to do so. Many of the systems that determine our access to jobs, insurance, credit, housing and the like are 'black boxes' that tell us little about what data is being used, how it is processed, which factors are taken into account or the various weightings that go into important decisions.[9] Algorithmic determinations are handed down like revelations. A job applicant might never know if the colour of her skin had anything to do with her rejection. A person denied parole might never know if the decision was based, directly or indirectly, on generalisations about her class or race.

A deeper issue is that machine learning algorithms often work in ways that even their human creators struggle to understand, still less explain. They do not apply neat rules. Instead, they look for patterns and relationships that can form the basis of predictions and decisions. But when millions of parameters are applied to billions of data points across hundreds of datasets, and processed thousands of times to form probabilistic outcomes, it can be hard for human engineers to answer simple questions like: *why did the system reach that answer?* In medicine, a system called Deep Patient looked for patterns in data from 700,000 patients and developed the ability to predict the early stages of several diseases, from liver cancer to schizophrenia. But the machine couldn't explain how it worked. Nor could its developers.[10] Systems which deal in predictions are often themselves highly unpredictable.

***

The truth is that most digital systems are not designed with transparency as a priority.[11] They are made to serve the ends of their creators as efficiently as possible. Sometimes that's fine. When a bank replaces its traders with algorithms, it matters little if the algorithms can't explain themselves, so long as the dollars flow and the bank takes the hit if things go wrong. But some types of decision *do* need an explanation, particularly where they affect the significant interests of others or involve an element of moral controversy.

Historically, the law has demanded a degree of rationality and procedural fairness from important decision-makers, including the provision of *reasons* so their logic can be understood and challenged if necessary. In the future, nonhuman systems will make countless determinations affecting individuals and society. What could be more offensive to the republican mind than blindly entrusting these decisions to processes we cannot even understand? The next chapter explores the nature and limits of transparency, looks at current practice in the tech sector and begins to explore what digital republicans might do differently.

THIRTY

Transparency about Transparency

Recent years have seen an explosion in new methods for making machine learning systems more understandable. Most, however, have been for engineers, not for ordinary people or regulators.[1] Nevertheless, three fields of inquiry show promise.

One involves trying to engineer systems that are more interpretable in the first place.[2] This usually means reducing the number of variables – the interplay of ten inputs is easier to understand than that of 1,000 – or using models that are easier to understand, like decision trees with a limited number of branches.[3] Critics, however, say that this approach introduces a trade-off between interpretability and efficacy. The easier to understand, the cruder the model.[4] This approach would rule out a number of the most powerful machine learning techniques.

A second approach is to use *other* computer models to generate *post hoc* interpretations of decisions that have already been taken.[5]

Such models can provide simple narrative descriptions, or diagrams, maps or other visualisations.[6]

A third approach involves using an interactive interface, so users can toggle inputs and see how the outputs change.[7] This is fun and intuitive, but the risk is that a small number of toggles could give a misleading impression of simple correlations that don't exist. (It may be that getting one credit card would improve your credit score, but you can't assume that getting fifty credit cards would improve your credit score fifty times over – or at all.)[8]

***

It's important to recognise, however, that transparency is not held back by technical constraints alone. Commercial factors matter too. Companies understandably want to protect their proprietary algorithms. They invest time and money in research and development, and it would be problematic for their workings to be laid bare for others to game or exploit. Moreover, time spent trying to be more transparent is time that cannot be spent doing other things.

There are also reasons of public policy that weigh against full transparency. It would not be in society's interest for the IRS to tell everyone how its algorithm decides which tax returns to audit.[9] And there is sometimes a tension between transparency and privacy: more visibility can sometimes mean a greater risk of exposing personal data to the outside world.

There are also republican reasons to be cautious about unlimited transparency. Too much information can lead to a *lowering* of trust in institutions. Sometimes it is easier to accept what seems like a fair decision without knowing how it was reached, just like it is easier to swallow a tasty sausage if you don't know how it was made. That might be why juries in England and Wales, for instance, are not asked to give reasons for their verdicts.[10] Another factor is that masses of raw information are useless to the ordinary person. Our working memory can only hold around seven items at a time, and 'information overload' is an obstacle to clear thinking.[11] Even with the best of intentions, we often fail to understand large bodies of data. Usually, we view problems through the distorting lens of our prior beliefs.[12]

For all these reasons, transparency is neither risk-free nor a panacea. So it has been interesting, in recent years, to see how the tech industry has responded to growing calls for transparency. The independent research group Ranking Digital Rights paints a picture of modest improvement across the board.[13] Several tech giants now publish reports with statistics relating to takedown requests from governments and copyright holders.[14] Some provide details of how platforms are assisting governments on surveillance and law enforcement.[15]

These reports are useful but they are also selective.[16] They leave out more than they include. For example, as Julie E. Cohen observes, platforms are happy to detail the number and nature of government takedown requests, but when it comes to their own algorithmic recommendation systems, 'the newfound commitment to openness ends'.[17] Search engines like Google remain highly secretive about the workings of their algorithms and what they do with user data.[18]

Some platforms have made their moderation processes marginally more visible. YouTube, for instance, now provides information about why it is removing a video and who requested it.[19] This is progress. But again the overriding incentive for platforms is to 'smooth and sanitise' the user experience, rather than keep users informed about the plumbing and mechanics of the systems.[20] We are still treated as consumers first, citizens second.

Moreover, most technology corporations still do not publish transparency reports at all, and their algorithms remain fundamentally mysterious. That's why some countries have started turning to the law. In US privacy enforcement proceedings, consent decrees increasingly require periodic reporting.[21] European data protection law provides a general commitment to data transparency, allowing people to request and access personal data that is held about them.[22] It also gives people the right to seek 'meaningful information about the logic involved' in automated decision-making and profiling systems.[23] But scholars disagree about whether European law actually contains a legally binding right to an explanation (and whether what is being explained can properly be described as 'logic' in any event).[24] If the 'right' does exist, it is exercisable only after a number of hurdles, limited in application and riddled with loopholes.[25]

***

What emerges from this survey is a more ambiguous picture than some might have imagined. Technologies are opaque but they can be made more transparent. Transparency is desirable but it has limitations. The industry is resistant to change but has started to do better. The law is inching in the right direction but there is more to be done. If this feels a little unsatisfying, the next chapter offers a tentative way forward.

THIRTY-ONE

A Duty of Openness

The law needs to strike a new balance. On the one hand, we need more transparency in digital technology and the tech industry. On the other hand, the law must respect the technical, commercial and public policy factors that weigh against total openness. This chapter imagines how that balance might be struck, in the form of a new *duty of openness* for tech firms.

***

Not every form of digital technology needs to be transparent, and some should be more transparent than others. Commercial software used for logistics, administration, inventory and the like does not need to operate in a transparent fashion, at least not for political reasons. A good rule of thumb is to ask whether there are legal rights or standards that apply to the use of a particular technology. If there are, then there will need to be a degree of transparency.

Where a republic has chosen to impose legal rights or standards in relation to a technology, what is the appropriate level of transparency? That will depend on the context. Sometimes, transparency will be the explicit goal of the law. So, if the aim is to reduce manipulation by bots, systems might have to make clear what they are.

For other technologies, there should be enough transparency to enable a concerned citizen or regulator to know whether a technology is compliant with the law.[1] But the duty to provide that transparency cannot be open-ended. What's needed is enough to enable a *reasonable* challenge.

What a reasonable challenge looks like will also differ according to the context. If the law gives individuals the right to appeal algorithmic *decisions*, then those decisions should be transparent enough for people to have a reasonable opportunity to tell whether they are compliant with the law. A regular person challenging a mortgage rejection is not going to benefit from 6,000 pages of code. She does not need disclosure of the algorithm and all its data. She just needs to know if her application was rejected for causes that are unlawful. Here, openness will mean simplicity and comprehensibility.

By contrast, if the law makes a regulator responsible for conducting heavy-duty audits of digital *systems* – systems-level transparency – then the duty will be broader. This is because a reasonable challenge by a regulator will naturally be more intensive than one by an ordinary person. The duty might require audit trails, checklists and impact assessments, as well as the source code and data itself.[2]

Pulling these threads together, the duty of openness would look something like this:[3]

> *Digital systems must be sufficiently transparent to be reasonably challenged according to the legal rights and standards which govern them.*

If this seems frustratingly vague, it is not because I want to make extra work for lawyers, but because the real world is sometimes too complex to be governed by long lists of rules for every possible scenario. That's why many governance regimes use tests like *reasonableness*, *practicability* and *fairness*, which differ from context to context. In common law systems at least, the law develops over time to create a more granular body of rules that can be applied to new circumstances. We currently lack such a body of law to guide us. The duty of openness would allow for development in the law.

In essence, the duty of openness would ask those who design and control digital technologies to apply some of the genius and resources that made their innovation *possible* to making that innovation *acceptable* in a free society. And it would reduce the need for the state to gather the information itself through more intrusive means of investigation.

Let's look a little closer at how it might work in practice.

***

A duty of openness would help citizens and public authorities answer two types of question:

Why?
What are you?

Let's start with the first. A person who is rejected for a job, declined a loan or booted off a social media platform will usually have one question: *why?* Where our rights or life chances are significantly affected by a decision, we expect to be given a reason. Judges, for instance, must explain their determinations. A world in which the powerful never had to justify their actions would be dark and confusing.

For *why* questions, the appropriate response is usually an *explanation*. With machine learning systems, that could take the form of 'information about the factors used in a decision and their relative weight', simplified if necessary.[4] Thus a high-stakes algorithmic system could provide a list of the factors that were taken into account, ordered by their significance.[5] That would enable people to understand whether race might have (even indirectly) played a part in the decision to charge them higher insurance premiums, or whether a credit algorithm inadvertently took into account the wrong data.

Another way of providing explanations would be to allow us to ask questions, factual or counterfactual:[6]

What are the characteristics of individuals who received similar treatment to me?
Are individuals similar to me classified erroneously more or less often than average?
Would the result have been different if I had been a man rather than a woman?
Would I have been granted parole if it were not for my earlier conviction for robbery?[7]

These kinds of question allow us to challenge decisions that have gone against us. And those who deploy powerful technologies should be able to answer them. If that poses technical or commercial difficulties, then it should be for the tech provider, not society, to bear the burden of overcoming them.

The good news is that we don't normally need to see the internal plumbing of a computer system any more than we need to inspect the brain tissue of a human decision-maker.[8] Explanations can generally be provided without the need for companies to disclose any proprietary algorithms or trade secrets.[9]

Explanations are useful not only when systems make apparently unfair decisions, but when they make weird decisions – like when a self-driving car stops in the middle of the motorway or a text-generation system uses a derogatory term. In cases like these, too, an *explanation* may well suffice for us to reasonably check that our rights have not been violated.

Sometimes the law will govern *systems* or *rules* rather than individual *decisions*. In these cases, a different kind of openness might be required, particularly if the systems or rules in question are to be challenged by professional regulators rather than lay citizens.

To test the safety of a self-driving car, regulators may well want to look beneath the bonnet at the nuts and bolts of the computer systems that make it run. They might need detailed information about the system's models and parameters, its training data and performance metrics, and its feedback mechanisms for evaluation and improvement.[10] It will sometimes be possible, however, for regulators to challenge a system without having to look under the bonnet. If the question is whether a recruitment algorithm discriminates against people of colour, then statistical evidence of how people of colour have fared against other groups might be enough to discharge the duty of openness. In that case, there might be no need for recourse to the algorithms or data itself.[11] Likewise, if the law required self-driving vehicles to be at least as safe as human-operated ones, then statistical evidence of comparative incident rates would go a long way to satisfying the duty of openness.[12]

***

In May 2018, Google unveiled a digital assistant that could make appointments and order pizza over the phone. Though harmless, the system was controversial because it adopted human mannerisms and affectations ('Mm hmm') that made it almost impossible for the recipient of the call to know they were speaking to a bot.[13] People found something unpleasant about this – but they only knew about it because Google's demo showed them.

Before long, we could be surrounded by nonhuman systems that do not identify themselves as bots – speaking to us over the phone, encouraging us to buy things, arguing with us about politics on the internet. Automated systems already take decisions about our lives without telling us. It's entirely possible that your most recent job application was rejected by a machine, not a person. And as digital technology improves and spreads, it will get harder to tell a chatbot from a human, a deepfake from a legitimate video, computer-generated text from the writing of a person, a self-driving car from a human-driven car. The duty of openness would enable us to ask *what are you?* and get a truthful answer. Of course, the answer we receive would depend on what the law required. A Turing Red Flag Law would require that 'an autonomous system should be designed so that it is unlikely to be mistaken for anything besides an autonomous system, and should identify itself at the start of any interaction with another agent.'[14] Promisingly, something similar is proposed in the EU's draft Artificial Intelligence Act.[15] A narrower version of this rule has already become law in California, where computer systems are prohibited from impersonating humans in certain circumstances.

***

The principle behind the duty of openness should not be controversial, in Europe at least. The GDPR already requires data controllers to put in place the measures needed to demonstrate that they are compliant with the regulation.[16] The idea is not that every digital system is challenged as a matter of course. But the onus should be on technology owners, designers and manufacturers not only to ensure that their systems are compliant, but to be able to show it – not out of charity but as their duty under the law.

What if a technology provider refuses to provide the transparency needed for a reasonable challenge? The answer is straightforward. If the technology is subject to a challenge by a citizen or regulator, that challenge should automatically succeed. A tech firm should not be able to hide behind its own lack of transparency to evade legal responsibility.

What about where a provider *cannot* provide the requisite transparency, for technical or commercial reasons? Well, if there are genuinely good reasons for a lack of transparency, such as concerns for privacy, then as far as possible the provider should be subject to confidential audit by a public authority. But if a particular system simply cannot be used in a way that enables anyone to reasonably check whether it is compliant with the law, then the problem is with the technology, not the law. Such a system should have no place in a free republic. A recruitment algorithm that rejects candidates for reasons that are literally incomprehensible is not worth using, not because its decisions are wrong, but because republican liberty requires us to be able to understand the logics of the systems that govern us. The ultimate surrender would be to entrust our future to machines we cannot comprehend, still less challenge. If that means some machine learning systems cannot be deployed in certain social contexts, then so be it. Technology must be led by politics, not the other way round.

PART IX

GIANTS, DATA AND ALGORITHMS

Tyranny is the wish to obtain by one means what can only
be had by another.
Blaise Pascal

THIRTY-TWO

Antitrust, Awakened

This part of the book considers two thorny challenges associated with the power of digital technology. The first is how to deal with the almost unimaginable size of the tech giants. The second is how to regulate personal data flows and high-stakes algorithms. Digital republicanism offers possible solutions for both.

Let's start with size.

***

Five years ago, calls to break up the tech giants were heard only on the margins. Now they are mainstream. In Congress, a new Antitrust Caucus is on the prowl.[1] Senate Democrats have called for a 'Trust Buster' to stop the abuse of market power.[2] Elizabeth Warren ran for president on a pledge to break up Amazon, Facebook and Google.[3] Influential voices on the right, too, assail the power of big tech.[4] In 2020, fifty attorneys-general unleashed a cannonade of antitrust actions against Google. Nearly as many did the same against Facebook.[5] One of Facebook's own billionaire co-founders has called for it to be broken up. [6] We have entered the age of the New Brandeis Movement, aimed at breaking up concentrations of corporate power.[7]

The clamour conceals a degree of sheepishness. American policymakers know that there have been no massive antitrust decisions in America since the 1990s and that this is a serious failure.[8] In the last decade or so, Facebook, Google and Amazon have happily hoovered up hundreds of smaller enterprises which could have bloomed into potential rivals.[9]

Across the pond, the European Commission has been confronting the tech giants with more gusto. It has slapped fines on Google for unfairly promoting its own services over rivals, for using its Android platform to cement its search dominance and for stifling competition by placing heavy requirements on third-party websites.[10] Facebook, Intel and Qualcomm have also fallen foul of the European antitrust regime. But according to one Commissioner, while 'structural separations' are on the table, they are not immediately on the cards.[11]

Why is antitrust – or as we call it in Europe, competition law – suddenly cool again? The facts speak for themselves. Five of the most valuable companies in the world are American tech companies.[12] The users of Apple and Facebook outnumber the populations of entire continents. Outside China, where it is banned, a majority of the world's population has an active Facebook profile.[13] Apple's market capitalisation alone is roughly the same as the GDP of Denmark, one of the richest countries in the world.[14] (Denmark, appropriately, is now onto its second 'digital ambassador' to the big tech companies. The first incumbent now works at Microsoft.[15]) Google commands 98 per cent of the market for mobile search.[16] More than 70 per cent of all internet referral traffic goes through sites owned or operated by Google and Facebook.[17] Nine in ten smartphone users use Google or Apple operating systems.[18]

Waves of mergers in the last three decades have seen a great concentration of wealth across the economy, and in the tech sector particularly.[19] A scramble is underway, not merely for market share but for our time and attention. The tech giants have become part of the social infrastructure. Without them many of our daily interactions, transactions and activities would be impossible or very different.[20] Amazon's CEO says that his goal to make sure that 'if you are not a Prime member, you are being irresponsible'.[21]

***

The growth of the tech giants can be explained in a few ways. One of the most important is the network effect. The more people who use a networked system, the more valuable it becomes. In fact, a network's value increases exponentially as its numbers increase, so doubling the number of users will quadruple the value of the network, and so on.[22] This makes it hard for users to leave a platform, because doing so would mean starting again with no friends or followers. It also makes it harder for rivals to elbow in on the action. You could design an online marketplace that was superior to Amazon in every technical respect, but if it only had a few hundred users, rather than a global network of millions of buyers and sellers, it would not pose much of a threat. Amazon's size, like Facebook's, is partly what makes it unassailable. That's why instead of looking to replace the tech giants, today's entrepreneurs can often hope merely to be acquired by them. As Zephyr Teachout notes, when another online merchant began to rival Amazon in the market for nappies, Amazon promptly bought it and shut it down. When Amazon and the publisher Hachette got into a dispute, Amazon downgraded Hachette's books on its website and started delivering them two weeks late.[23] Venture capitalists are wary of investing in start-ups that compete against the core offerings of the tech giants. These are known in the industry as 'innovation kill zones'.[24]

On top of the network effect, tech giants benefit from economies of scale. As giant businesses grow, it costs them less and less to add extra users to their services.[25] In turn, they accumulate more and more data, which allows for the development of ever-more capable machine learning systems. This, too, makes competition harder. We are witnessing a wealth cyclone, sucking in everything around it and destroying stragglers in its path.[26]

***

Large corporations have always derived power from the wealth under their command.[27] That's not new. Their size gives them the ability to shunt capital around the economy, employ armies of workers and choose which regions will prosper or decline. When Amazon announced that it was seeking a location for its new headquarters, American cities scrabbled desperately to win its favour, like pimply suitors chasing a debutante at a nineteenth-century ball. New York alone offered more than a billion dollars in tax breaks. The same mating ritual plays out on the international stage, as corporations chivvy governments into offering the lowest taxes and lightest regulation.

Tech firms translate their economic heft into political influence. An astonishing fact is that 94 per cent of members of Congress with jurisdiction over privacy and antitrust issues have received money from a 'Big Tech corporate PAC or lobbyist'.[28] The two biggest corporate lobbyist spenders in the US are Facebook and Amazon.[29] In 2017, Google's lobbyists occupied an office space in Washington DC as large as the White House.[30] The tech giants raise funds for political candidates on both sides of the aisle[31] and lend employees to political campaigns to nurture political relationships.[32] They fund think tanks and academic research, giving their works the sheen of scholarly respectability.[33] They plant op-eds in newspapers to cultivate public opinion.[34]

There are thus two basic ways that size translates into power. One is through command of economic resources. Another is influence on the conventional political process. Then, as argued in this book, a third source of power is ownership and control of digital technologies that exert power directly. It's the combination of the three that makes the tech industry unusual.

***

The republican tradition is hostile to concentrations of economic power. As enterprises grow bigger, they are inherently more likely to dominate the lives of the citizenry. And supermassive entities can destroy or swallow up their rivals, leaving people with no alternative but submission. These challenges, of concentration and (lack of) competition, have led republicans to argue that power should be dispersed around society rather than allowed to accumulate in the hands of a few. More than 2,000 years ago, Polybius noted that the Roman Republic divided power among the Consuls, the Senate and the People.[35] Machiavelli in the fifteenth century, and Montesquieu in the eighteenth, also argued that power could not safely be left to gather in the hands of a few.[36] In Montesquieu's time, republican thinkers regarded England's 'mixed' constitution as attractive because of its 'peculiar capacity to balance and check the basic forces within society'.[37] The Founders of the American republic took this lesson to heart. No thinker was quoted more than Montesquieu in the revolutionary era.[38] James Madison pronounced him an oracle.[39] Ironically, many of the American patriots fighting the British redcoats believed they were fighting for the lost English ideal of the freeman, protected by the 'balanced counterpoise of social and governmental forces'.[40] In the *Federalist Papers*, Alexander Hamilton counselled for the 'distribution of power into distinct departments'.[41] Madison argued that 'the accumulation of all powers, legislative, executive, and judiciary, in the same hands . . . may justly be pronounced the very definition of tyranny'.[42] In the economic arena, Jefferson wanted a 'restriction against monopolies' to be written into the Bill of Rights.[43]

As American capitalism gathered pace, these threads of republican idealism were woven into a seam of ideas usually associated with Theodore Roosevelt and Louis Brandeis. These men supported the aggressive break-up of trusts and corporate interests that had grown too large.[44] In the economic realm, as well as the political, they saw concentrated power as a recipe for unfreedom. What made their ideas republican was the insistence that the mere *capacity* to abuse power was enough to cause concern. For as long as corporate entities were able to upend the political order (even if they chose not to), the republic was not free. As Brandeis put it:[45]

> benevolent absolutism . . . is an absolutism all the same; and it is that which makes the great corporation so dangerous. There develops within the State a state so powerful that the ordinary social and industrial forces existing are insufficient to cope with it.

***

It is one thing to say that excessive corporate power is undesirable. It is another to say what should be done. We already have a mature legal regime for governing the abuse of economic power – the one described at the start of this chapter. It is known as antitrust law, or competition law (for simplicity, I refer to it as antitrust).

Yet antitrust law has not done a terribly good job of constraining the might of the tech giants in recent years. This is partly because the existing law has been enforced too limply, particularly in America. But the problem also lies with the law itself, at least as it is currently interpreted.

The first issue is technical. Antitrust law uses economic terms and models that can be hard to apply in the digital context. To take just one example, in working out whether a company is abusing a dominant position, the orthodox approach is to start by identifying the specific market in which it is dominant. But tech companies are more amorphous than companies of the past. Markets are many-sided and overlapping, companies span more than one at the same time, and often they provide their services without upfront charge.[46] It can be hard to find a technical hook on which to hang an allegation of abuse.

A second challenge is that the law is simply not tough enough, or at least not applied with enough force.[47] Too many cosy settlements, too many slaps on the wrist. Some critics call for greater scrutiny of prospective mergers.[48] Others demand more proactivity in breaking up businesses. Tim Wu, for instance, asks what the social cost would be if Facebook were forced to divest WhatsApp and Instagram from its portfolio of companies.[49] For that matter, as Zephyr Teachout asks, is it right that Amazon is not only the Everything Store but a credit lender, a publisher, a television producer and a major supplier of cloud-computing services, able to decide whether to pull entire social media platforms from the internet, as it did in 2021?[50] As Teachout points out, America has a record of breaking up companies according to function. The 1933 Banking Act banned investment banks from taking deposits. The 1934 Air Mail Act divorced airlines from aircraft manufacturers. Why not decide that 'a search company cannot also be an ad company, or a store, or a map provider'?[51]

The third and most fundamental challenge to antitrust law, however, goes beyond the need to beef it up. The New Brandeis Movement maintains that under the prevailing orthodoxy, the law cannot adequately protect against the harms of digital dominance, no matter how stringently it is applied. Since the 1970s, American antitrust law has proceeded on the basis that only one kind of harm is relevant: 'consumer welfare'.

In fact, this orthodoxy, known as the Chicago School, prizes one aspect of consumer welfare above all others: prices. For the last half-century, prices have been the principal lens through which concentrations of power have been examined. Will this merger lead to higher prices? Will breaking up this conglomerate lead to lower prices? Has this dominant player charged customers more than it would have in a more competitive market? Even in economic terms, these are fairly narrow tests.[52] And they seem dated in a world where many of the tech giants provide their services without upfront charge. As a matter of political and legal history, it's far from clear that the Chicago School is right in its central contention that Congress in 1890 and 1914 intended for antitrust law to be limited to the scrutiny of consumer welfare, as reflected in prices.[53] Certainly that was not how Theodore Roosevelt saw it, as he and his successors sought to dismantle Standard Oil, US Steel and AT&T in the 1910s.[54] The government, he said, 'must be freed from the sinister influence or control of special interests':[55]

> The citizens of the United States must effectively control the mighty commercial forces which they have called into being.

Roosevelt was right. Antitrust should not just be about prices, or even just about the consumers that use a particular product or service. Concentrations of power affect all of us. If the Brandeis critique is right, and antitrust law needs a reboot, what form should it take? That's the subject of the next chapter.

THIRTY-THREE

Republican Antitrust

We cannot reform antitrust law without first understanding its inherent limitations. For one thing, size is sometimes what makes the tech giants useful. Consider Facebook's core social media platform. It is useful precisely because so many other people are on it. When you log on, you can be pretty sure that almost everyone you know can be found there too. Few want to break up this offering. It would harm users as well as Facebook.

That is not, however, a prescription for doing nothing. On the contrary, as Theodore Roosevelt argued in 1910, the alternative to antitrust is regulation: 'The way out lies, not in attempting to prevent such combinations, but in completely controlling them in the interest of the public welfare.'[1] The new forms of democratic oversight and counterpower set out earlier in this book could adequately regulate businesses whose size is what makes them useful.

If increasing the amount of competition is the aim, antitrust law is not the only way of achieving it. Sometimes lack of competition is caused by the difficulty of moving between alternatives rather than a lack of alternatives. You might be tempted to join a new start-up platform, but if all your treasured photos are on Instagram and there's no easy way of shifting them over, that's a strong disincentive. To meet this problem, the GDPR introduced a *data portability* right. It gives users the right to have their personal data transmitted directly from one entity to another (where it is technically feasible).[2] Thus if you want to move from one streaming service to a rival, it should be possible for your personal preferences to be transferred to a new service.[3] It's the same logic that lets you keep your mobile phone number when you switch to a new provider.[4] To their credit, Microsoft, Google, Facebook and Twitter have begun collaborating on their own data portability initiative, which could enable, for instance, Facebook photos to be moved to Google Videos without hassle.[5]

There are, however, loopholes in the GDPR and the data portability right does not apply in all circumstances.[6] As Lilian Edwards points out, we could go further and require *interoperability* between platforms. This would allow members of different platforms to interact directly with each other – imagine sending a Twitter direct message into a friend's Facebook Messenger inbox.[7]

Of course, interoperability would require more cooperation between tech providers, not less.[8] Which leads to a third reason to be cautious about antitrust: sometimes we want companies to work together. If someone is being harassed by a troll on one platform and moves to another, only to be trolled by the same person again, they should not have to seek fresh aid from the second platform. Companies should be encouraged to cooperate to prevent bullying or the spread of misinformation.[9] Haydn Belfield and Shin-Shin Hua identify fourteen areas in which cooperation between AI developers might be desirable.[10] Yet the general effect of antitrust law is to encourage rivalry, not cooperation, and so it can work against the public good if not wielded judiciously.

A final consideration, and perhaps the most important, is that unaccountable power is not always caused by monopoly. Often it is caused by a combination of economic, social, legal and technical factors. Imagine that today's giants were broken up into a thousand smaller companies. This would lead to a reduction in the concentration of power. But as we saw in chapter ten, making power less concentrated is not the same as increasing freedom for the rest of us. People will still be unfree if they are the passive subjects of rules written by others, or subject to moral codes that are alien to them. They will be unfree if they are made to live under constant scrutiny, or depend on others for the quality of their public deliberation. They will be unfree if they live at the mercy of hidden algorithms for their access to social goods.

Domination of this kind is caused by systemic factors, not just large corporations.[11] It would be foolish to break up a handful of leviathans only to see them replaced by a thousand smaller oppressors, like replacing a tyrannosaurus rex with a pack of velociraptors. Antitrust on its own will never be enough.

***

Having sounded these notes of caution, we can sketch out the principles of a new system of structural regulation. Call it republican antitrust law.

The starting point is that a republican antitrust regime would recognise its own limits. 'Break them up' is just a slogan, and populist sentiment alone does not guarantee liberty. Antitrust is just one tool in the republican toolbox. The parsimony principle dictates that the state should do no more than necessary, and breaking up firms should not be done lightly. Thus, where it is possible to counter domination using other forms of counterpower, the state should presumptively favour those.[12]

With that note of caution, it is clear that antitrust law cannot be geared towards consumer welfare alone. It should be aimed, first and foremost, at preventing corporations from becoming too powerful in social and political terms, not just economic. When scrutinising an acquisition or contemplating a break-up, power should be among the first criteria for regulators to consider. This is 'political antitrust', a lost American tradition.[13] In my last book, I called it *the new separation of powers*.[14]

Some scholars have already called for 'citizen welfare' to replace 'consumer welfare' as the test for mergers.[15] There is wisdom in this. In fact, there is no limit to the factors that could become the goals of antitrust policy: system resilience, labour standards, wealth equality and media diversity could all factor into determinations by republican antitrust regulators.[16] Taking the latter as an example, if a large social media platform proposed to swallow up a smaller one, that acquisition might be blocked in the name of media diversity even if it had no adverse economic implications. Antitrust authorities in the digital republic would thus consider the broader welfare of the republic when they review market activity, not just the merits of a particular transaction. And they would consider the *potential* for abuses of power in the current economic arrangements, taking the necessary action to break them up without waiting for disaster to strike.

A republican antitrust policy would acknowledge that consumer welfare and citizen welfare are not just alternatives, but have the potential to be rivals too. From a consumer perspective, a free digital service might seem marvellous. But it might also be utterly corrosive to democracy, and thus undesirable from the perspective of the citizenry. Current law and practice would prioritise the consumer. A republican antitrust regime would strike a balance between the two, but prioritise the citizen over the consumer, and the republic over both. Republican antitrust law must also make space for cooperation where it is in the interests of the republic. That might mean exempting certain categories of behaviour – cooperating on security, for instance – from the regime altogether.

Traditionalists will say that this approach would upend fifty years of practice, turning antitrust law into something different from the limited economic instrument they imagine it to be. They would be right. We cannot tackle the political problems of big tech using the parched vocabulary of twentieth-century economics. Under a republican antitrust regime, the process of scrutinising a merger or pursuing a break-up would undoubtedly look different.

Unlike the present model, a republican antitrust regime would involve the citizenry as far as possible. Antitrust should not be the preserve of a narrow clique of lawyers and economists.[17] Why not convene deliberative mini-publics – perhaps citizen juries – to judge the acceptability of particularly sensitive mergers alongside the professionals?

***

When we speak of republican antitrust or the new separation of powers, we imagine a regulatory regime that prevents the concentration of social and political power in corporate hands, balances economic considerations with broader public policy goals, involves the citizenry in big decisions and recognises its own limits. It would break up concentrations of social and political power, as well as economic power. Whether you see this as a reversion to tradition or as a new departure is ultimately less relevant than whether this system of antitrust would better serve the needs of the next century. I believe it would.

THIRTY-FOUR

Beyond Privacy

Data protection law and *privacy* law are often spoken of as if they were the same thing. But governing data is about more than ensuring personal privacy. Without suitable controls, data can be used by one group in society to dominate others – whether by distortion of public debate, manipulation of individuals or algorithmic determinations that affect people's life chances. For republicans, the purpose of governing data is to reduce this dominating effect, not merely to preserve the privacy of the individual. The next two chapters sketch out what republican data governance might look like.

***

We should start by acknowledging that the European system of data protection, the GDPR, already offers an advanced and sophisticated body of rules. A natural first step for the US would be to introduce a comprehensive data protection regime of the kind that exists in Europe, staunching the unbridled flow of personal data around the American economy. Lawmakers, policymakers, activists and even tech firms have called for a 'US GDPR'.[1]

An even braver approach, however, would be to move beyond the GDPR to a new way of understanding and governing the collection and use of personal data. This means challenging some fundamental assumptions.

Despite their differences, most data governance regimes share a common intellectual heritage. From the US to the EU (and its cousin regimes in Canada, Singapore, Australia and Japan), most are derived from the 'Fair Information Practices' (FIPs), a set of principles devised by the US government in the 1960s.[2] The FIPs have been reformulated many times over the years.[3] The general idea is that data collectors should gather as little personal data as possible, and that data should not be used for ancillary purposes without good cause. In America, the modern approach to the FIPs boils down to *notice* (consumers should be given clear notice of what is done with their data), *choice* (consumers should be able to choose how their personal data is used), *access* (consumers should be able to see what data is held about them, and correct or delete it if necessary) and *security* (personal data should be reasonably protected).[4] The European approach includes the additional idea that data collection should be limited to certain lawful purposes.

The FIPs were developed in an era when the very idea of data processing 'seemed revolutionary'.[5] In the 1960s and 1970s, there was not much electronic data around, personal computers were still in the future, and the commercial internet did not exist.[6] People were impressed by the potential of data analysis, but they feared that intimate or sensitive information would be seen by those who had no right to see it. This led to calls for 'privacy' measures. The FIPs struck a balance between the extraordinary potential of data processing on the one hand, and personal privacy on the other. Overall, and perhaps counterintuitively, the FIPs accelerated the commercialisation of personal data.

By introducing limited restrictions on conduct, and effectively okaying the rest, they cleared the ground for ever more gathering and processing.[7]

Today's data scientists, unlike those of forty or fifty years ago, usually want as much data as possible, regardless of quality or even pertinence. This is because machine learning systems often function better when trained on large but messy datasets rather than small clean ones.[8] Data scientists don't know what the data will tell them, or which aspects will prove to be valuable, until it has been processed. They might use it for purposes that were impossible to know at the time of collection.

This poses a problem for the FIPs, even in their beefed-up form under the GDPR. The biggest issue is what I have called the consent trap (chapter fourteen). To recap, 'notice' and 'choice' are both illusions. Meaningful consent is impossible when individuals cannot know what their data will reveal in different contexts, or combined with older data, or the data of countless others.

The FIPs regime is afflicted by the same fundamental flaw as market individualism: it leaves us to fend for ourselves. The FIPs were supposed to balance the rights of individuals with those of data processors, but in recent years the balance has become lopsided. Giving people control over their data is attractive in the abstract but makes little sense when the harms of data analytics are so often caused and felt collectively, not to mention well hidden. The FIPs treat data collection as a dyadic process involving two players: the data subject and the gatherer. But data law should also weigh a third factor in the balance: the interests of society.

If the FIPs approach is too individualised, then reforms that only inject more individualism into the system are unlikely to help. One popular idea, making companies pay people for collecting or processing their data, falls into precisely this error. 'If companies are profiting from it, you should get paid for it', runs the tagline.[9] But this is answering the wrong question. The problem with data processing is not that consumers get a bad commercial deal. On the contrary, exchanging personal data for free stuff is often a good bargain. Tech firms take our data, which is of little financial value to us, and turn it into something profitable. There's nothing commercially objectionable about that. The real problem is that data can be used to monitor, exploit or manipulate people. For the republican, the data dividend cannot compensate for that loss of liberty.

***

How, then, can we move past the FIPs into a new era of republican data protection? We should start by acknowledging that the most pressing area for reform is the *use* of data, rather than its collection. That is where the principal harms lie.[10]

A hardline strategy would be to ban companies from buying, selling or sharing personal data for profit. This is favoured by the Oxford scholar Carissa Véliz.[11] It has the benefit of clarity. And it would wipe out the data market overnight. But it might also entrench the dominance of companies that can gather their own data, as against smaller rivals that can only purchase it. It would also deprive society of many of the benefits of the data trade – of course, there are some.

More attractive is Véliz's proposal that the definition of personal data should include 'inferred sensitive information' – i.e. what the data *tells us* about people – as well as raw data itself.[12] Thus all the restrictions that apply to personal data would apply to the inferences from that data too.

Véliz and others also argue for an expiry date on personal data. A rule that required firms to destroy personal data after five years would prevent it from being reused time and again.[13] That said, once raw data has been used to train a machine learning system, it is pretty much impossible to withdraw it – like trying to remove an ingredient from a pot of soup after it's been cooked.[14]

A promising option would be to supplement *individual* consent with new forms of *collective* consent. Imagine placing your data rights under the management of a trusted third party who could make informed decisions on your behalf about how your data could be used, along with the data of thousands or millions of others. These collective bodies – call them data trusts, data coalitions or data unions – would also be in a stronger position to negotiate with industry players than individuals acting alone.[15] If a platform used data in a way that was unacceptable, it might face the threat of a data trust withdrawing the consent of thousands or millions of users all at once, not just a few disgruntled activists.[16] *That* would be a check on digital power.

A collective approach would also make it easier for citizens to decide which social causes should benefit from their personal data. Many of us are comfortable sharing our personal data for scientific and public interest projects, but less keen on corporate surveillance. Citizens could choose from a suite of packages, each representing a different set of values or priorities.[17]

We would need some new technical infrastructure to introduce a system of collective consent. But work is already underway. Researchers are hunting for alternatives to the 'data silo' model, which results in huge, centralised troves of personal data. So-called 'edge computing', 'personal data containers' and 'distributed machine learning analytics' are all promising fields of inquiry.[18]

***

Let's go further. There is a strong case for limiting what can be done with people's data, even with their consent.[19] One of the most important decisions a society can make is about which rights and freedoms may be signed away, and which cannot. In ancient Athens, the legislator Solon made it unlawful for citizens to pledge their bodies – to consent themselves into slavery – as securities for debts.[20] Nowadays, a worker cannot agree to earn less than the minimum wage. A debtor cannot promise a pound of flesh in the event of default.

The GDPR gestures at *some* limitations on what rights can be waived,[21] but on the key question – whether data collected for one purpose may be used for an entirely different purpose – it does appear to let individuals sign their rights away.[22] The GDPR also generally allows businesses to process data if they have a 'legitimate interest' in doing so – a term which has been applied and interpreted broadly.[23]

A political community is entitled to say that certain kinds of data should never be used for particular purposes, even if individual people have agreed to them.[24] Lilian Edwards suggests forbidding companies from requiring data when they don't need it.[25] Under such a rule, platforms would not be allowed to request, or require, consent for data gathering or processing where it wasn't necessary for the service being provided. This idea is attractive in principle, although it would be a legal minefield: companies that rely on gathering and selling data for their business model might say (with some justification) that data is necessary to the service they provide because without it, the business would be unprofitable.

Alternatively, in some contexts it might be appropriate to treat data processors as fiduciaries, or to impose tortious duties beyond what the parties have agreed. Say an applicant for credit or insurance has consented to personal data being used to assess their application. The law might say that the algorithms may only use specified categories of personal data where doing so would lead to a more favourable outcome for the applicant than would have been achieved without it. The less powerful party would be able to benefit from the wonders of machine learning without worrying about being punished. Alternatively, a policy of 'forbidden variables' would hold that certain factors X should play no part in determination Y. For instance, whether Pete's friends have paid their debts should play no part in whether Pete is able to get a loan. Or the law could specify what data may be used. Instead of letting recruitment algorithms comb the internet for the social connections, sexual history or shopping habits of candidates, perhaps they should only be permitted to process data like qualifications or work experience.[26]

Whatever approach is adopted, what matters is that a future regime places responsibility on data processors to comply with the law, rather than leaving individuals to fend for themselves.[27]

***

The FIPs began as general guidelines for the gathering and use of data. But as technology grows more sophisticated, we will need a more contextual way of thinking about data.[28] An acceptable data practice in the workplace may not be right in healthcare, or in banking, or insurance, credit, advertising, or criminal justice. Society has developed complex norms for governing these fields of activity, so why should data be governed largely in the abstract, divorced from its actual social functions?

THIRTY-FIVE

Acceptable Algorithms

Most of the world's algorithms are inconsequential. They perform unglamorous tasks quietly and with little fanfare. They matter only to the businesses that use them. A small number of algorithms, however, carry more social importance. They are used to determine who gets a job, or a college place, or a mortgage. They frame and shape our perception of the world. When these algorithms go wrong, it matters.

But what does it really mean to say an algorithm has gone wrong?

It might mean there's been a technical error, like faulty programming or data corruption. More commonly, it will mean that an algorithm has been trained using data that is incomplete or skewed. Voice recognition systems that have been trained only on the voices of people with typical speech patterns will not be able to process the speech of people who stutter, or speak with slow or slurred speech. This is significant – around 7.5 million Americans have difficulty using their voices.[1]

Even algorithms trained on data that is supposed to be representative of the real world, however, can produce repugnant results. The text-generating algorithm GPT-3 generates violent and terroristic stereotypes of Muslims because the vast language dataset on which it is trained – taken from actual human speech – contains a lot of speech associating Muslims with bombs and murder.[2] The real world contains patterns of injustice. These patterns are reflected in data. Algorithms reproduce and amplify them.

Thus while important algorithms should be engineered with precision, they cannot be judged by *technical* standards alone. A well-designed algorithm, like a well-oiled chainsaw, is not inherently objectionable but nevertheless capable of gruesome results if wielded carelessly.[3]

***

The *democracy principle* (chapter eighteen) requires that certain algorithms – like other vectors of digital power – should be brought into broad alignment with the moral standards of the community. That sounds simple enough, except for two factors.

Firstly, moral standards are *contextual*. Every sphere in which algorithms are used – employment, credit, insurance, healthcare, welfare, the military, education, criminal justice and others – has its own moral code. Thus a norm that applies in a courtroom (*every defendant must receive a fair trial before suffering punishment*) would not necessarily apply on the battlefield (*in the theatre of war, there is no need for a trial before taking the life of a hostile assailant*). The norm that underpins public education (*provision should be made for every child to have a place in school*) does not apply in the job market (*jobs are finite in number, and they should go to the candidates who deserve them most*). Just as expectations about human behaviour differ by context – a swimsuit is OK at the beach, but not in the office – an algorithm that is acceptable in one context may be repugnant in another.

Secondly, moral standards are *contested*. In each context, reasonable people disagree about right and wrong. Imagine an algorithm used to distribute welfare payments to unemployed people. Everyone in the community might want the algorithm to work justly, but their conceptions of what justice requires may differ sharply. One group might say that *everyone must have enough to get by, but no one should get more than necessary*. An algorithm engineered on this basis would make sure that the very poorest received enough to survive, but that those with assets or savings get nothing. Another group might retort that *welfare payments should be earned*. Their algorithm would only distribute payments to those who were actively job-hunting, giving nothing to those who idle at home (no matter how needy). Same context, different visions of justice.

Meaty moral questions are always going to divide opinion. But if reasonable disagreement is inevitable, how can we decide the moral standards that apply to algorithms in different contexts?

The first step is to decide which contexts are apt for governance at all. Remember the *parsimony principle*, which requires that the law should never extend further than necessary. Algorithms that are morally uncontroversial or socially unimportant should not attract the attention of the law. There is no need for the community to fret about the moral nuances of logistics or accountancy software.

But as we have seen, there are some algorithms that are too important to be left to the market. These algorithms share a few important characteristics. First, they are likely to have a significant impact on the lives of those they affect. Second, they make decisions that are moral or political in character, in that they invariably optimise for some values over others.[4] Third, they tend to be used in scenarios where the decision-maker is in a position of relative power to the person who is the subject of the decision. We have been calling these kinds of algorithms *high-stakes algorithms*.

High-stakes algorithms ought to be governed according to the moral standards of the community. How? Through the political process. It is for the citizenry to decide the moral standards by which algorithms should be governed, just as they decide other big moral questions like the law governing abortion, climate change policy or media regulation. A sensible way to handle moral disagreement is to subject it to the rigour of deliberative democracy and find a decision that can legitimately bind the community.

***

Stepping back, then, republican governance of algorithms means ensuring that:

> High-stakes algorithms must be (a) technically sound, and (b) consistent with the moral standards of the community in the context in which they are used, as determined in democratic processes.

This is a deliberately broad formulation. It moves the debate on from endless discussions about algorithmic discrimination. In recent years, there has been an explosion of scholarship about whether existing anti-discrimination regimes can be applied to new algorithmic scenarios. (The general consensus is: sometimes, but not always.[5]) But legal analysis can only take us so far. For one thing, knowing what the law *is* and how it applies is not the same as knowing what the law *ought to be*. A facial-recognition system that persistently misidentified people with unusual facial features (short of severe disfigurement) would probably not be considered discriminatory in a court of law, as it would not implicate any legally protected characteristic. The question is not merely whether this is unlawful, but *should it be?* If the algorithm offends the moral standards of the community in the context in which it is deployed, then the answer might be yes.

The problems caused by algorithms are not limited to discrimination, so governance should not be so limited either. A social media algorithm that degraded the integrity of the electoral process would be objectionable, but not because it is discriminatory. To call an algorithm discriminatory, like calling it *biased*, tells you nothing about whether it is morally objectionable. A university admissions algorithm weighted in favour of students from deprived areas would be both biased and discriminatory – but in the context of widening access to higher education, it might also be morally acceptable.

***

One morally vexing aspect of high-stakes algorithms is that they find patterns without understanding or explaining them. Indeed, most 'learning' algorithms are utterly uninterested in the *why* of their predictions. (There is a 94.7 per cent correlation between cheese consumption and fatal bedsheet tangling accidents. But why?[6])

Some people see no problem with using systems like these to decide important questions about life and society. If the data says that people who use Hotmail for their personal email address are more likely to crash their car, then make them pay more in insurance premiums. Who cares if we don't understand why?

This is a somewhat stunted moral attitude, and it reflects the *computational ideology* described in chapter nine. To the republican mind, a society in which important decisions are made for incomprehensible reasons would not be a free society. It would leave us utterly in thrall to technology, trusting without understanding.

There is a consistent and reasonably straightforward way to determine whether a particular algorithm is objectionable. Consider a problem from earlier in the book. Statistically, customers who enter their names on online forms using all lower case letters (*jamie susskind* rather than *Jamie Susskind*) are more than twice as likely to default on their loans. Should mortgage lenders be able to decline credit or charge higher interest rates on this basis?

A morally serious response requires answers to three questions.

The first is whether there is a known *causal* connection between lower-case typing and mortgage default. Do we know if the former *causes* the latter? If the machine can help answer this question, great. But it may not be able to.[7]

If no causal connection can be found, is there a *common sense* way of drawing a link between the two variables, even if it's not the actual basis for the correlation? Perhaps those who are lackadaisical with their typing are also less punctilious in their financial management.

If it's possible to find either a causal or common sense connection, the final question is whether that connection is *morally acceptable*, having regard to the values and priorities of the community. For example, there may be an overriding concern that charging poor typists more would disproportionately affect people who don't type in their jobs, or the less educated.[8] A different political community might decide that those who fill out their forms carefully are morally entitled to the fruits of their fastidiousness. Either way, if there is no causal or common sense connection between data and a decision, or no acceptable moral connection either, then the data should not be used in that context.

Machines have a dazzling ability to find patterns. But we do not need to reorganise society on the basis of those patterns. Some things are more important than the data.[9]

***

A final necessity when governing algorithms is the need to allow for exceptions. Centuries ago, Aristotle observed that any decision-making process based on generalised rules would occasionally give rise to bad decisions. Any such process, he argued, had to be able to provide special treatment for special cases.[10] In the context of laws, as Lord Sales, a UK Supreme Court judge, has shown, Aristotle's solution was that judges and magistrates should have wiggle-room to make exceptions.[11] He called this *epeikeia*, usually translated as equity. Republican jurists seized upon the concept of *epeikeia* with gusto. Cicero cited it as a means to soften Roman law, which was 'so formal and so precise' that even well-founded claims would fail on technicalities.[12] As Frederick Schauer notes, similar considerations guided the emergence of equity in English law.[13]

Aristotle's logic applies to machine learning algorithms. Even if the data points a certain way, certain decisions ought to be subject to human oversight to make sure that injustice has not been done in the circumstances. Take, for example, the automated systems used on social media platforms to remove inauthentic content. Ninety-five out of every hundred deepfakes might breach a platform's terms of service, but the other five per cent might be legitimate satire.[14] Likewise, flesh-coloured images in a particular arrangement may well be pornographic, but a small number will be tasteful advertisements for life-drawing classes or close-up photos of peaches. But these, too, may be taken down by a pattern-finding algorithm – with potentially disastrous consequences for the businesses affected. In these cases, the algorithm is functioning properly, but the exceptions give rise to injustice. That's why we cannot blindly follow their decisions. The tech tribunals described in chapter twenty-three would allow for equity to be done.

***

It is not easy to identify and distil our moral standards before applying them to algorithms. The process demands unusual levels of clarity about what really matters to us.[15] But that's because algorithms do not have ethical intuitions of their own. Humans design them (even if they design them to work in unpredictable ways) and those entrusted with that power can't just be left to make important decisions on behalf of everyone else.

Some say it would be wrong to govern algorithms more strictly than we govern humans who perform the same tasks. Humans, after all, make irrational and unfair decisions all the time. A juror might find a defendant guilty because he doesn't like the set of his jaw. A manager might reject a job applicant because she doesn't like the colour of his tie. A judge might hand down a longer prison sentence because he stubbed his toe shortly before hearing legal argument. People justify their decisions with plausible-sounding reasons which bear little resemblance to the truth.[16] So why should algorithms be exposed, probed and held to a higher standard?

There is force to this objection, but it rests on a misunderstanding. While it would be impossible to require people to disclose their secret reasons for acting, algorithms are not people. They lose nothing when we expose and tinker with their inner workings. They have no dignity to be violated, no conscience to leave unmolested, no feelings to be hurt, no ego to be dented, no self-esteem to be diminished. Algorithms can be closely regulated without a concomitant loss of human freedom. Yes, governing algorithms means that those who design and control them must ensure that they are in conformity with the law. But that is a modest price to pay, particularly for those algorithms whose effects on society are truly systemic.

Looked at this way, algorithms present a social opportunity, not just a threat. The important ones can be engineered to conform to, and thus help deliver, our shared priorities, like liberty and democracy. Code need not be a threat to the republic.

PART X

GOVERNING SOCIAL MEDIA

The problem of maintaining a system of freedom of
expression in a society is one of the most complex any
society has to face. Self-restraint, self-discipline, and
maturity are required . . . The members of the society must
be willing to sacrifice individual and short-term advantage
for social and long-range goals.

Thomas Emerson

THIRTY-SIX

The Battlefield of Ideas

How should social media be regulated, if at all? The world has begun to grapple with this puzzle. There are no easy answers. Reasonable people disagree on almost every element of the debate. Not many would say they are happy with the ragged state of online discourse, but there is also unease about proposals which would allow the state to trespass in the forbidden territory of free speech.

It's time to find a path forward. That begins with a diagnosis of the problems associated with social media. The next chapter considers the regulation of media in the past, and what we can learn from it. Then we examine the rival philosophies at play. Then, finally, a tentative plan of action.

But first, four hard truths.

First, doing nothing is a choice in itself — a choice of surrender. Social media platforms are already drawing the borders of free expression and deciding the quality of democratic deliberation.

They order, filter and present information. They set rules about what may be said and who may say it. They approve and ban speakers and ideas. And when they do these things, they necessarily apply their own rules, principles, biases and philosophies.[1] None of this is a criticism. But it does mean that the choice is not between governing speech and leaving it alone. It is already being governed – by the platforms themselves. The question is whether platforms should be left to their own devices or bound by rules set by the political community.

Second, 'free speech absolutism' is not an option. It is neither desirable in theory nor possible in practice to protect every utterance equally. Every democracy recognises that certain forms of speech – threats, bribes, perjury, sedition, fraud – are not deserving of the law's protection. Even societies that sanctify free speech, like the US, legislate against false advertising, workplace harassment and defamation.[2] The Securities and Exchange Commission (SEC) controls what people may say when they sell financial products. The Food and Drug Administration (FDA) lays down what must (and must not) be said about food and drug products. The Federal Trade Commission (FTC) restricts 'unfair and deceptive' speech relating to trade.[3] Why? Because some restrictions are necessary to preserve civilised society. Permitting or protecting *all* speech would permit criminality and sedition of the kind that could destabilise the republic itself, undermining the system of laws that enables social order and stability. That's why every mature democracy has some kind of speech hierarchy, with precious political speech at the top and criminal speech at the bottom.[4]

Third, governing online speech inevitably involves trade-offs between quality and quantity of speech, between laxity and over-censorship, and between the rights of speakers and the rights of those harmed by their speech. Any balance will inevitably please some and disappoint others. As one headline put it, 'Everybody Hates Mark Zuckerberg—But Can't Agree Why'.[5]

Finally, regulating online speech is extremely hard as a technical exercise. For small platforms, burdensome regulation or open-ended liability would make survival impossible. For larger platforms, the scale of the challenge is mind-boggling. Every day, Facebook hosts billions of new posts and receives millions of review requests. Hundreds of thousands of hours of new video are uploaded to YouTube.[6] After a British teenager took her own life in 2017, Facebook and Instagram took down around 35,000 posts relating to self-harm and suicide *every day*.[7] While individuals experience online harms at a personal level, the work of moderation is necessarily 'industrial, not artisanal'.[8] At the planetary scale, mistakes are inevitable. As Monika Bickert, Facebook's Head of Global Policy Management, fairly puts it: 'A company that reviews a hundred thousand pieces of content per day and maintains a 99 per cent accuracy rate may still have up to a thousand errors.'[9]

In the face of these truths, caution and humility are advisable. Before ploughing ahead, therefore, let's start by considering the way social media platforms are currently organised and regulated.

***

In his celebrated *Abrams* dissent of 1919, Oliver Wendell Holmes argued for 'free trade in ideas'. The 'best test of truth,' he wrote, 'is the power of the thought to get itself accepted in the competition of the market'. This mercantile mindset has infiltrated Silicon Valley at three levels. First, the market governs rivalries between platforms, which compete for the attention of their users. Second, most platforms are ultimately intended as engines of profit for their owners. Third – and perhaps most faithful to Holmes's ideal – online speech itself has come to be treated as a commodity, like sugar or steel, which sells or flops according to the laws of supply and demand.[10] These factors have heavily influenced platform design, and through that design, the shape of public discourse. It is a triumph of market individualism.

Unfortunately, despite its prevalence, it is doubtful whether the market was ever a sensible guiding metaphor for free speech. It has allowed commercial logic to distort a sphere where it doesn't properly apply. And some of the most troubling failures of social media have been market failures in disguise.[11]

Let's begin with the business model. For most major platforms, data is integral. It allows platforms to build machine learning systems. And it allows them to show users content and advertisements that are personalised to their taste.[12] For these reasons, platforms are generally incentivised to gather as much data about as many people as possible.[13] This is part of what Shoshana Zuboff calls 'surveillance capitalism'.[14]

Advertisements are particularly important. Facebook makes 98 per cent of its revenue from advertising.[15] You've probably seen thousands of its ads, even if you didn't notice them at the time. Facebook uses 2 million features to predict how people will respond to a digital advertisement, from the last place they 'ate a hamburger' to the 'percentage of battery life left' on their phone.[16] The data allows Facebook to group users into categories that can be efficiently targeted by advertisers, who bid for slots in real-time instantaneous auctions.[17]

The pursuit of revenue through advertisements, however, can lead to perversities. Facebook's systems have targeted advertisements at people based on their sexuality and race, and at teenagers who feel 'worthless' and 'insecure'.[18] Ads for military gear were shown alongside posts about the 2021 storming of the US Capitol.[19] Google suggested to advertisers that their ads be placed next to search results for 'evil Jew' and 'Jewish control of banks'.[20] Twitter enabled advertisements targeted at users interested in keywords such as 'anti-gay' and 'white supremacists'.[21]

Platforms show people more of what they like and less of what they dislike.[22] In commercial terms this makes sense. But it can also have regrettable consequences. Instagram's desire to grab and secure its users' attention led it to steer paedophiles towards the accounts of young children.[23] Searches for 'Holocaust' have yielded pages promoting Holocaust denial.[24] It's not that the people working at these platforms are hostile towards Jews or partial towards paedophiles. It's simply that a major purpose of their systems is to maximise advertising revenue. And sometimes that leads to unpleasant results.

Business sense does not always make moral sense, particularly at scale. Platforms compete to grab people's attention and hook them for as long as they can.[25] The *virality* principle promotes content that is popular, as measured by clicks, shares and likes. But the most popular content is not always safe, or true, or conducive to a functioning democracy. Clickbait headlines like 'This Porpoise Has a Face Like Donald Trump And People Are Freaking Out' and 'You Won't Believe What Meghan Markle Has Said About 9/11' are more likely to get attention than actual journalism. Salacious gossip is more interesting than boring old facts.[26] The market is blind to the truth or validity of ideas so long as demand is high. Online 'rumour cascades' accelerate the circulation of false claims many times faster than true ones. Political falsehoods spread the fastest.[27]

As a result, misinformation is rife and shows no sign of abating. Since the first major panic about online misinformation in 2016, Facebook likes, comments and shares from outlets that regularly publish falsehoods have roughly *tripled*.[28] During the Covid-19 epidemic, millions of people saw social media posts claiming that eating garlic cures the virus, that the Pope had caught it and that it was caused by 5G technology.[29] The top-performing link on Facebook in the first three months of 2021 was a bogus article about the fatal effects of the coronavirus vaccine.[30] One study estimates that 45 per cent of tweets about the virus came from bots spreading false information.[31] This is an issue with medical information generally, as three-quarters of people use the internet as their first port of call for health issues. Many consult YouTube, where the most engaging medical videos are often medically unsound.[32] Likewise, many of those searching Amazon for basic information about Covid-19 were recommended books of quack (but sadly popular) 'science'.[33]

Platforms have tried to clamp down on fake news in recent years, with some success. But this will be an uphill struggle as long as they are fighting their own business model. And moderation and fact-checking are expensive. Platform owners are incentivised to do the absolute minimum that the market requires. In 2019,

Mark Zuckerberg boasted that Facebook would spend $3.7 billion on platform safety – an impressive figure until you consider that Facebook's revenue for the year was about $70 billion, with $24 billion in profit.[34]

Even as platforms try to tackle the problem of misinformation, they are awash with information that is not necessarily false but selective, misleading or exaggerated. The internet is home to what Alexander Hamilton would have recognised as a 'torrent of angry and malignant passions'.[35] This makes civility difficult and compromise rare. Of course, technology did not cause the 'cankerworm of faction'.[36] In America, much of the polarisation in recent years was caused by cable news channels, not social media.[37] One authoritative study identifies the 'right-wing media ecosystem' as 'the primary culprit in sowing confusion and distrust'.[38] But platforms exacerbate social divisions, particularly among those who are not generally interested in politics or who rely on a narrow range of media sources.[39] Conflict gets more clicks than consensus. Bare-knuckle brawls get more views than collegiate debates.[40] Appeals to 'tribal belonging' fare better than neutral headlines.[41] Every word of moral outrage is said to increase the rate of retweets by 17 per cent.[42] And extremists somehow always manage to find each other. 'Hateful users are seventy-one times more likely to retweet other hateful users.'[43]

YouTube's algorithmic recommendation system, responsible for nearly three-quarters of time spent on the platform,[44] has been criticised for leading users to content that gets steadily more outrageous the longer they spend watching.[45] Clips about recreational jogging lead to videos about ultramarathons. Vegetarian recipes lead to videos about militant veganism.[46] A video of a right-leaning politician might lead, clip-by-clip, into a vipers' nest of alt-right extremism.[47] YouTube would object that its algorithm does not favour any particular ideology.[48] Perhaps so, but is that an answer to the charge of irresponsibility? It's just the *neutrality fallacy* again (chapter eight). A policy of neutrality towards extreme, racist and violent speech only empowers those with extreme, racist and violent things to say. Most platforms implicitly acknowledge this in their community guidelines, but guidelines are little match for algorithmic virality tugging in the other direction.

**

Political advertisements bring their own problems, particularly when personal data is used to target them at those who are likely to agree, a practice known as microtargeting. The main issue is that journalists, fact-checkers and political opponents have less opportunity to rebut them.[49] Traditionally, political debates were held in the open to force politicians to appeal to a range of voters, not just whoever they happened to be speaking to.[50] Microtargeting, by contrast, allows politicians to whisper things that could never be said in the town hall: 'Vote for me and things will be good for you – don't worry about anyone else.'[51] Instead of exposing ideas to the sunlight of public criticism, platforms create thousands of mini-markets for specific ways of thinking. Politicians can lurk in the ones in which they receive the warmest reception. (The creation of openly accessible, searchable archives by some platforms has gone some way to addressing this problem, but enabling post hoc criticism is no substitute for preventing the harm in the first place.)

The New Yorker has argued that the problem with political microtargeting is that advertisements 'bypass' the 'marketplace of ideas'.[52] This is a misdiagnosis. In a truly public forum, speakers must show their faces and defend their ideas openly. But parties to a commercial transaction have no such duty. There is no norm of subjecting ordinary market activity to public scrutiny. The problem is not that platforms bypass the marketplace of ideas. Instead, they take the concept to its logical endpoint: a privatised system of discourse that operates between consumers rather than citizens, according to mercantile rather than democratic norms.

**

If social media platforms are markets, then they are globalised markets. Information flows across national borders. In theory, this means that people might be exposed to benign influences from around the world. But it also means that foreign actors can inject harmful content into the bloodstream of other countries. Famously, in 2016 a firm engaged by the Kremlin launched advertisements against Hillary Clinton which reached 150 million internet users.[53] Tens of millions of tweets and thousands of YouTube videos were disseminated by hostile powers seeking to sow political and social discord.[54] But these kinds of operations did not stop after 2016.[55] The Covid-19 crisis saw a 'significant disinformation campaign' by Russian outlets on social media platforms, including claims, started in Iran, that coronavirus was a US biological weapon.[56]

We find it natural that commodities are traded across national borders, but political communities have not yet globalised in the same way. There is no sensible reason why Chinese propagandists should be able to influence elections in Europe, or why Macedonian workers should be able to spread pro-Trump propaganda in the American rustbelt,[57] or why the largest Black Lives Matter group on Facebook should be under the secret control of a white man living in Australia.[58] By the same token, there is probably no good reason why a European court should be able – as it is under current European law – to order platforms to take down content in other countries, where that content is perfectly legal.[59]

***

Here's the rub. A marketplace without rules eventually becomes a kind of battlefield,[60]

> Swept with confused alarms of struggle and flight,
> Where ignorant armies clash by night.

Being right matters less than being strong. Finding truth matters less than being forceful. Reason matters less than marshalling resources and concentrating them on the enemy. On the battlefield of ideas, people can enjoy the almost erotic sensation of 'spouting shit, the joy of pure emotion, often anger, without any sense'.[61] Anything goes.

In the digital age, we have reverted to an earlier and more primitive way of thinking about speech. Before the market metaphor, free speech advocates drew on the language of war. In the seventeenth century, John Milton described ideas as wrestlers in a brawl: 'Let [truth] and falsehood grapple; who ever knew Truth put to the worse, in a free and open encounter?'[62] In the nineteenth century, John Stuart Mill argued that truth would emerge from 'a struggle between combatants fighting under hostile banners'.[63]

As Nathaniel Persily has pointed out, social media platforms might be subject to a double standard.[64] By some they are accused of over-censoring. By others they are accused of under-moderating. For the republican, however, there is no contradiction. The republican critique is structural: under the present system, the flow of information in society is too heavily conditioned by private commercial incentives rather than the public good, and citizens have little or no say in the rules that govern them. Following democratic reflection, some republics might lean towards a more controlled information environment. Others might opt for more of a free-for-all. Either way, the information infrastructure should not be abandoned to the whims of corporations, dictated by market norms rather than democratic ones.

Introducing laws to govern social media would not be our first attempt at regulating the media. The next chapter looks at the approaches taken by democratic societies in the past, the mixed results that have followed, and the lessons that can be learned.

THIRTY-SEVEN

Toasters with Pictures

The regulation of speech does not have an auspicious history. For thousands of years, political and religious authorities have tried, usually for insalubrious reasons, to decide what may be uttered and what must remain unspoken. The Athenian statesman Solon outlawed 'evil speaking' in 594 BC. Socrates was executed in 399 BC for his controversial views.[1] In medieval Europe there were strict limits on what could be said and written. In 1275, the English Parliament forbade 'any false news or tales' whereby 'discord or slander' may 'grow between the king and his people'.[2] Religious authorities, too, kept a lid on heretical ideas. In 1500, Pope Alexander VI grandly declared that 'Rome is a free city; everyone may write and speak as he pleases.' A year later he proclaimed that unlicensed printing was to be forbidden.[3] In the next century, the Inquisition found Galileo 'vehemently suspect' of heresy and required him to 'abjure, curse, and detest' the views that came to mark the start of modern physics.[4] These were not enlightened times. Books and other publications were licensed, censored and sometimes burned.

In England, matters improved after the 1689 Bill of Rights and the lapsing of the Licensing Act in 1695. Subjects gained the freedom to petition political authorities openly. Newspapers that had begun life as government mouthpieces evolved into the free and burly press for which Britain is now famed. This marked the emergence of what we now call 'the public sphere'.[5]

The First Amendment to the American Constitution, ratified in 1791 and extended to the states in 1868, was unusual in that it gave constitutional protection to a specific industry: the press. It was also remarkable in that rather than seeking to interfere with speech, it sought to restrict such interference by prohibiting any law 'abridging the freedom of speech'.[6] Perhaps surprisingly, however, the First Amendment did not stop the enactment and enforcement of the Sedition Act in 1798, which made it a crime 'to write, print, utter or publish . . . any false, scandalous, and malicious writing or writings against the government' for defamatory purposes. (The constitutionality of the Sedition Act was never directly challenged.[7]) Nor did it prevent the Espionage Act of 1917, or the Sedition Act of 1918, which forbade any anti-government speech that interfered with the success of the military. Famously, the Supreme Court did not interfere with the prosecution, under the Espionage Act, of men who had authored a pamphlet describing conscription as 'slavery'.

The US Supreme Court and generations of scholars have interpreted the First Amendment in different ways, some of which we will look at in the next chapter. It may surprise some readers to know that the modern First Amendment tradition, with its broad conception of speech and severe suspicion of government interference, is largely a creation of the twentieth century.

***

It is easy to forget that for nearly half of the twentieth century, American broadcasters were subject to federal oversight. Since the early 1930s, the Federal Communications Commission (FCC) has had power to regulate commercial broadcasters in pursuit of 'the public interest, convenience, and necessity'.[8] Until the Reagan years, the FCC required broadcasters to spend time on issues of public importance, treat opposing views in a balanced way and cover diverse viewpoints. This was known as the fairness doctrine.[9] The airwaves were deemed a natural resource, held in common by all Americans, with broadcasters as its public trustees. Broadcasting licences were granted only to those who pledged to fulfil public obligations.[10]

The FCC's work was not to everyone's liking. In its heyday, it was criticised from the right for being overly intrusive, and from the left for being a toothless patsy of big business.[11] The right eventually won out. The dial swung fiercely against media governance in the 1970s and 1980s, with the FCC itself endorsing the view that the free market provided quite enough regulation on its own. The FCC's view at this time was that there were now so many broadcasters, enabled by satellite and cable technology, that consumers could vote with their feet if they didn't like what they saw. Savvy consumers would demand quality. The necessary incentives were already in place.

Eventually the FCC abandoned the idea of public service broadcasting altogether. Its chairman declared that 'it was time to move away from thinking about broadcasters as trustees. It was time to treat them the way almost everyone else in society does – that is, as businesses.' On this view, television was 'just another appliance. It's a toaster with pictures.'[12] In 1987, the FCC decided that the fairness doctrine was irreconcilable with the First Amendment and jettisoned it. This marked the end of federal efforts to enforce a standard of fairness on American broadcasters.[13] The FCC formally struck the fairness doctrine from its books in 2011.[14]

Commentators have very strong views about the fairness doctrine. It is often discussed, however, without much consideration of how the American media might have developed without the doctrine. Indeed, American commentators sometimes forget that other democracies have tried their own versions of broadcasting regulation, with some success. The UK is a good example. Broadcasters there have been subject to state regulation since the earliest days of the medium.[15] To broadcast on commercial radio or television, a person must hold a licence from the statutory regulator, Ofcom. To secure and maintain a broadcast licence, a person must be declared 'fit and proper'. In deciding whether someone is fit and proper, Ofcom will consider applicants' historic compliance with media law, as well as other aspects of their conduct that might render them unfit to broadcast. (In 1998, for instance, a radio licence was removed from a man convicted of rape.[16])

As it once did in America, the 'public interest' plays a prominent role in broadcast regulation in the UK. The legal regime is designed to ensure that the media 'performs the functions that are necessary to a democratic society'.[17] Ofcom has a Broadcasting Code that contains various rules.[18] Factual television programmes 'must not materially mislead the audience'. Depictions of self-harm and suicide must not feature 'except where they are editorially justified and are also justified by the context'. Hate speech is prohibited unless justified by the context. Broadcasters must not treat individuals or organisations unjustly or unfairly.

The Code contains specific provisions for news broadcasters, requiring them to report with 'due accuracy' and 'due impartiality'. Significant mistakes should 'normally be acknowledged and corrected on air quickly'. Any personal interest of a reporter or presenter that would call 'due impartiality' into question 'must be made clear to the audience'. For matters of 'major political and industrial controversy' and 'major matters relating to current public policy', broadcasts must include an 'appropriately wide range of significant views', each 'given due weight'. During elections, due weight must also be given to parties and independent candidates. The law prohibits almost all paid political advertising.[19] A small number of major broadcasters are subject to specific 'public service obligations' that include quotas for independent and original productions and regional programming.[20]

To back up these rules, Ofcom has powers of enforcement. More often than not, a breach of the Code will result only in a rap over the knuckles – the publication of an adverse finding. But Ofcom can also fine broadcasters, direct them to broadcast a correction, forbid them from repeating certain programmes and shorten or revoke their licences.[21] Under current government proposals, Ofcom is to be the new regulator for social media, as well as the broadcast media.

To the American eye, this regulatory regime may appear somewhat authoritarian – just as to the British eye, some American broadcasters look partisan to the point of lunacy. But it can scarcely be said that the UK is not a functioning democracy or that its citizens do not enjoy freedom of speech. On the contrary, as we will see in the next chapter, the British approach simply reflects a different philosophy of free expression from the one favoured by First Amendment purists.

***

A common question is: *if you're going to regulate Facebook, are you going to regulate the New York Times as well?* The answer to this question is 'no'. To understand why, it's first worth noting the difference between the treatment of the broadcast media and the print media, both in America and the UK. Whereas the broadcast media has been closely regulated, newspapers have traditionally been left to themselves, at least in the last couple of centuries. In the UK, where broadcasters are under strict duties, the newspapers are politically partisan and gleefully offensive.

There are, in fact, at least seven meaningful differences between newspapers and social media platforms.

First, as we have seen, social media platforms gather immense amounts of data about their users. The newspaper – at least in its paper form – never listens to our conversations, nor reads our private messages. It doesn't watch us back.

Second, when we read a physical newspaper, we know we are getting the same content as everyone else. On platforms, however, the flow of information differs depending on the preferences of the user. What I see is unlikely to be the same as what you see.

Third, when you read the *Guardian* or *Wall Street Journal*, you know what you are getting. These outlets wear their allegiances openly and their priorities can easily be inferred. Platforms, by contrast, operate in the dark. Their algorithms are occluded from public scrutiny. Even the most vigilant citizens cannot really know if platform moderation systems are working properly, or even what 'working properly' would mean in any given context.

Fourth, a newspaper (even an online one) is a relatively controlled medium in which a comparatively small amount of content is generated and presented to the outside world. By contrast, the domain of online platforms is fissile and vast. Facebook has more adherents than Christianity. Ninety-four per cent of Americans aged eighteen to twenty-four use YouTube.[22] These platforms host millions of voices clamouring to be heard all at once. The online public sphere is inherently more volatile.

Fifthly, the newspaper industry is self-governed according to norms that are long established and widely understood. There are good journalists and bad journalists, and we generally know what distinguishes a good one from a bad one.[23] A good journalist is suitably detached, protects her sources, fact-checks allegations, gives subjects the opportunity to comment and so forth. No shared system of norms yet exists for social media.

Sixthly, compared with broadcast media, the newspaper industry is more decentralised and competitive.[24]

Finally, and crucially, platforms do not just edit their own content. They edit, sort, rank and rate the rest of us. No mass medium has ever had such substantive influence on the conversations that take place *between* citizens.

For all these reasons, it is perfectly sensible to argue that social media platforms should be governed more intensively than newspapers. A different tack, however, would be to *welcome* the inconsistency. Taking a light-touch approach to print media and a more substantive approach to broadcast and social media could offer an ideal balance. A thriving republic needs a fearless print media to hold the government to account, but it might also need spaces in which deliberation can be conducted in a more democratic fashion. Why not build a media ecosystem in which both are available?[25]

***

'Toasters with pictures' was never a clever way to think about the broadcast media, and it is equally unfit to describe social media.[*]

A toaster never exerted power on anyone, threatened their liberty or damaged democracy. Given social media can do all these things, there is a prima facie case for governance. But what philosophical principles should underpin a regulatory regime? That's the question for the next chapter.

---

[*]As an aside, an internet search for the phrase 'toaster with pictures' shows advertisements for the 'SCOTUS Toaster', which 'toasts the face of Justice Ruth Bader Ginsburg on your toast'. There is a lesson in there somewhere.

THIRTY-EIGHT

A System of Free Expression

Speaking and listening are prerequisites of civilized life. Without freedom to express ourselves, we could not achieve autonomy or personal growth. We could not pursue the truth through public debate. And we could not enjoy the other rights and freedoms that depend on free expression. The right to a fair trial, for example, presupposes a right to speak in one's own defence. The right to an education presupposes that teachers may impart ideas without molestation. The right to practise religion requires space to preach and pray. A meaningful system of free expression must allow us to say things that are disgusting, heretical and repugnant. 'Freedom only to speak inoffensively,' says the common law of England and Wales, 'is not worth having.'[1]

Within the West today, however, there are two broad approaches to the governance of expression. For simplicity, I call them the American approach and the European approach, though neither is a complete portrait.

The American approach begins (and arguably ends) with the First Amendment: Congress 'shall make no law . . . abridging the freedom of speech, or of the press'. Many Americans regard this as the gold standard for the protection of speech. One scholar calls America 'the most speech protective of any nation on Earth, now or throughout history'.[2] Speaking one's mind is regarded as 'part of the national identity'.[3]

The American approach takes a somewhat literal approach to the First Amendment. It sees the primary function of a system of free expression as to exclude *Congress* from making *laws* that restrict freedom of speech. It is, above all, a formidable shield against state censorship. It reflects a deep and justified suspicion of governments – even democratic ones – that use crises or other pretexts to suppress speech they don't like.[4]

But does the American approach protect free expression in the twenty-first century? In the past, the working assumption was that speech was a scarce resource in need of conservation. Any legal restrictions would shrink a finite reservoir. Now, as many authors have carefully documented, the world is deluged with information.[5] The worry is not that speech dries up, but that *meaningful* communication is lost in a torrent of fake news, trollogenic invective, bot-generated misdirection and viral misinformation.[6] The First Amendment offers little or no protection against these threats. Milan Kundera once described 'the madness of quantity' in the production of information, with culture 'perishing in overproduction, in an avalanche of words'.[7] And that was *before* the internet.

These days, we need platforms and other digital systems to filter and sort the world's information for us. But as we have seen, this is an inherently political task, fraught with controversy and risk. It gives platforms – private enterprises, rather than Congress – the power to abridge free expression, or to suppress certain forms of expression while promoting others. Yet the First Amendment asks nothing of platforms. It places no limitation on speech-suppressing or speech-promoting activities. Social media platforms in America have no duty to permit any form of expression, still less to moderate discourse in a way that is democratic.

The problem runs deeper. It's not just that the First Amendment places no obligations on tech firms. It places no obligation on the state to introduce laws that would encourage free expression. In fact, the First Amendment is arguably hostile to such laws, on the grounds that they might be construed as abridging freedom of speech.[8]

As we saw in chapter thirty-six, Holmes said that the 'theory of our Constitution' is that 'the best test of truth is the power of the thought to get itself accepted in the competition of the market'. But he qualified this with a caution: 'It is an experiment, as all life is an experiment'.

Has the experiment been a success?

Whichever way you look at free expression – as a vital means of self-actualisation, as a generator of truths, or as the bedrock of democracy – it is hard to say that the First Amendment, as currently interpreted, is a match for the challenge of digital technology. It may have made perfect sense in the late eighteenth century, when the principal threat to free expression was government censorship. But it is starting to look old-fashioned.

***

What about the European approach? (I include within this shorthand the approach taken in Canada, New Zealand, South Africa and elsewhere.) Europe's counterpart to the First Amendment is Article 10 of the European Convention on Human Rights (ECHR).

Like the First Amendment, Article 10 protects freedom of expression. But its protection is not absolute. For one thing, it allows governments to impose 'formalities, conditions, restrictions or penalties' on free expression in a few limited circumstances. Any restrictions must be prescribed by law – and not simply the whim of whoever is in government. Restrictions must also be 'necessary in a democratic society', i.e. a proportionate way of meeting a pressing need. That need can only include national security, territorial disorder, the protection of health or 'morals', the protection of the reputation or rights of others, and maintaining the authority and impartiality of the judiciary. These categories are developed and explored in the case law of the European courts.

The upshot is that some forms of speech that are permitted in America are forbidden in Europe. Holocaust denial, for instance, is illegal in France and Germany. Burning a cross in the garden of a black family would not be protected by a European court, as it was by the US Supreme Court (in the absence of evidence of intent to intimidate).[9] In Europe, governments can lawfully demand that fascist symbols be removed from social media; the US government would seldom even try to do the same.

Looked at in isolation, the European approach plainly allows governments to censor more speech activity than the American approach. In this sense, the American approach is indeed 'the most speech protective of any nation on Earth'. There is always a risk that European governments, even democratic ones, might try to intervene repressively in the information ecosystem.

A crucial difference between the European and American approaches, however, is that the European approach does not merely prohibit states from doing things, but requires them to do certain things. European governments have a positive duty to take reasonable steps to ensure that the right to free expression can be enjoyed in practice.[10] Article 10 requires them to create a favourable environment for participation in public deliberation.[11] This is profoundly important. It places an obligation on governments to ensure that their system of free expression is conducive to democratic self-rule. Free expression is treated as a collective good, not just an individual one – and sometimes the two are balanced against each other.

***

Although there is a division between the American and European approaches, they are not as far apart as they seem. Some American jurists have argued that a correct interpretation of the First Amendment would be closer to the approach taken in Europe than the present orthodoxy. They tend to make three points.

First, there is the argument from history. The First Amendment has been interpreted in different ways over time. As we saw in the previous chapter, the Sedition Act became law while Framers were still alive – and many saw no problem with it.[12] Indeed, American jurisprudence from before the 1930s can be searched in vain for the idea that free speech was 'a core constitutional commitment, let alone a formidable constraint on government power'.[13] In 1969, the Supreme Court held that the fairness doctrine – allowing for federal oversight of the broadcast media – was consistent with the First Amendment. And although that case has been doubted and occasionally sidelined, it has not been overturned.[14] Today's orthodoxy, which recognises no threat to free speech other than that of the government, is neither eternal nor universal.

Secondly, the Supreme Court's own decisions suggest that reinterpretation may be possible within existing lines of case law. The Court famously justified the fairness doctrine, in part, on the basis of the distinctive capabilities of the broadcast media. So why not revisit the First Amendment considering the distinctive capabilities of digital technology?[15] Similarly, the Court has equated the internet with 'public streets and parks' and held that 'cyberspace' and 'social media in particular' have become 'the most important places (in a spatial sense) for the exchange of views'. Shouldn't that finding prompt a broader reconsideration of the rights of people who want to express themselves within that public space?[16]

Finally, there is a powerful intellectual tradition in America that holds that the First Amendment does more than merely prohibit government interference. Found in the writings of James Madison and Louis Brandeis, and often associated with Alexander Meiklejohn, it holds that the principle of free expression cannot be divorced from the project of self-government.[17] On this view, a system of free expression must, above all, facilitate the 'central constitutional goal' of deliberative democracy.[18] Meiklejohn points out that the First Amendment forbids the abridging of *freedom of speech*, not the abridgment of speech itself. Thus, he argues that legislation to 'enlarge and enrich' freedom of speech is perfectly constitutional.[19] If Meiklejohn is right, the First Amendment is not meant to safeguard 'unregulated talkativeness' but rather to prevent the 'mutilation of the thinking process of the community'.[20] This way of interpreting the First Amendment chimes with the republican philosophy. For republicans, free expression matters principally because it is the prerequisite to self-government. Without laws safeguarding free expression, *deliberation* and *counterpower* – the two pillars of republican freedom – would be under constant threat. Republicans recognise that politics cannot be separated from the communicative acts that make it up: arguing, discussing, listening, reasoning, declaring, declaiming. '[There] can be no government by the people,' held one of the most distinguished English judges, 'if they are ignorant of the issues to be resolved, the arguments for and against different solutions and the facts underlying those arguments.'[21]

***

Our generation, whether in the American or European tradition, will need to think anew about the problem of free speech. Any system of social media governance must reckon with the fact that some of the most harmful online speech is not necessarily illegal. Consider, for instance, content that glorifies eating disorders among young girls, surprisingly common on popular platforms. Many would like the law to require platforms to reduce the prevalence of such material for young female users, and in the UK there are plans to make it happen. But (runs the counterargument) the state has no business in telling anyone what to do with speech that is perfectly lawful.

This dilemma arises because forms of speech which were mildly problematic in the past now have the potential to be mortally dangerous. Material advocating self-harm and eating disorders was hard to find thirty years ago, but is now available to any little girl with an iPad. Likewise, misinformation can be disseminated to every corner of the planet in a few seconds and targeted at the most vulnerable and credulous. This obviously makes it more dangerous than it used to be, particularly if platforms filter out alternative views.

Another new puzzle is whether the same rules should apply to the individual speaker who *utters* misinformation as to the platform that *amplifies* it. Again, in the past it made sense to treat these both as forms of 'speech' deserving equal protection under the law. But they are not the same, either in social function or effect.

The challenge requires us to think like philosophers, not lawyers. The lawyer asks: *Is this lawful?* The philosopher asks: *Should it be?* Speech that was perfectly acceptable in the past might not be today. Speech norms change − but they aren't changing fast enough. The laws we have inherited should not be treated with Talmudic reverence. It's possible that the First Amendment is simply obsolete, incapable of offering a governing philosophy that is fit for our times. Some distinguished scholars, like Daphne Keller, doubt that popular measures to regulate social media platforms can be constitutional.[22] Or perhaps, as others argue, the First Amendment can be revitalised or reinterpreted.

This book does not seek to resolve the question of what the US Supreme Court should or will decide on such matters in the future. Instead it tries to imagine a system of free expression that would be suitable for the twenty-first century, putting legal precedent aside and combining the best of the American and European approaches. That's the subject of the last chapter.

THIRTY-NINE

Governing Social Media

Speech is the citizen's first and final defence against domination. In a free society, even the mightiest powers can come in for criticism and challenge. For a republic to survive, its citizens must be able to question any person or policy without fear. 'Give me the liberty,' wrote John Milton, 'to know, to utter, and to argue freely according to conscience, above all liberties.'[1]

Freedom of expression must be protected not only from the *imperium* of the state, but the *dominium* of private companies. My aim in this chapter is to describe a republican scheme for governing social media (in reality a republican scheme for protecting free expression) and explain how it could be realised in practice. The regulatory model defended in this chapter is closest to the system presently being considered in the UK. This chapter also draws on the work of a number of important scholars – Danielle Citron, Daphne Keller, Kate Klonick, Nathaniel Persily, Will Perrin, Lorna Woods, Renee DiResta and many more – who have been thinking carefully about these issues for years, even if they wouldn't necessarily agree with what I have to say.

The starting point is that not all platforms are equal. A local chatroom for dachshund enthusiasts does not have the same systemic importance as Facebook, 'the world's largest censorship body'.[2] A sensible framework for governing social media would start by ranking platforms according to their level of social risk. In the lowest risk tier would be modest online spaces like community forums, hobbyist groups and fansites. These should mostly be left alone and subject only to minimal regulation. This is not because they are always pleasant places – some are cesspits – but because they are not especially powerful. They are easy to leave, easy to replace and the harms they generate do not usually spill over into wider society. Moreover, they tend to be run by volunteers or modest organisations without extensive resources. Too much regulation would be crushing. Sometimes people are going to be horrible to each other on the internet and there's not much that can be done – at least without doing more harm than good.

At the other end of the spectrum, in the highest tier, would be very large, open platforms like Facebook and Twitter, which I will refer to as major platforms. Because of their size and social function, major platforms have the capacity to frame the political agenda, rapidly disseminate content and shape the opinions and behaviour of millions of people. They are difficult for users to leave, and for rivals to challenge. They are important spaces for civic and commercial life. When they block or exclude people, it matters. Because major platforms pose greater risks to society, they should be regulated more intensively than minor ones.[3]

Of course, sorting platforms by size alone would be unsatisfactory. It would mean drawing arbitrary lines to distinguish between the tiers. But in fairness, that is how regulation works in many other sectors, including banking, in which larger institutions usually bear a heavier regulatory burden than smaller ones. And it is increasingly how the tech industry is governed too. In German law the threshold for hate speech regulations is 2 million users per platform, and the EU is already planning to regulate platforms according to size, distinguishing between 'micro or small enterprises' and 'very large online platforms' that pose 'systemic risks'.[4]

That said, size should not be the only factor that determines which tiers platforms are assigned to for regulatory purposes. As the UK government appears to recognise, platforms that (deliberately or otherwise) become hotbeds of child pornography, sexual harassment or extreme political activity should expect to be placed in a higher-risk tier than others their size.[5] They (like groups on Facebook and subreddits on Reddit) can pose a serious threat to individuals and the republic despite only having a few thousand users.

***

Once platforms have been sorted into tiers, they should be governed accordingly. Platforms in the lowest tiers should remain largely immune from liability for the content they host. A local chatboard should not be held responsible for all the fruity content posted on it by users. But that immunity should not be absolute. Even small platforms should be expected to have reasonable systems (proportionate to their size and resources) for the removal of material that is highly injurious to identifiable people, or that falls into the most serious categories of prohibited speech. Revenge porn would be an example.

Platforms in higher tiers should expect more legal responsibility. Instead of unconditional immunity from liability for the content on their platforms (as they currently enjoy in the US under Section 230 of the Communications Decency Act – see chapter sixteen), immunity should be conditional on certification of their critical systems. By this I mean, in line with proposals in the UK, that major platforms should be regulated at the *system* or *design* level.[6] The people, acting through the democratic organs of the state, would decide the legal standards that apply. These would differ from place to place, but they might look like this (drawing inspiration from proposals in the UK):

- Major platforms should have reasonable systems in place to reduce the flow and/or visibility of harmful misinformation.
- Major platforms should have reasonable systems in place to prevent online harassment.
- Major platforms should have reasonable systems in place to prevent foreign interference in the political process.
- Major platforms should have reasonable systems in place to mitigate the harmful effects of synthetic or inauthentic coordinated activity.

In general, legal standards for social media should be strict liability: good intentions are not enough. But they should also be *outcome*-based, placing the burden on platforms to work out how best to fulfil their obligations. Finally, they should be risk-based, aimed at reducing risks rather than eradicating harm. Richard Allan, who has worked in both politics and tech, proposes Harm Reduction Plans along these lines:[7]

> Each plan would provide any of us who are interested in these matters with a clear and easily intelligible description of how a platform has assessed the risk of a particular harm and what specifically it is doing to combat this harm.
>
> These plans would form the basis of the engagement between the regulator and platforms, with the regulator able to test each aspect of the platform's understanding of the problem and proposed response.
>
> There should be published versions of the plans that contain sufficiently detailed information for us all to understand what is happening, while some information may be provided only to the regulator if it is especially sensitive.

There is a growing body of empirical research about what works and what doesn't.[8] Platforms should be given space and incentives to develop that field of knowledge and apply it in their own domains. And there is no reason why the work should not be outsourced to companies whose *raison d'être* is to create and licence systems that are compliant with the law.

Legal standards for major platforms need not be limited to the reduction of harm. They could include duties to promote certain positive goals too. Take, for example, the following:

- Major platforms should have reasonable systems in place to encourage civil deliberation on matters of public importance.

At first glance, this might seem like an onerous and overly complex thing for the state to require. What is meant by *reasonable*, or *civil*, or *public importance* for that matter? But these terms are no more difficult to define than many others used in the law. To define them with increasing precision, over time if necessary, would be the job of parliament and the courts. The English law of negligence, for instance, has been built over many years around the concept of what a reasonable person would do in a given situation. It is not beyond us to develop the law in new ways.

Any number of measures could be used to satisfy the standard of reasonableness in this instance: systems to monitor bullying and limit abuse; algorithms that promote reputable news sources, or check the spread of misleading viral posts, or show people different viewpoints; design features that encourage people to read articles before sharing them, or encourage them to change their minds; independent procedures for fact-checking and labelling; terms of service that are fit for purpose, enforced fairly and promptly. Some commentators have suggested that platforms ought to place higher duties on accounts with high numbers of followers – an idea with merit.[9]

This is not rocket science. Nor is it an exact science. Schemes for the governance of speech should never strive for perfection, only the reduction of imperfection. But we can ask a little more of major platforms, and we ought to be able to hold them to account when they fall short.

***

Systems-level, outcome-based, risk-based governance has three advantages in this context.

First, it keeps the state out of individual decisions about content moderation. That is important. It is desirable to avoid a regulatory system in which the state is encouraged to interfere with individual speech acts. Much more useful would be a broad focus on the prevalence of harmful information on a platform, which might be down to the proclivities of its users, the work of bad actors, algorithmic amplification, or all three.

Second, platforms and third-party service providers will usually be in the best position to find ways of improving their systems. We can expect content moderation and fact-checking, for instance, to continue to grow into big businesses.[10]

Third, major platforms operating at scale should be allowed to make small mistakes if they can show that their systems are fit for purpose. When it comes to misinformation, the state should not be penalising decisions to remove or restore particular items of content. But it *should* be able to ask: are the platform's algorithms worsening this problem or improving it? Are there adequate procedures for detecting or flagging false content? Is there a certified system, run by certified people, for separating fake and real news? Is there an adequate complaints system for users who flag misinformation? Are proper safeguards in place to prevent foreign interference around the time of elections?[11]

Systems-level governance is common in other industries. Major-hazard oil rigs, for example, submit safety plans to regulators for approval. Those plans are carefully inspected – tested against the requirements of the law, checked for auditability, run past stakeholders – and then certified. The regulator monitors whether the plans are then properly implemented, with sanctions for failure.[12] Similarly, in the financial sector, instead of prosecuting individual failures, regulators tend to require banks to put in place *systems* for combatting money laundering and terrorist financing. The emphasis is on risk management and harm reduction – not the elimination of all harm.

Many platforms have been working to improve their systems for years. But these improvements do not always reflect public values and priorities. And they tend to move at the pace that suits the platforms, not the public need. What's more – and putting it gently – platforms' existing systems do not always perform to a reasonable standard. Some are poorly designed, others insufficiently resourced, others not properly implemented. Whatever the reasons, there needs to be a shift in incentive. In the digital republic, maintaining reasonable standards would be a matter of law, not goodwill.

***

The duty of openness (chapter thirty-one) would apply to social media platforms in the digital republic. Depending on the law, we would expect major platforms to disclose the content of their moderation policies, the algorithms and human processes used to implement those policies, the principles and practices guiding the ordering of content on the platform, statistics concerning regulated areas (including the number and nature of takedowns), and details about how data is gathered and used.[13] Germany already has an extensive public reporting regime in place, and it has yielded valuable insight for scholars and activists.[14]

As for individual rights, in the higher tiers (but not in the lower ones) individuals would have some protection against being arbitrarily thrown off or censored. Major platforms would be required to apply their community standards consistently, transparently and promptly. In the most serious cases affecting individuals – highly defamatory statements, harassment, revenge porn and the like – individuals should be able to appeal unsatisfactory decisions to tech tribunals that can order the platforms to take action.[15]

Once platforms' systems are certified, then that should be a defence to claims arising out of the content posted by third parties. Immunity would only be forfeited where a regulator has found a systemic failing or a failure to comply with the duty of openness.

The republican approach would thus abolish the unconditional liability currently granted to American platforms under Section 230 of the Communications Decency Act. And it would go further than most European systems that grant immunity to intermediaries if they have no notice or knowledge of illegal content.[16] Egregious or repetitive failures could result in criminal sanction or the revocation of licences.

***

The scheme designed in this chapter would still leave space for a universe of different platforms. But it would require each of them to meet certain minimum standards, in line with their size and social importance. It rests on a theory of free expression that holds that the primary aim of governance is to promote and protect the social conditions necessary for deliberative self-government, as well as individual autonomy.

This does not mean trying to sanitise political debate. People must have the right to articulate challenging or even repugnant ideas without being shut down, either by governments or platforms. But when taken to the extreme, that right must also be balanced against the other needs of deliberation: the need to reduce misinformation, harassment and foreign interference.

The scheme described here would be concerned with *all* the threats posed to free expression, whether from the state, platforms, or malevolent third parties.[17] It would *prevent* any platform from dominating the speech environment, while taking reasonable steps to *promote* the conditions necessary for democratic deliberation. And it would seek to meet these goals with as little state intrusion as possible.

This scheme reflects a republican understanding of freedom that is older by many centuries than the First Amendment and the European Convention on Human Rights. It would combine the best elements of the American and European approaches, while adapting to the needs of the century unfolding before us.

Conclusion:
The Digital Republic

The great republican thinker Montesquieu wrote that a state may change in two ways: either because it is corrected or because it is corrupted.[1]

Which will it be?

No matter where we are or what we are doing – even in the sanctum of our own minds – it is getting harder to escape the power of digital technology. With each passing year, we grow more vulnerable to the whims and prerogatives of those who design and control the most capable systems.

It is hard to pin down what makes the power of technology so difficult to govern. The conventional explanation is that it is because it is concentrated in the hands of a small number of governments and mega-corporations. This is certainly part of the answer. But as we have seen, technology's power is present even when it is not concentrated in the hands of a few. It is dispersed throughout society, emanating from countless lines of code and bits of hardware, enveloping us without any particular site of responsibility. Often technology's power is invisible. Often it is subtle. Often it doesn't look like power at all. And as it swells and spreads, the 'fabric of freedom', as Abraham Lincoln called it, is stretched ever thinner.

In this book I have tried to persuade you that the real issue facing the tech industry is not the personal failings of CEOs, or the venality of corporations, or the size of the tech giants, or even the potency of the technologies themselves. It is that powerful digital technologies – and those who design and control them – are not adequately governed. The law does too little to redress the imbalances in power that arise where technology is present. It gives too much protection to industry, and not enough to the rest of us. It expects individuals to stand alone, when what's needed is collective action.

This tattered framework of laws did not come about by accident. Nor did the pervasive assumption – sometimes explicit, often implicit – that the market mechanism can *regulate* innovation as well as it generates it. The underlying philosophy of the status quo, which I have called market individualism, has not only contributed to the challenges we face, but clouded our thinking and hindered our desire to do things differently.

My hope is that digital republicanism offers an alternative. Instead of turning a blind eye to unaccountable power, it seeks to identify and reduce it at every turn. Whether it is the unaccountable power of corporations over citizens, of one group of citizens over others, or of the government over everyone else, there are always ways to redress the imbalances. We can do better.

***

Technological advance should be a blessing for humanity. Innovation is inseparable from social progress. Tech can make life safer, more vibrant, more dignified, more fun and interesting.

Properly deployed, it can even make societies more democratic and just. Technological capitalism, for all its flaws, has enabled sustained growth in the standard of living for billions of people – albeit with a fair bit of help from the state along the way.

For these reasons, I know that some readers may worry that the wrong kind of regulatory burden could stifle innovation, depriving humanity of all the benefits of digital technology.[2] This would indeed be a tragedy. But the assumption here is that regulation and innovation are antithetical, whereas the reality is more complicated. Sometimes regulation and innovation go hand in hand.

Regulated industries plainly benefit from the public trust that comes with proper oversight. All other things being equal, customers will usually prefer to go to a café with a food hygiene certification in its window than one without.[3] The law gives consumers 'the confidence to try something new' because they know they can trust regulated people, products and services.[4] And it makes space for responsible producers to offer high-quality wares without fear that competitors will cut their knees off with cheap, dodgy alternatives.[5] Other big industries like healthcare and pharma are closely regulated, but they are certainly still innovative. And we'd be pretty concerned if they were governed in the same way as the tech industry.

Stable governance regimes also bring the economic benefits of harmonisation. Rather than inefficient competing standards – like VHS and Betamax in the 1970s[6] – the consolidation of regulatory standards allows businesses to compete on a large and level playing field. The idea of the EU data governance regime, for instance, is not merely to govern the use of data, but to facilitate 'data-driven business across national borders'.[7] The Cambridge economist Diane Coyle attributes the ubiquity of mobile phones, which we now take for granted, to an EU technical standard enforced in 1987, which created a continent-wide (and eventually worldwide) market.[8]

Contrary to myth, governance regimes do not destroy the incentive to innovate. But they can change the nature of the incentives for the better. As Julie E. Cohen explains, the precautionary approach to innovation in other sectors has generated new incentives for 'lean manufacturing and energy production, safe drug delivery and the like'.[9] With good governance, it's not that innovation withers; it's that some of the genius, energy and investment that was previously funnelled into the blind pursuit of growth is diverted into making products and services that cohere with society's values, as laid down in the law. I, for one, would not object to a race to create the safest social network, the securest database or the fairest algorithm. There are already whole industries dedicated to servicing the growing need for social media moderation, algorithmic oversight, data management and the like.[10] These industries would flourish in the digital republic. That is to be welcomed.

All this being said, we should also reject the assumption that the only purpose of economic activity is to generate growth, rather than to produce a society in which life is worth living. The ultimate surrender to market individualism would be to subordinate the scope of our freedoms or the strength of our democracy to the need for economic growth or even technological advance. There is, on occasion, a tension between the logic of capitalist innovation and the public good. And we should not be afraid to say that the public good must sometimes be given priority.

Once we are clear about our values – about what really matters to us – the challenge becomes less about political controversy, and more about regulatory design. We need systems of governance that are suitable for the sectors they regulate; that bring out the best and curtail the worst; and that channel market forces into powerful streams of progress.

***

In the dying pages of the eighty-fifth (and final) Federalist essay of 1877, written to persuade Americans to adopt their fledgling constitution, Alexander Hamilton quoted the Scottish philosopher David Hume:

'The judgments of many must unite in the work: EXPERIENCE must guide their labour: TIME must bring it to perfection: And the FEELING of inconveniences must correct the mistakes which they inevitably fall into, in their first trials and experiments.'

They were far from perfect, but there is much we can learn from the institution-builders of the American revolution: their indignant spirit, their sense of possibility, their willingness to make mistakes in pursuit of a higher goal. Our task seems modest by comparison. But building the digital republic will not be easy. There will be errors and missteps along the way. There will be rival approaches, rival philosophies. There will be doubters and haters. Yet the great task that has fallen to our generation is to be welcomed and not feared: the task of harnessing the awesome power of technology and binding it tightly to the shared hopes and aspirations of humanity. That's worth fighting for.

Acknowledgements

My grandfather, Werner Susskind (1933–2021), passed away before I could send him a copy of this book. He escaped Nazi Germany as a little boy and arrived in Glasgow as a refugee. He grew up to become a doctor and beloved family man. He never had much regard for politics or politicians, but the story of his life spoke powerfully enough about the fragility of freedom and democracy (although he would never have put it so pompously). His unwavering support for my work meant a great deal to me while he was alive. His memory will always be a blessing and an inspiration.

This book builds on the work of hundreds of others, and in writing it I have racked up many intellectual debts. I owe a great deal to Philip Pettit, the foremost republican thinker of our time. Since we met a few years ago, he has become a cherished friend and mentor.

In the early stages of writing, I benefitted from the guidance of Mike Kenny, delivered over a tasty pie at the Cambridge Blue. I was also greatly assisted by the work of my researchers Imre Bard, Roberta Fischli, Joe Robinson, Benjamin Slingo and George Tarr.

As work progressed, I was fortunate to receive feedback from Claire Benn, Simon Caney, Ashley Casovan, Tim Clement-Jones, Diane Coyle, Julie E. Cohen, Kate Dommett, Yaël Eisenstat, Nathan Gardels, Sam Gilbert, Nils Gilman, Mireille Hildebrandt, Mike Kenny, Carly Kind, Chris Hoofnagle, Julian Huppert, Seth Lazar, Barry Lynn, Matthijs Maas, Helen Margetts, Roger McNamee,

Martin Moore, Geoff Mulgan, Chi Onwurah, Peter Pomerantsev, Rob Reich, Bruce Schneier, Tom Upchurch, Shannon Vallor and John Zerilli, all of whom were kind enough to read the book in draft.

I would also like to record my particular thanks to Julie E. Cohen, Mireille Hildebrandt, Nils Gilman, Seth Lazar and Rob Reich, who went far beyond the call of duty in helping me to improve *The Digital Republic*.

Many friends were good enough to provide careful, and at times withering, feedback. Alex Canfor-Dumas, Kim FitzGerald, Josh Glancy, Laurence Mills, Matt Orton (who read the book twice – a proper friend), Fred Popplewell, Owain Williams and Tom Woodward made many corrections and suggestions. I also benefitted from conversations with Richard Allan, Roberta Katz, Jan Middendorp, Yascha Mounk, Jonnie Penn, Will Perrin, Nathaniel Persily, Carissa Véliz, Jimmy Wales and Lorna Woods. I owe a debt of gratitude to Christy O'Neil for helping to get the manuscript over the line.

I would like to thank the Leverhulme Centre for the Future of Intelligence at the University of Cambridge, and the Berggruen Institute in Los Angeles, for hosting me as a Visiting Fellow. I am also grateful to the Bennett Institute of Public Policy at the University of Cambridge for bringing me in as an Affiliated Researcher.

Before the Covid-19 crisis I was able to test and clarify some of my ideas at the Berggruen Institute's *Future of Democracy* gathering at the Bellagio Center in July 2019, the Stanford *Regulating Cyberspace* conference in October 2019, and the Stanford conference on *AI Ethics, Policy, and Governance* and the political philosophy working group which followed it, also in October 2019.

In early 2021, the Berggruen Institute and the Bennett Institute kindly convened two workshops that focused on early drafts of this book. Many busy people – some of whom I knew only by their considerable reputation – were kind enough to offer their thoughts on an early draft. They were, as well as those mentioned above: Kenneth Cukier, Ken Goldberg, Julian Huppert, Toomas Ilves, Barry Lynn, Viktor Mayer-Schönberger, Natasha McCarthy,

Claire Melamed, Dawn Nakagawa, Reema Patel, Eleonore Pauwels, Philip Sales, Amelia Sargent, Elizabeth Seger, Jo Stevens, Zephyr Teachout and Meredith Whitaker. The book is much better as a result of their extensive input. Thank you to Jennifer Bourne, Rebecca Leam, Sarah Rosella and Amelia Sargent for organising so expertly.

As ever, I am hugely thankful to have Caroline Michel as my agent, along with her immensely talented team at Peters Fraser + Dunlop, including Rebecca Wearmouth who has handled the foreign rights. And I am lucky to be able to lean on Dominic Byatt, my former editor at OUP and a mensch of the first order.

The team at Bloomsbury have been brilliant. It has been a pleasure to work with Alexis Kirschbaum and Jasmine Horsey. I am likewise enormously grateful to have Jessica Case as my editor at Pegasus Books. Joe Hall copy-edited the draft with great care and skill. An author could not hope for a more supportive and expert group of publishing professionals.

There is much for which to thank my family. I often fire ideas at my siblings, Ali Susskind and Daniel Susskind. Both provide me with limitless support and help. As ever, my mum, Michelle, has been encourager-in-chief. And no one gives more selflessly than my dad, Richard. Many times during the writing of this book – indeed, throughout my life – I have turned to him for advice. He never lets me down.

The Digital Republic would have been completed much sooner were it not for the attentions of my miniature dachshund, Mr Pickle. At crucial moments he has halted the writing process to demand food, present me with a rancid chew-toy, or volunteer his belly for a rub. As a political philosopher he has limitations, but as a writing partner he is hard to beat.

Finally, my wife Joanna. My hero, my constant companion, the only critic that matters, and my best friend. This book is dedicated to you with all my love.

Jamie Susskind
London
March 2022

NOTES

INTRODUCTION: UNACCOUNTABLE POWER

1 See Lawrence Lessig, *Code Version* 2.0, Basic Books, New York, 2006; Jamie Susskind, *Future Politics: Living Together in a World Transformed by Tech*, Oxford University Press, Oxford, 2018.

2 Nathaniel Popper, 'Lost Passwords Lock Millionaires Out of Their Bitcoin Fortunes', *The New York Times*, 12 January 2021 <https://www.nytimes.com/2021/01/12/technology/bitcoin-passwords-wallets-fortunes.html> (accessed 19 August 2021).

3 Adi Robertson, 'Facebook and Twitter are restricting a disputed New York Post story about Joe Biden's son', *The Verge*, 14 October 2020 <https://www.theverge.com/2020/10/14/21515972/facebook-new-york-post-hunter-biden-story-fact-checking-reduced-distribution-election-misinformation> (accessed 4 October 2021).

4 Susskind, *Future Politics*.

5 David Reinsel, John Gantz and John Rydning, 'The Digitization of the World: From Edge to Core' (2019) IDC White Paper No. US44413318.

6 Ryan Mac, 'Facebook Apologizes After A.I. Puts "Primates" Label on Video of Black Men', *The New York Times*, 14 September 2021. <https://www.nytimes.com/2021/09/03/technology/facebook-ai-race-primates.html> (accessed 3 October 2021).

7 I am grateful to Sam Gilbert and Simon Caney for helping to clarify this point.

8 Evan Osnos, 'Can Mark Zuckerberg Fix Facebook Before It Breaks Democracy?', *The New Yorker*, 17 September 2018 <https://www.newyorker.com/magazine/2018/09/17/can-mark-zuckerb

erg-fix-facebook-before-it-breaks-democracy> (accessed 19 August 2021).

9 Eric Bradner and Sarah Mucha, 'Biden campaign launches petition lambasting Facebook over refusal to remove political misinformation', CNN, 11 June 2020 <https://edition.cnn.com/2020/06/11/politics/joe-biden-facebook-open-letter/index.html> (accessed 19 August 2021).

10 Lauren Feiner and Megan Graham, 'Pelosi says advertisers should use their "tremendous leverage" to force social media companies to stop spreading false and dangerous information', CNBC, 16 June 2020 <https://www.cnbc.com/2020/06/16/pelosi-says-adve rtisers-should-push-platforms-to-combat-disinformation.html> (accessed 19 August 2021).

11 Marcus Tullius Cicero, The Republic and The Laws, Oxford University Press, Oxford, 1998, p. 49.

12 Cass Sunstein, Democracy and the Problem of Free Speech, Free Press, New York, 1995, p. 37.

13 See, for an eloquent modern exposition, K. Sabeel Rahman, Democracy Against Domination, Oxford University Press, New York, 2017.

14 Richard Dagger, 'Republicanism and the Foundations of Criminal Law', in R. A. Duff and Stuart P. Green (eds), Philosophical Foundations of Criminal Law, Oxford University Press, Oxford, 2013, p. 47.

15 Alexander Hamilton, James Madison and John Jay, The Federalist Papers, Penguin Classics, New York, 2012, p. 96.

16 Alex Gourevitch, From Slavery to the Cooperative Commonwealth: Labor and Republican Liberty in the Nineteenth Century, Cambridge University Press, Cambridge, 2015, p. 10.

17 I am grateful to Julie E. Cohen for suggesting this term, and to Simon Caney, Rob Reich and others for coaxing me away from some of my earlier ideas.

18 I make reference throughout to the draft Artificial Intelligence Act and Digital Services Act, among others.

19 Peter Cihon, Matthijs M. Maas and Luke Kemp, 'Should Artificial Intelligence Governance be Centralised? Design Lessons from History', Proceedings of the AAAI/ACM Conference on AI, Ethics and Society (2020), 228–234.

20 See Gary E. Marchant, Kenneth W. Abbott and Braden Allenby (eds), Innovative Governance Models for Emerging Technologies, Edward Elgar Publishing, Cheltenham, 2013, p. 3.

21   K. Sabeel Rahman, 'Regulating Informational Infrastructure: Internet Platforms as the New Public Utilities', *Georgetown Law and Technology Review*, Vol. 2, No. 2 (2018), 234–251.

22   Diane Coyle, *Markets, State, and People: Economics for Public Policy*, Princeton University Press, Princeton, 2020, p. 267.

23   See Paul Tucker, *Unelected Power: The Quest for Legitimacy in Central Banking and the Regulatory State*, Princeton, Princeton University Press, 2018, p. 66; Stigler Committee on Digital Platforms, 'Final Report', Stigler Centre for the Study of the Economy and the State, 2019 <https://www.chicagobooth.edu/-/media/research/stigler/pdfs/digital-platforms---committee-report---stigler-center.pdf> (accessed 20 August 2021)

24   Rahman, 'Regulating Informational Infrastructure'.

25   Coyle, *Markets*, p. 262.

26   Diane Coyle, 'Three Cheers for Regulation', *Project Syndicate*, 17 July 2018 <https://www.project-syndicate.org/commentary/positive-effects-market-regulation-by-diane-coyle-2018-07> (accessed 20 August 2021).

27   Mark Scott, Laurens Cerulus and Steven Overly, 'How Silicon Valley gamed Europe's privacy rules', *Politico*, 22 May 2019 <https://www.politico.eu/article/europe-data-protection-gdpr-general-data-protection-regulation-facebook-google/> (accessed 20 August 2021).

28   Frank Pasquale, 'The Automated Public Sphere', *University of Maryland Legal Studies Research Paper* No. 2017-31 (2017).

29   Michael Walzer, *Exodus and Revolution*, Basic Books, New York, 1985.

CHAPTER ONE

1   Donna Rose Addis, Alana T. Wong and Daniel L. Schacter, 'Remembering the Past and Imagining the Future: Common and Distinct Neural Substrates During Event Construction and Elaboration,' *Neuropsychologia*, 45, No. 7 (2007), pp. 1363–1367.

2   Philip Pettit, *Just Freedom: A Moral Compass for a Complex World*, W. W. Norton and Company, New York, 2014, p. 3.

3   There is some debate as to whether the Greek city-states fall into the same republican tradition as the Roman Republic. Plato's *Republic* is one of the most influential books of all time; while Aristotle had called it *Politeia* ('citizenship, constitution, government, way of life') Cicero mischievously translated it as *De Republica* – 'ensuring that for the next two millennia important political theorists would derive their view of the "republic" from a Greek philosopher who had

never even heard the term.' Eric Nelson, *The Greek Tradition in Republican Thought*, Cambridge University Press, Cambridge, 2006, p. 1.

4   Melissa Lane, *Greek and Roman Political Ideas*, Penguin Books, London, 2014, p. 12; Marcus Tullius Cicero, *The Republic and The Laws*, Oxford University Press, Oxford, 1998, editor's note at p. 181.

5   Lane, *Political Ideas*, pp. 245–246.

6   Of course, the Roman Republic was also home to slavery, the oppression of women and violent conquest. But the basic idea, that the state and those within it must never be in *potestate domini* (under the power of another), is ripe for rehabilitation. Philip Pettit, 'Law and Liberty', in Samantha Besson and José Luis Martí (eds), *Legal Republicanism: National and International Perspectives*, Oxford University Press, Oxford, 2009, p. 44.

7   The Roman historian Sallust attributed its downfall, in part, to the loss of republican spirit among the people. The majority did not actually want to be free. What they wanted, Sallust said with disgust, was 'nothing more than fair masters.'

8   Iseult Honohan, *Civic Republicanism*, Routledge, Abingdon, 2002, p. 43; Quentin Skinner, 'Machiavelli's *Discorsi* and the pre-humanist origins of republican ideas', in Gisela Bock, Quentin Skinner and Maurizio Viroli (eds), *Machiavelli and Republicanism*, Cambridge University Press, Cambridge, 1993, p. 122; David Held, *Models of Democracy*, Polity Press, Cambridge, 2006, p. 32.

9   Pettit, *Just Freedom*, p. 8.

10   Nicolai Rubenstein, 'Machiavelli and Florentine republican experience', in Bock et al., *Machiavelli and Republicanism*, p. 4.

11   Quentin Skinner, 'Classical Liberty and the Coming of the English Civil War', in Martin van Gelderen and Quentin Skinner (eds), *Republicanism: A Shared European Heritage (Volume II)*, Cambridge University Press, Cambridge, 2006, p. 10.

12   See Pettit, *Just Freedom*; Quentin Skinner, 'Freedom as the Absence of Arbitrary Power' in Cécile Laborde and John Maynor (eds), *Republicanism and Political Theory*, Blackwell Publishing, Oxford, 2008, p. 85; Quentin Skinner, *Hobbes and Republican Liberty*, Cambridge University Press, Cambridge, 2008; Quentin Skinner, 'On trusting the judgement of our rulers', in Richard Bourke and Raymond Geuss (eds), *Political Judgement: Essays for John Dunn*, Cambridge University Press, Cambridge, 2009, pp. 117–118.

13   Skinner, 'Classical Liberty', p. 16.

14   Skinner, 'On trusting', p. 116.

15 John Milton, 'A Defence of the People of England', *Political Writings* (ed. Martin Dzelzainis), Cambridge University Press, Cambridge, 1998, p. 51.

16 Blair Worden, 'Milton's republicanism and the tyranny of heaven', in Bock et al., *Machiavelli and Republicanism*, p. 226.

17 Skinner, 'Classical Liberty', p. 16.

18 Skinner, 'On trusting', p. 122.

19 Roger Scruton, *England: An Elegy*, Bloomsbury, London, 2006, p. 121.

20 *R (Miller) v Prime Minister & ors* [2019] UKSC 41; [2020] AC 373, per Baroness Hale of Richmond PSC at [32].

21 Lea Campos Boralevi, 'Classical Foundational Myths of European Republicanism: The Jewish Commonwealth', in Martin van Gelderen and Quentin Skinner (eds), *Republicanism: A Shared European Heritage* (Volume I), Cambridge University Press, Cambridge, 2006, pp. 250–260; Anna Grześkowiak-Krwawicz, 'Anti-monarchism in Polish Republicanism in the Seventeenth and Eighteenth Centuries', in van Gelderen and Skinner (eds) *Republicanism* (Volume I), pp. 43–45; Vittorio Conti, 'The Mechanisation of Virtue: Republican Rituals in Italian Political Thought in the Sixteenth and Seventeenth Centuries', in van Gelderen and Skinner (eds), *Republicanism (Volume II)*, p. 73; Pettit, *Just Freedom*, p. 9.

22 The Declaratory Act, 18 March 1766.

23 See F. A. Hayek, *The Constitution of Liberty*, Routledge, Abingdon, 2006, p. 155.

24 Alexander Hamilton, James Madison and John Jay, *The Federalist Papers*, Penguin Classics, New York, 2012, no. 1.

25 Philip Pettit, *On the People's Terms: A Republican Theory and Model of Democracy*, Cambridge University Press, Cambridge, 2014, p. 7; Bernard Bailyn, *The Ideological Origins of the American Revolution*, Harvard University Press, Cambridge, MA, 1992, p. 34.

26 Philip Pettit, *Republicanism: A Theory of Freedom and Government*, Oxford University Press, Oxford, 201, p. 34; see also Pettit, *Just Freedom*, p. 11.

27 Alex Gourevitch, 'Liberty and Democratic Insurgency: The Republican Case for the Right to Strike', in Yiftah Elazar and Geneviève Rousselière (eds), *Republicanism and the Future of Democracy*, Cambridge University Press, Cambridge, 2019, p. 173.

28 Mary Wollstonecraft, *A Vindication of the Rights of Women*, Penguin Classics, London, 2004, p. 5.

29 Cited in Frank Lovett, *A General Theory of Domination and Justice*, Oxford University Press, Oxford, 2012, pp. 45–46.

30   See generally Pettit, *People's Terms*; *Republicanism*, p. 5.

31   Shaila Dewan and Serge F. Kovaleski, 'Thousands of Complaints Do Little to Change Police Ways', *The New York Times*, 8 June 2020 <https://www.nytimes.com/2020/05/30/us/derek-chauvin-george-floyd.html> (accessed 20 August 2021).

32   Pettit, *Republicanism*, p. 6; Frank Lovett, 'Algernon Sidney, Republican Stability, and the Politics of Virtue', *Journal of Political Science*, Vol. 48, No. 1, Article 3 (2020), 59–83.

33   Cited in Fania Oz-Salzberger, 'Scots, Germans, Republic and Commerce', in van Gelderen and Skinner (eds), *Republicanism (Volume II)*, p. 202.

CHAPTER TWO

1   See generally Julie E. Cohen, *Between Truth and Power: The Legal Constructions of Informational Capitalism*, Oxford University Press, Oxford, 2019.

2   Philip Pettit, *Republicanism: a Theory of Freedom and Government*, Oxford University Press, Oxford, p. 9.

3   Quentin Skinner, *Liberty before Liberalism*, Cambridge University Press, Cambridge, 1998, p. 10; Pettit, *Republicanism*, pp. 37–45

4   Skinner, *Liberty before Liberalism*, p. 10.

5   Quentin Skinner, 'Classical Liberty and the Coming of the English Civil War', in Martin van Gelderen and Quentin Skinner (eds), *Republicanism: A Shared European Heritage (Volume II)*, Cambridge University Press, Cambridge, 2006, p. 27.

6   Nathaniel Persily, *The Internet's Challenge to Democracy: Framing the Problem and Assessing Reforms*, Kofi Annan Foundation, Stanford, CA, 2019.

7   Stephen Mulhall and Adam Swift, *Liberals & Communitarians* (2nd ed.), Blackwell, Oxford, 1996, pp. 15; 54; Charles Taylor, *Philosophy and the Human Sciences: Philosophical Papers 2*, Cambridge University Press, Cambridge, 1999, pp. 187-189.

8   Michael Walzer, 'The Communitarian Critique of Liberalism', *Political Theory*, Vol. 18, No. 1 (Feb 1990), 6–23.

9   Hannah Arendt, *The Human Condition*, University of Chicago Press, Chicago, 1998, p. 8.

10   Samuel Bowles and Herbert Gintis, *Democracy and Capitalism: Property, Community, and the Contradictions of Modern Social Thought*, Basic Books, London, 1986, p. 67; Philip Pettit, *On the People's Terms: A Republican Theory and Model of Democracy*, Cambridge University Press, Cambridge, 2014, p. 11.

11   Pettit, *Republicanism*, p. 45.

12 Alexander Hamilton, James Madison and John Jay, *The Federalist Papers*, Penguin, New York, 2012, p. 115.

13 Besson and Martí, 'Law and Republicanism', p. 35; Philip Pettit, 'Law and Liberty', in Besson and Martí, *Legal Republicanism*, p. 49; Martin Loughlin, *Foundations of Public Law*, Oxford University Press, Oxford, 2014, p. 174; Nicolas P. Suzor, *Lawless: The Secret Rules that Govern Our Digital Lives*, Cambridge University Press, Cambridge, 2019, p. 8.

14 James Harrington, *The Commonwealth of Oceana and A System of Politics*, Cambridge University Press, Cambridge, 2008, p. 41.

15 Skinner, *Liberty before Liberalism*, ix–x.

16 Julie E. Cohen, 'How (Not) to Write a Privacy Law', *Knight First Amendment Institute*, 23 March 2021 <https://knightcolumbia.org/content/how-not-to-write-a-privacy-law> (accessed 20 August 2021).

CHAPTER THREE

1 See Tim Wu, *The Master Switch: The Rise and Fall of Information Empires*, Atlantic, London, 2010; Jamie Susskind, *Future Politics: Living Together in a World Transformed by Tech*, Oxford University Press, Oxford, 2018.

2 See Lawrence Lessig, *Code Version 2.0*, Basic Books, New York, 2006.

3 Bruce Schneier, *Click Here to Kill Everybody: Security and Survival in a Hyper-Connected World*, W. W. Norton & Company, New York, 2018, p. 5.

4 See e.g. Amy Webb, *The Big Nine: How the Tech Titans & Their Thinking Machines Could Warp Humanity*, Hachette, New York, 2019, p. 90; Laura Denardis, *The Internet in Everything: Freedom and Security in a World with No Off Switch*, Yale University Press, New Haven, 2020; Susskind, *Future Politics*.

5 Seung-min Park et al., 'A mountable toilet system for personalized health monitoring via the analysis of excreta', *Nature Biomedical Engineering*, Vol. 4, No. 6 (2020), 624–635.

6 Uri Bram and Martin Schmalz, *The Business of Big Data: How to Create Lasting Value in the Age of AI*, Amazon, UK, 2019, pp. 32–33.

7 Cited in Clive Thompson, *Coders: Who They Are, What They Think And How They Are Changing Our World*, Picador, London, 2019, p. 189.

8 Colin Lecher, 'How Amazon Automatically Tracks and Fires Warehouse Workers for 'Productivity'', *The Verge*, 25 April 2019 <https://www.theverge.com/2019/4/25/18516004/amazon-warehouse-fulfillment-centers-productivity-firing-terminations> (accessed 20 August 2021). Article 88 of the GDPR has effect here: see Robert Jeffrey, 'Would you let AI recruit for you?' *People*

Management, 12 December 2017 <https://www.peoplemanagem ent.co.uk/long-reads/articles/recruiting-algorithms> (accessed 20 August 2021).

9   Michèle Finck, *Blockchain Regulation and Governance in Europe*, Cambridge University Press, Cambridge, 2019, p. 40.

10   McKenzie Raub, 'Bots, Bias, and Big Data: Artificial Intelligence, Algorithmic Bias and Disparate Impact Liability in Hiring Practices', *Arkansas Law Review*, Vol. 71, No. 2 (2018), 529–570.

11   Charles Hymas, 'AI used for first time in job interviews in UK to find best applicants', *The Telegraph*, 27 September 2019 <https:// www.telegraph.co.uk/news/2019/09/27/ai-facial-recognit ion-used-first-time-job-interviews-uk-find/> (accessed 20 August 2021); Raub, 'Bots, Bias, and Big Data'.

12   See *Belong.co* <https://belong.co/hireplus/> (accessed 20 August 2021); *Mighty Recruiter* <https://www.mightyrecruiter.com> (accessed 20 August 2021); *deepsense.ai* https://deepsense.ai> (accessed 20 August 2021).

13   Shirin Ghaffary, 'Facebook is taking down some, but not all, quarantine protest event pages', *Vox*, 20 April 2020 <https:// www.vox.com/recode/2020/4/20/21228224/facebook-coro navirus-covid-19-protests-taking-down-content-moderation-free dom-speech-debate> (accessed 20 August 2021).

CHAPTER FOUR

1   Bernard E. Harcourt, *Exposed: Desire and Disobedience in the Digital Age*, Harvard University Press, Cambridge, MA, 2012, pp. 198–199.

2   Harcourt, *Exposed*, pp. 199–204. After it became public that Medbase2000 was selling such information, it removed these offers from its website.

3   See e.g. Justin Sherman, 'Data Brokers Are a Threat to Democracy', *Wired*, 13 April 2021 <https://www.wired.com/story/opin ion-data-brokers-are-a-threat-to-democracy/> (accessed 3 October 2021); Justin Sherman, 'Data Brokers and Sensitive Data on U.S. Individuals: Threats to American Civil Rights, National Security, and Democracy', Duke University Sanford Cyber Policy Program, 2021 <https://sites.sanford.duke.edu/techpolicy/report-data-brokers-and-sensitive-data-on-u-s-individuals/> (accessed 3 October 2021).

4   Stuart Russell, *Human Compatible: AI and the Problem of Control*, Allen Lane, London, 2019, p. 75.

5   Bruce Schneier, *Click Here to Kill Everybody: Security and Survival in a Hyper-Connected World*, W. W. Norton & Company, New York, 2018, p. 58.

6   Dan Robitzski, 'Ex-Googler: Company Has "Voodoo Doll, Avatar-Like Version of You"', Futurism, 2 May 2019 <https://futurism.com/google-company-voodoo-doll-avatar/amp> (accessed 20 August 2021).

7   See Mireille Hildebrandt, 'Profile transparency by design? Re-enabling double contingency' in Mireille Hildebrandt and Katja de Vries (eds), *Privacy, Due Process and the Computational Turn: The Philosophy of Law Meets the Philosophy of Technology*, Routledge, Abingdon, 2013, p. 227; Julie E. Cohen, *Between Truth and Power: The Legal Constructions of Informational Capitalism*, Oxford University Press, Oxford, 2019, p. 67.

8   See generally Carissa Véliz, *Privacy is Power: Why and How you Should Take Back Control of Your Data*, Transworld Publishers, London, 2020.

9   Steven Pinker, 'Tech Prophecy and the Underappreciated Causal Power of Ideas' in John Brockman (ed.), *Possible Minds: 25 Ways of Looking at AI*, Penguin Press, New York, 2019, p. 107.

10   Rana Faroohar, *Don't be Evil: The Case Against Big Tech*, Allen Lane, London, 2019, p. 238.

11   Ella Fassler, 'Here's How Easy It Is for Cops to Get Your Facebook Data', OneZero, 17 June 2020 <https://onezero.medium.com/cops-are-increasingly-requesting-data-from-facebook-and-you-probably-wont-get-notified-if-they-5b7a2297df17> (accessed 20 August 2021).

12   Zoom has since ceased this practice. Kaitlyn Tiffany, 'No, the Internet Is Not Good Again', The Atlantic, 16 April 2020 <https://www.theatlantic.com/technology/archive/2020/04/zoom-facebook-moderation-ai-coronavirus-internet/610099/> (accessed 20 August 2021).

13   Joseph Cox, 'Zoom iOS App Sends Data to Facebook Even if You Don't Have a Facebook Account', Vice, 26 March 2020 <https://www.vice.com/en/article/k7e599/zoom-ios-app-sends-data-to-facebook-even-if-you-dont-have-a-facebook-account> (accessed 20 August 2021).

14   The Pillar, 'Location-based apps pose security risk for Holy See', The Pillar, 27 July 2021 <https://www.pillarcatholic.com/p/location-based-apps-pose-security> (accessed 3 October 2021).

15   Philip Pettit, 'Is Facebook Making Us Less Free?' The Institute of Art and Ideas, 26 March 2018 <https://iai.tv/articles/the-big-brotherhood-of-digital-giants-is-taking-away-our-freedom-auid-884>

(accessed 20 August 2021). See also Michael Walzer, *Spheres of Justice: A Defense of Pluralism and Equality*, Basic Books, New York, 1983, p. 17.

16   See e.g. David Owen, 'Should We Be Worried About Computerized Facial Recognition?' *The New Yorker*, 10 December 2018 <https://www.newyorker.com/magazine/2018/12/17/should-we-be-worried-about-computerized-facial-recognition> (accessed 20 August 2021); Davey Alba, 'Facial Recognition Moves Into a New Front: Schools', *The New York Times*, 6 February 2020 <https://www.nytimes.com/2020/02/06/business/facial-recognition-schools.html > (accessed 20 August 2021).

17   Stuart A. Thompson and Charlie Warzel, 'Twelve Million Phones, One Dataset, Zero Privacy', *The New York Times*, 19 December 2019 <https://www.nytimes.com/interactive/2019/12/19/opinion/location-tracking-cell-phone.html> (accessed 20 August 2021); 'How to Track President Trump', *The New York Times*, 20 December 2019 <https://www.nytimes.com/interactive/2019/12/20/opinion/location-data-national-security.html> (accessed 20 August 2021).

18   Bruce Schneier, 'We're Banning Facial Recognition. We're Missing the Point', *The New York Times*, 20 January 2020, <https://www.nytimes.com/2020/01/20/opinion/facial-recognition-ban-privacy.html> (accessed 20 August 2021).

19   Sonia Fernandez, 'WiFi System Identifies People Through Walls By Their Walk', *Futurity*, 1 October 2019 <https://www.futurity.org/wifi-video-identification-through-walls-2173442/> (accessed 20 August 2021).

20   Amber Marks, Benjamin Bowling and Colman Kennan, 'Automatic Justice? Technology, Crime, and Social Control' in Roger Brownsword, Eloise Scotford and Karen Yeung (eds), *The Oxford Handbook of Law, Regulation, and Technology*, Oxford University Press, Oxford, 2017, p. 712.

21   See e.g. Daphne Leprince-Ringuet, 'Facial Recognition: This New AI Tool Can Spot When You Are Nervous or Confused', ZDNet, 21 October 2019 <https://www.zdnet.com/article/this-new-ai-tool-can-spot-if-you-are-nervous-or-confused/> (accessed 20 August 2021).

22   See Andrew McStay, *Emotional AI: The Rise of Empathic Media*, Sage Publications, London, 2018.

23   Terrence J. Sejnowski, *The Deep Learning Revolution*, MIT Press, Cambridge, MA, 2018, p. 184; Ben Dickson, 'Your Next Car Will

Be Watching You More Than It's Watching the Road', *Gizmodo*, 28 November 2019 <https://gizmodo.com/your-next-car-will-be-watching-you-more-than-its-watchi-1840055386> (accessed 20 August 2021).

24 Matt Simon, 'This Robot Can Guess How You're Feeling by the Way You Walk', *Wired*, 18 May 2020 <https://www.wired.com/story/proxemo-robot-guesses-emotion-from-walking/> (accessed 20 August 2021).

25 Jonas Rauber, Emily B. Fox and Leon A. Gatys, 'Modeling patterns of smartphone usage and their relationship to cognitive health', *arXiv*: 1911.05683 (2019).

26 Johannes C. Eichstaedt et al., 'Facebook Language Predicts Depression in Medical Records', *Proceedings of the National Academy of Sciences of the United States of America*, Vol. 115, No. 44 (2018), 11203–11208.

27 Michal Kosinski, David Stillwell and Thore Graepel, 'Private traits and attributes are predictable from digital records of human behaviour', *Proceedings of the National Academy of Sciences of the United States of America*, Vol. 110, No. 15 (2013), 5802–5805.

28 See Quentin Skinner, 'On trusting the judgement of our rulers', in Richard Bourke and Raymond Geuss (eds), *Political Judgement: Essays for John Dunn*, Cambridge University Press, Cambridge, 2009, pp. 123–124. See Alex Marthews and Catherine Tucker, 'The Impact of Online Surveillance on Behavior', in David Gray and Stephen E. Henderson (eds), *The Cambridge Handbook of Surveillance Law*, Cambridge University Press, New York, 2019; Jamie Susskind, *Future Politics: Living Together in a World Transformed by Tech*, Oxford University Press, Oxford, chapter 7.

29 Anthony Cuthbertson, 'Google Admits Workers Listen to Private Audio Recordings From Google Home Smart Speakers', *The Independent*, 11 July 2019 <https://www.independent.co.uk/life-style/gadgets-and-tech/news/google-home-smart-speaker-audio-recordings-privacy-voice-spy-a9000616.html> (accessed 20 August 2021); 'Apple 'sorry' that workers listened to Siri voice recordings', BBC, 28 August 2019 <https://www.bbc.co.uk/news/technology-49502292> (accessed 20 August 2021).

30 Sarah Marsh, 'Councils let firms track visits to webpages on benefits and disability', *The Guardian*, 4 February 2020 <https://www.theguardian.com/technology/2020/feb/04/councils-let-firms-track-visits-to-webpages-on-benefits-and-disability> (accessed 20 August 2021).

31 Ifeoma Ajunwa, 'Algorithms at Work: Productivity Monitoring Applications and Wearable Technology as the New Data-Centric Research Agenda for Employment and Labor Law', *Saint Louis University Law Journal*, Vol. 63, No. 1, Article 4 (2018), 21–54.

32 Sarah O'Connor, 'When your boss is an algorithm', *Financial Times*, 8 September 2016 <https://www.ft.com/content/88fdc 58e-754f-11e6-b60a-de4532d5ea35> (accessed 20 August 2021).

33 Lora Jones, "I monitor my staff with software that takes screenshots", BBC, 29 September 2020 <https://www.bbc.co.uk/news/busin ess-54289152> (accessed 3 October 2021).

34 Chris Stokel-Walker, 'If You're a Remote Worker, You're Going to Be Surveilled. A Lot', *OneZero*, 23 April 2020 <https://onezero.med ium.com/if-youre-a-remote-worker-you-re-going-to-be-surveil led-a-lot-f3f8d4308ee> (accessed 20 August 2021).

35 Lauren Kaori Gurley and Joseph Cox, 'Inside Amazon's Secret Program to Spy On Workers' Private Facebook Groups', *Vice*, 1 September 2020 <https://www.vice.com/en/article/3azegw/ amazon-is-spying-on-its-workers-in-closed-facebook-groups-inter nal-reports-show> (accessed 3 October 2021).

36 This idea permeates the work of the republican philosopher Philip Pettit.

CHAPTER FIVE

1 See generally the work of Neil Richards and Woodrow Hartzog.

2 See John Danaher, 'The Ethics of Algorithmic Outsourcing in Everyday Life', in Karen Yeung and Martin Lodge, *Algorithmic Regulation*, Oxford University Press, Oxford, 2019, pp. 102–103.

3 Julie E. Cohen, *Between Truth and Power: The Legal Constructions of Informational Capitalism*, Oxford University Press, Oxford, 2019, p. 83.

4 See e.g. Richard H. Thaler and Cass Sunstein, *Nudge: Improving Decisions about Health, Wealth and Happiness*, Penguin, London, 2009.

5 Eliza Mik, 'Persuasive Technologies: From Loss of Privacy to Loss of Autonomy', in Kit Barker, Karen Fairweather and Ross Grantham (eds), *Private Law in the 21st Century*, Hart Publishing, Oxford, 2017, p. 375.

6 See Richard Susskind, *The Future of Law: Facing the Challenges of Legal Technology*, Oxford University Press, Oxford, 1996.

7 Sam Biddle, Paulo Victor Ribeiro and Tatiana Dias, 'Invisible Censorship: TikTok Told Moderators to Suppress Posts by "Ugly" People and the Poor to Attract New Users', *The Intercept*, 16 March

2020 <https://theintercept.com/2020/03/16/tiktok-app-mod erators-users-discrimination/> (accessed 20 August 2021).

8   Carissa Véliz, *Privacy is Power: Why and How you Should Take Back Control of Your Data*, Transworld Publishers, London, 2020, pp. 103–104.

9   Robert Epstein and Robert E. Robertson, 'The search engine manipulation effect (SEME) and its possible impact on the outcomes of elections', *Proceedings of the National Academy of Sciences of the United States of America*, Vol. 112, No. 33 (2015), E4512–E4521.

10 Robert Epstein, Roger Mohr Jr. and Jeremy Martinez, 'The Search Suggestion Effect (SSE): How Search Suggestions Can Be Used to Shift Opinions and Voting Preferences Dramatically and Without People's Awareness', *98th Annual Meeting of the Western Psychological Association*, Portland, OR, 26 April 2018. See also Yuji Develle, 'Why we cannot trust Big Tech to be apolitical', *Wonk Bridge*, 4 May 2020 <https:// medium.com/wonk-bridge/why-we-cannot-trust-big-tech-to-be-apolitical-f031af9386cf> (accessed 20 August 2021).

11 Epstein et al., 'Search Suggestion Effect'.

12 Emily Bell, 'The Unintentional Press: How Technology Companies Fail as Publishers', in Lee C. Bollinger and Geoffrey R. Stone, *The Free Speech Century*, Oxford University Press, New York, 2019, p. 237.

13 Jamie Susskind, 'What we need from social media is transparency, not apologies', *The New Statesman*, 6 September 2018 <https://www.newst atesman.com/science-tech/2018/09/what-we-need-social-media-transparency-not-apologies> (accessed 24 September 2021).

14 Melissa Lane, *Greek and Roman Political Ideas*, Penguin Books, London, 2014, p. 12; see also the editor's note at Marcus Tullius Cicero, *The Republic and The Laws*, Oxford University Press, Oxford, 1998, p. 181.

CHAPTER SIX

1   The Queen's message to the Commonwealth on the coronavirus pandemic, 6 April 2020.

2   Steven Morris, 'Facebook apologises for flagging Plymouth Hoe as offensive term', *The Guardian*, 27 January 2021 <https://www.theg uardian.com/uk-news/2021/jan/27/facebook-apologises-flagg ing-plymouth-hoe-offensive-term> (accessed 20 August 2021).

3   'Mark Zuckerberg Testimony Transcript: Zuckerberg Testifies on Facebook Cryptocurrency Libra', *Rev*, 23 October 2019 <https:// www.rev.com/blog/transcripts/mark-zuckerberg-testimony-tra nscript-zuckerberg-testifies-on-facebook-cryptocurrency-libra> (accessed 20 August 2021).

4   Jillian C. York and Ethan Zuckerman, 'Moderating the Public Sphere', in Rikke Frank Jørgensen (ed.), *Human Rights in the Age of Platforms*, MIT Press, Cambridge, MA, 2019, p. 140.

5   York and Zuckerman, 'Moderating', p. 156.

6   Timothy Karr, 'Why Facebook Filtering Will Ultimately Fail', *Start It Up*, 15 November 2019 <https://medium.com/swlh/why-facebook-filtering-will-ultimately-fail-90606ec98c11> (accessed 20 August 2021).

7   Chloe Hadavas, 'Why We Should Care That Facebook Accidentally Deplatformed Hundreds of Users', *Slate*, 12 June 2020 <https://slate.com/technology/2020/06/facebook-anti-racist-skinheads.html> (accessed 20 August 2021).

8   Julia Carrie Wong and Hannah Ellis-Petersen, 'Facebook planned to remove fake accounts in India – until it realized a BJP politician was involved', *The Guardian*, 15 April 2021 <https://www.theguardian.com/technology/2021/apr/15/facebook-india-bjp-fake-accounts> (accessed 3 October 2021).

9   Nicolas P. Suzor, *Lawless: The Secret Rules that Govern Our Digital Lives*, Cambridge University Press, Cambridge, 2019, p. 51; Tarleton Gillespie, *Custodians of the Internet: Platforms, Content Moderation, and the Hidden Decisions that Shape Social Media*, Yale University Press, New Haven, 2018, p. 38.

10 Gillespie, *Custodians*, p. 38.

11 Alex Hern, 'Revealed: how TikTok censors videos that do not please Beijing', *The Guardian*, 25 September 2019 <https://www.theguardian.com/technology/2019/sep/25/revealed-how-tiktok-censors-videos-that-do-not-please-beijing> (accessed 20 August 2021).

12 Rikke Frank Jørgensen, 'Rights Talk: In the Kingdom of Online Giants', in Jørgensen (ed.), *Human Rights*, p. 174.

13 Daphne Keller, 'Who Do You Sue? State and Platform Hybrid Power Over Online Speech' (2019) Hoover Institution, Aegis Series Paper No. 1902.

14 Alex Hern, 'TikTok's local moderation guidelines ban pro-LGBT content', *The Guardian*, 26 September 2019 <https://www.theguardian.com/technology/2019/sep/26/tiktoks-local-moderation-guidelines-ban-pro-lgbt-content> (accessed 20 August 2021).

15 Tanya Basu, 'How a ban on pro-Trump patterns unraveled the online knitting world', *MIT Technology Review*, 6 March 2020 <https://www.technologyreview.com/2020/03/06/905

472/ravelry-ban-on-pro-trump-patterns-unraveled-the-online-knitting-world-censorship-free/> (accessed 20 August 2021).

16 Julia Jacobs, 'Will Instagram Ever "Free the Nipple"?' *The New York Times*, 22 November 2019 <https://www.nytimes.com/2019/11/22/arts/ design/instagram-free-the-nipple.html> (accessed 20 August 2021); Instagram Inc, 'Community Guidelines', 2021 <https://help. instagram.com/477434105621119> (accessed 20 August 2021).

17 Kim Lyons, 'Twitter removes tweets by Brazil, Venezuela presidents for violating COVID-19 content rules', *The Verge*, 30 March 2020 <https://www.theverge.com/2020/3/30/21199845/twit ter-tweets-brazil-venezuela-presidents-covid-19-coronavirus-jair-bolsonaro-maduro> (accessed 20 August 2021).

18 Lizzie Dearden, 'Iran's Supreme Leader claims gender equality is "Zionist plot" aiming to corrupt role of women in society', *The Independent*, 21 March 2017 <https://www.independent.co.uk/news/world/middle-east/iran-supreme-leader-ayatollah-khame nei-gender-equality-women-zionist-plot-society-role-islamic-lea der-theocracy-a7641041.html> (accessed 20 August 2021).

19 Hanna Kozlowska, 'Each platform's approach to political ads in one table', *Quartz*, 13 December 2019 <https://qz.com/1767 145/how-facebook-twitter-and-others-approach-political-advertis ing/> (accessed 20 August 2021).

20 Sue Halpern, 'The Problem of Political Advertising on Social Media', *The New Yorker*, 24 October 2019 <https://www.newyorker.com/tech/annals-of-technology/the-problem-of-political-advertis ing-on-social-media> (accessed 20 August 2021).

21 Lauren Jackson and Desiree Ibekwe, 'Jack Dorsey on Twitter's Mistakes', *The New York Times*, 7 August 2020 < https://www.nyti mes.com/2020/08/07/podcasts/the-daily/Jack-dorsey-twit ter-trump. html > (accessed 20 August 2021).

22 Jeff Horwitz, 'The Facebook Files: Facebook Says Its Rules Apply to All. Company Documents Reveal a Secret Elite That's Exempt', *The Wall Street Journal*, 13 September 2021 <https://www.wsj. com/articles/the-facebook-files-11631713039> (accessed 3 October 2021). It is understood that the case against Neymar was dropped for lack of evidence.

23 Carl Schmitt, *Political Theology: Four Chapters on the Concept of Sovereignty*, University of Chicago Press, Chicago, 2005.

CHAPTER SEVEN

1 Thomas Macaulay, 'Someone let a GPT-3 bot loose on Reddit – it didn't end well', The Next Web, 7 October 2020 <https://thenextweb.com/news/someone-let-a-gpt-3-bot-loose-on-reddit-it-didnt-end-well> (accessed 3 October 2021).

2 Jamie Susskind, 'Chatbots Are a Danger to Democracy', The New York Times, 4 December 2018 <https://www.nytimes.com/2018/12/04/opinion/chatbots-ai-democracy-free-speech.html> (accessed 20 August 2021).

3 Susskind, 'Chatbots'. See also Davey Alba, 'Fake "Likes" Remain Just a Few Dollars Away, Researchers Say', The New York Times, 6 December 2019 <https://www.nytimes.com/2019/12/06/technology/fake-social-media-manipulation.html> (accessed 20 August 2021).

4 Toby Walsh, Android Dreams: The Past, Present and Future of Artificial Intelligence, C. Hurst & Co (Publishers), London, 2017, p. 218.

5 Ryan Calo, 'Against Notice Skepticism in Privacy (and Elsewhere)', Notre Dame Law Review, Vol. 87, No. 3 (2012), 1027–1072.

6 Jane Croft, 'Chatbots join the legal conservation', Financial Times, 7 June 2018 <https://www.ft.com/content/0eabcf44-4c83-11e8-97e4-13afc22d86d4> (accessed 20 August 2021).

7 Will Douglas Heaven, 'IBM's Debating AI Just Got a Lot Closer to Being a Useful Tool', MIT Technology Review, 21 January 2020 <https://www.technologyreview.com/2020/01/21/276156/ibms-debating-ai-just-got-a-lot-closer-to-being-a-useful-tool/> (accessed 20 August 2021); IBM AI Research, 'Project Debater', IBM, 2021 <https://www.research.ibm.com/artificial-intelligence/project-debater/> (accessed 20 August 2021).

8 Tom Simonite, 'The AI Text Generator That's Too Dangerous to Make Public', Wired, 14 February 2019 <https://www.wired.com/story/ai-text-generator-too-dangerous-to-make-public/> (accessed 20 August 2021).

9 Nick Statt, 'Google expands AI calling service Duplex to Australia, Canada, and the UK', The Verge, 8 April 2020 <https://www.theverge.com/2020/4/8/21214321/google-duplex-ai-automated-calling-australia-canada-uk-expansion> (accessed 20 August 2021).

10 Centre for Data Ethics and Innovation, Snapshot Paper – Deepfakes and Audiovisual Disinformation, 12 September 2019 <https://www.gov.uk/government/publications/cdei-publihes-its-first-series-of-three-snapshot-papers-ethical-issues-in-ai/snapshot-paper-deepfakes-and-audiovisual-disinformation> (accessed 20 August

2021). See also Nina Schick, *Deepfakes:The Coming Infocalypse*, Monoray, London, 2020.

11   Gregory Barber, 'Deepfakes Are Getting Better, But They're Still Easy to Spot', *Wired*, 26 May 2019 <https://www.wired.com/story/deepfakes-getting-better-theyre-easy-spot/> (accessed 20 August 2021). See Robert Chesney and Danielle Keats Citron, 'Deep Fakes: A Looming Challenge for Privacy, Democracy, and National Security', 107 *California Law Review* 1753 (2019).

12   Chesney and Citron, 'Deep Fakes'.

13   Mark Zuckerberg, 'Mark Zuckerberg: Big Tech needs more regulation', *Financial Times*, 16 February 2020 <https://www.ft.com/content/602ec7ec-4f18-11ea-95a0-43d18ec715f5> (accessed 20 August 2021).

CHAPTER EIGHT

1   Jamie Susskind, *Future Politics: Living Together in a World Transformed by Tech*, Oxford University Press, Oxford, 2018, p. 290.

2   Nicolas P. Suzor, *Lawless: The Secret Rules that Govern Our Digital Lives*, Cambridge University Press, Cambridge, 2019, p. 29.

3   April Glaser, 'Is a Tech Company Ever Neutral?' *Slate*, 11 October 2019   <https://slate.com/technology/2019/10/apple-chinese-government-microsoft-amazon-ice.html> (accessed 20 August 2021); Alexi Mostrous and Peter Hoskin, 'Foreign policy: the great game', *Tortoise Media*, 7 January 2020 <https://www.tortoisemedia.com/2020/01/07/tech-states-apple-foreign-policy/> (accessed 20 August 2021).

4   Dave Gershgorn, 'Amazon's "holy grail" recruiting tool was actually just biased against women', *Quartz*, 10 October 2018 <https://qz.com/1419228/amazons-ai-powered-recruiting-tool-was-biased-against-women/> (accessed 20 August 2021). See also Cathy O'Neil, *Weapons of Math Destruction: How Big Data Increases Inequality and Threatens Democracy*, Crown, New York, 2016.

5   See Gary Marcus and Ernest Davis, *Rebooting AI: Building Artificial Intelligence We Can Trust*, New York, Pantheon Books, 2019, p. 36; Solon Barocas and Andrew D. Selbst, 'Big Data's Disparate Impact', *California Law Review*, Vol. 104, No. 3 (2016), 671–732, p. 691; Talia B. Gillis and Josh Simons, 'Explanation < Justification: GDPR and the Perils of Privacy', *Pennsylvania Journal of Law and Innovation*, Vol. 2 (2019), 71–99.

6   Ziad Obermeyer et al., 'Dissecting racial bias in an algorithm used to manage the health of populations', *Science*, Vol. 366, No. 6464 (2019), 447–453.

7   Katy Cook, *The Psychology of Silicon Valley: Ethical Threats and Emotional Unintelligence in the Tech Industry*, Palgrave Macmillan, London, 2020, p. 53.

8   See e.g. Natasha Singer and Cade Metz, 'Many Facial-Recognition Systems Are Biased, Says U.S. Study', *The New York Times*, 19 December 2019 <https://www.nytimes.com/2019/12/19/technology/facial-recognition-bias.html> (accessed 20 August 2021).

9   'In 1947, Alan Turing told the London Mathematical Society that "what we want is a machine that can learn from experience." Nine years later, scholars at Dartmouth College coined the term "artificial intelligence".' McKenzie Raub, 'Bots, Bias, and Big Data: Artificial Intelligence, Algorithmic Bias and Disparate Impact Liability in Hiring Practices', *Arkansas Law Review*, Vol. 71, No. 2 (2018), 529–570.

10   Aylin Caliskan-Islam, Joanna J. Bryson and Arvind Narayanan, 'Semantics derived automatically from language corpora necessarily contain human biases', *arXiv*: 1608.07187v2 (2016).

11   Peter Birks, *The Roman Law of Obligations*, Oxford University Press, Oxford, 2014, p. 221.

12   Francis Fukuyama, *Identity: Contemporary Identity Politics and the Struggle for Recognition*, Profile Books, London, 2018, xiii.

CHAPTER NINE

1   Will Hazell, 'A-level results 2020: 39% of teacher predicted grades downgraded by algorithm amid calls for U-turn', *inews*, 13 August 2020 <https://inews.co.uk/news/education/a-level-results-2020-grades-downgraded-algorithm-triple-lock-u-turn-result-day-578194> (accessed 20 August 2021).

2   Louise Amoore, 'Why "Ditch the algorithm" is the future of political protest', *The Guardian*, 19 August 2020 <https://www.theguardian.com/commentisfree/2020/aug/19/ditch-the-algorithm-generation-students-a-levels-politics > (accessed 20 August 2021).

3   Cited in Scott R. Peppet, 'Regulating the Internet of Things: First Steps Toward Managing Discrimination, Privacy, Security, and Consent', *Texas Law Review*, Vol. 93, No. 83 (2014), 85–176.

4   Hannah Arendt, *The Human Condition*, University of Chicago Press, Chicago, 1998, p. 43. See, seminally, Alain Desrosières, *The Politics of*

Large Numbers: A History of Statistical Reasoning, Harvard University Press, Cambridge, MA, 1998.

5   Arendt, *Human Condition*, p. 43.

6   Sophia Moreau, 'Equality Rights and Stereotypes', in David Dyzenhaus and Malcolm Thorburn (eds), *Philosophical Foundations of Constitutional Law*, Oxford University Press, Oxford, 2019, p. 293.

7   Tyler Vigen, *Spurious Correlations*, Hachette Books, New York, 2015, p. 13.

8   Pavlo Blavatskyy, 'Obesity of politicians and corruption in post-Soviet countries', *Economics of Transition and Institutional Change*, Vol. 29, No. 2 (2020), 343–356.

9   Brendan McGurk, *Data Profiling and Insurance Law*, Hart Publishing, London, 2019, pp. 12–16; Peppet, 'Regulating', 85–176.

10 Mikella Hurley and Julius Adebayo, 'Credit Scoring in the Era of Big Data', *Yale Journal of Law and Technology*, Vol. 18, No. 1 (2016), 148–216.

11 Charles Randell, 'How can we ensure that Big Data does not make us prisoners of technology?' *Reuters Newsmaker Event*, Reuters News & Media, London, 11 July 2018 <https://www.fca.org.uk/news/speeches/how-can-we-ensure-big-data-does-not-make-us-prisoners-technology> (accessed 20 August 2021).

12 Hurley and Adebayo, 'Credit Scoring', 148–216.

13 Graeme Paton, 'Admiral charges Hotmail users more for car insurance', *The Times*, 23 January 2018 <https://www.thetimes.co.uk/article/admiral-charges-hotmail-users-more-for-car-insurance-hrzjxsslr> (accessed 20 August 2021).

14 Uri Bram and Martin Schmalz, *The Business of Big Data: How to Create Lasting Value in the Age of AI*, Amazon, UK, 2019, p. 61.

15 Bram and Schmalz, *Business of Big Data*, p. 61; Tobias Berg et al., 'On the Rise of FinTechs – Credit Scoring using Digital Footprints', *National Bureau of Economic Research Working Paper*, No. 24551 (2018).

16 Frederick Schauer, *Profiles, Probabilities, and Stereotypes*, Harvard University Press, Cambridge, MA, 2003, p. 4.

17 Ibid., p. 216.

18 Ibid., p. 269.

19 Ibid., p. 67.

20 Ibid., p. 69.

CHAPTER TEN

1   Derek Parfit, *Reasons and Persons*, Oxford University Press, Oxford, 1987, p. 216.

2   Roger Brownsword, *Law, Technology and Society: Re-Imagining the Regulatory Environment*, Routledge, Abingdon, 2019, p. 40.

3   Thomas Hobbes, *Leviathan*, Cambridge University Press, Cambridge, 1996, p. 147.

4   Marcus Tullius Cicero, *The Republic and The Laws*, Oxford University Press, Oxford, 1998, p. 24.

5   See Lawrence Lessig, *Republic, Lost: How Money Corrupts Congress – and a Plan to Stop It*, Hachette Book Group, New York, 2011.

6   Megan Slack, 'From the Archives: President Teddy Roosevelt's New Nationalism Speech', *The White House: President Barack Obama*, 6 December 2011 <https://obamawhitehouse.archives.gov/blog/2011/12/06/archives-president-teddy-roosevelts-new-nationalism-speech> (accessed 20 August 2021).

7   See Jeremy Waldron, *Political Political Theory: Essays on Institutions*, Harvard University Press, Cambridge, MA, 2016, p. 64.

8   Brownsword, *Law, Technology and Society*, vii.

9   Roger Brownsword and Morag Goodwin, *Law and the Technologies of the Twenty-First Century: Texts and Materials*, Cambridge University Press, Cambridge, 2012, p. 447.

10   F. A. Hayek, *The Constitution of Liberty*, Routledge, Abingdon, 2006, p. 27.

CHAPTER ELEVEN

1   Bernard E. Harcourt, *The Illusion of Free Markets: Punishment and the Myth of Natural Order*, Harvard University Press, Cambridge, MA, 2012, pp. 50–51.

2   'The dominant narrative within Silicon Valley is that technology is inseparable from capitalism.' Wendy Liu, *Abolish Silicon Valley: How to Liberate Technology from Capitalism*, Repeater Books, London, 2020, p. 4.

3   Cited by Fania Oz-Salzberger, 'Scots, Germans, Republic and Commerce', in Richard Vetterli and Gary Bryner (eds), *Republicanism: A Shared European Heritage (Volume II)*, Cambridge University Press, Cambridge, 2006, p. 199.

4   See Robert S. Taylor, *Exit Left: Markets and Mobility in Republican Thought*, Oxford University Press, Oxford, 2017.

5   Alexi Mostrous and Peter Hoskin, 'Part II: The constitution', *Tortoise Media*, 7 January 2020 <https://www.tortoisemedia.com/2020/01/07/tech-states-apple-constitution/> (accessed 20 August 2021).

6   Brad Smith and Carol Ann Browne, *Tools and Weapons: The Promise and the Peril of the Digital Age*, Hodder & Stoughton, London, 2019, p. 147.

7   Joint Committee on Human Rights, 'The Right to Privacy (Article 8) and the Digital Revolution' (HC 122).

8   See James Grimmelmann, 'Saving Facebook', *NYLS Legal Studies Research Paper* No. 08/09-7 (2008).

9   Geoffrey A. Fowler, 'You downloaded FaceApp. Here's what you've just done to your privacy', *Washington Post*, 17 July 2019 <https://www.washingtonpost.com/technology/2019/07/17/you-downloaded-faceapp-heres-what-youve-just-done-your-privacy/> (accessed 20 August 2021).

10 Cass Sunstein, *Free Markets and Social Justice*, Oxford University Press, New York, 1997, pp. 153–154.

11 Edward J. Watts, *Mortal Republic: How Rome Fell Into Tyranny*, Basic Books, New York, 2018, p. 18.

12 Adam Ferguson, *An Essay on the History of Civil Society*, Cambridge University Press, Cambridge, 2007, p. 173.

13 Cited in Sheera Frenkel and Cecilia Kang, *An Ugly Truth: Inside Facebook's Battle for Domination*, The Bridge Street Press, London, 2021, p. 124.

14 Cited in Marco Geuna, 'Republicanism and Commercial Society in the Scottish Enlightenment: The Case of Adam Ferguson', in Vetterli and Bryner (eds), *Republicanism*, p. 185.

CHAPTER TWELVE

1   Reid Hoffman, *Blitzscaling: The Lightning-Fast Path to Building Massively Valuable Companies*, HarperCollins, London, 2018, pp. 2, 9, 12.

2   Ibid., p. 21.

3   Ibid., p. 5.

4   Clive Thompson, *Coders: Who They Are, What They Think And How They Are Changing Our World*, Picador, London, 2019, p. 335.

5   Hoffman, *Blitzscaling*, p. 14.

6   Milton Friedman, 'The Social Responsibility of Business is to Increase its Profits', *The New York Times Magazine*, 13 September 1970.

7   Tarleton Gillespie, *Custodians of the Internet: Platforms, Content Moderation, and the Hidden Decisions that Shape Social Media*, Yale University Press, New Haven, 2018, pp. 11–12.

8   See e.g. Naomi Nix, 'Facebook Ran Multi-Year Charm Offensive to Woo State Prosecutors', *Bloomberg*, 27 May 2020 <https://www.bloomberg.com/news/articles/2020-05-27/facebook-ran-multi-year-charm-offensive-to-woo-state-prosecutors> (accessed 20 August 2021); Shira Ovide, 'Facebook and Its Secret Policies', *The New York Times*, 28 May 2020 <https://www.nytimes.

com/2020/05/28/technology/facebook-polarization.html> (accessed 20 August 2021).

9   Richard Susskind and Daniel Susskind, The Future of the Professions: How Technology will Transform the Work of Human Experts, Oxford University Press, Oxford, 2015, p. 15.

10 For a comprehensive comparison, see Brent Mittelstadt, 'Principles alone cannot guarantee ethical AI', Nature Machine Intelligence, Vol. 1 (2019), 501–507.

11 Robert Baldwin, Martin Cave and Martin Lodge, Understanding Regulation: Theory, Strategy, and Practice, Oxford University Press, Oxford, 2012, p. 138.

12 Ian Tucker, 'Yaël Eisenstat: "Facebook is ripe for manipulation and viral misinformation", The Guardian, 26 July 2020 <https://www.theguardian.com/technology/2020/jul/26/yael-eisenstat-facebook-is-ripe-for-manipulation-and-viral-misinformation> (accessed 20 August 2021).

13 Ross LaJeunesse, 'I Was Google's Head of International Relations. Here's Why I Left', Ross LaJeunesse, 2 January 2020 <https://medium.com/@rossformaine/i-was-googles-head-of-international-relati ons-here-s-why-i-left-49313d23065> (accessed 20 August 2021).

14 Richard Waters, 'Google scraps ethics council for artificial intelligence', Financial Times, <https://www.ft.com/content/6e291 2f8-573e-11e9-91f9-b6515a54c5b1> (accessed 20 August 2021).

15 Karen Hao, 'We read the paper that forced Timnit Gebru out of Google. Here's what it says.' MIT Technology Review, 4 December 2020 <https://www.technologyreview.com/2020/12/04/1013 294/google-ai-ethics-research-paper-forced-out-timnit-gebru/> (accessed 3 October 2021); Khari Johnson, 'Google employee group urges Congress to strengthen whistleblower protections for AI researchers', VentureBeat, 8 March 2021 <https://venturebeat. com/2021/03/08/google-employee-group-urges-congress-to-str engthen-whistleblower-protections-for-ai-researchers/> (accessed 3 October 2021).

CHAPTER THIRTEEN

1   Nathan Benaich and Ian Hogarth, 'State of AI Report', 28 June 2019 <https://www.stateof.ai/2019> (accessed 20 August 2021).

2   Amy Webb, The Big Nine: How the Tech Titans & Their Thinking Machines Could Warp Humanity, Hachette Book Group, New York, 2019, p. 64.

3   Ibid.

4   Benaich and Hogarth, 'State of AI'.

5   For an overview of what is taught in these classes, see Casey Fiesler, 'What do we teach when we teach tech & AI ethics?' *CUInfoScience*, 17 January 2020 <https://medium.com/cuinfoscience/what-do-w e-teach-when-we-teach-tech-ai-ethics-81059b710e11> (accessed 20 August 2021).

6   Webb, *Big Nine*, p. 60.

7   Ibid.

8   McKenzie Raub, 'Bots, Bias, and Big Data: Artificial Intelligence, Algorithmic Bias and Disparate Impact Liability in Hiring Practices', *Arkansas Law Review*, Vol. 71, No. 2 (2018), 529–570.

9   Katrina Lake, 'Stitch Fix's CEO on Selling Personal Style to the Mass Market', in Harvard Business Review, *HBR's 10 must reads on AI, analytics, and the new machine age*, Harvard Business Publishing Corporation, Cambridge, MA, 2019, p. 23; Clive Thompson, *Coders: Who They Are, What They Think And How They Are Changing Our World*, Picador, London, 2019, p. 211.

10   Mary Ann Azevedo, 'Untapped Opportunity: Minority Founders Still Being Overlooked', *crunchbase*, 27 February 2019 < https://news. crunchbase.com/news/untapped-opportunity-minority-found ers-still-being-overlooked/ > (accessed 20 August 2021).

11   Thompson, *Coders*, p. 23.

12   Ibid., pp. 21, 22, 103.

13   Katy Cook, *The Psychology of Silicon Valley: Ethical Threats and Emotional Unintelligence in the Tech Industry*, Palgrave Macmillan, London, 2020, p. 40.

14   Thompson, *Coders*, p. 24.

15   Tarleton Gillespie, *Custodians of the Internet: Platforms, Content Moderation, and the Hidden Decisions that Shape Social Media*, Yale University Press, New Haven, 2018, p. 12.

16   Yelena Dzhanova, 'Facebook did not hire Black employees because they were not a "culture fit," report says', *Business Insider*, 6 April 2021 <https://www.businessinsider.com/facebook-workplace- hiring-eeoc-black-employees-culture-fit-2021-4> (accessed 3 October 2021).

17   See Thompson, *Coders* e.g. pp. 35, 236.

18   Fred Turner, *From Counterculture to Cyberculture: Stewart Brand, the Whole Earth Network, and the Rise of Digital Utopianism*, University of Chicago Press, Chicago and London, 2008.

19   Carissa Véliz, 'Three things digital ethics can learn from medical ethics', *Nature Electronics*, Vol. 2, No. 8 (2019), 1–3.

20   Rodrigo Ochigame, 'The Invention of "Ethical AI"', *The Intercept*, 20 December 2019 <https://theintercept.com/2019/12/20/mit-ethical-ai-artificial-intelligence/> (accessed 20 August 2021). See also Brent Mittelstadt, 'Principles alone cannot guarantee ethical AI', *Nature Machine Intelligence*, Vol. 1 (2019), 501–507.

21   Jacob Metcalf, Emanuel Moss and danah boyd, 'Owning Ethics: Corporate Logics, Silicon Valley, and the Institutionalization of Ethics', *Social Research: An International Quarterly*, Vol. 82, No. 2 (2019), 449–476.

22   Jessica Fjeld et al., 'Principled Artificial Intelligence: Mapping Consensus in Ethical and Rights-Based Approaches to Principles for AI', *Berkman Klein Center Research Publication*, No. 2020-1 (2020).

23   Gary M. Fleischman et al., 'Ethics Versus Outcomes: Managerial Responses to Incentive-Driven and Goal-Induced Employee Behavior', *Journal of Business Ethics*, Vol. 158, No. 4 (2019), 951–967.

24   Mittelstadt, 'Principles alone'.

25   Michael Veale calls this the 'framing problem'. 'A Critical Take on the Policy Recommendations of the EU High-Level Expert Group on Artificial Intelligence', *European Journal of Risk Regulation* (2020), pp. 1–10.

26   For a particularly sceptical take, see Ochigame, 'The Invention of "Ethical AI"'.

27   Leonard Haas and Sebastian Gießler, with Veronika Thiel, 'In the realm of paper tigers – exploring the failings of AI ethics guidelines', *Algorithm Watch*, 28 April 2020 <https://algorithmwatch.org/en/ai-ethics-guidelines-inventory-upgrade-2020/> (accessed 20 August 2021).

28   Ochigame, 'The Invention of "Ethical AI"'.

29   Ibid.

30   Ben Wagner, 'Ethics as an escape from regulation: From "ethics-washing" to ethics-shopping?' in Emre Bayamlıoğlu et al. (eds), *Being Profiled: Cogitas Ergo Sum: 10 Years of Profiling the European Citizen*, Amsterdam University Press, Amsterdam, 2018.

CHAPTER FOURTEEN

1   See Julie E. Cohen, *Between Truth and Power: The Legal Constructions of Informational Capitalism*, Oxford University Press, Oxford, 2019; Solon Barocas and Helen Nissenbaum, 'Big Data's End Run around

Anonymity and Consent', in Julia Lane, Victoria Stodden, Stefan Bender and Helen Nissenbaum (eds), *Privacy, Big Data, and the Public Good: Frameworks for Engagement*, Cambridge University Press, New York, 2014; Elettra Bietti, 'Consent as a Free Pass: Platform Power and the Limits of the Informational Turn', *Pace Law Review*, Vol. 40, No. 1 (2020), 307–397.

2   I naïvely thought I was the first to use the term 'consent trap' in this field. An early reader, Seth Lazar, pointed out that Sascha Molitorisz got there first.

3   Samuel D. Warren and Louis D. Brandeis, 'The Right to Privacy', *Harvard Law Review*, Vol. 4, No. 5 (1980), 193–220.

4   Christophe Lazaro and Daniel Le Métayer, 'Control over Personal Data: true Remedy or Fairy Tale?' *Scripted*, Vol. 12, No. 1 (June 2015), 3–34

5   Daniel J. Solove, 'Introduction: Privacy Self-Management and the Consent Dilemma', *Harvard Law Review*, Vol. 126, No. 7 (2013), 1880–1903; Nicholas LePan, 'Visualizing the Length of the Fine Print, for 14 Popular Apps', *Visual Capitalist*, 18 April 2020 <https://www.visualcapitalist.com/terms-of-service-visualizing-the-length-of-internet-agreements/> (accessed 20 August 2021).

6   Gillian K. Hadfield, *Rules for a Flat World: Why Humans Invented Law and How to Reinvent it for a Complex Global Economy*, Oxford University Press, Oxford, 2017, p. 170.

7   Shoshana Zuboff, *The Age of Surveillance Capitalism: The Fight for a Human Future at the New Frontier of Power*, Profile Books, London, 2019, p. 237.

8   Kevin Litman-Navarro, 'We Read 150 Privacy Policies. They Were an Incomprehensible Disaster', *The New York Times*, 12 June 2019 <https://www.nytimes.com/interactive/2019/06/12/opinion/facebook-google-privacy-policies.html> (accessed 20 August 2021).

9   Zuboff, *Surveillance Capitalism*, p. 50.

10   Neil Richards and Woodrow Hartzog, 'The Pathologies of Digital Consent', *Washington University Law Review*, Vol. 96 (2019), 1461–1503, p. 1480. Sandra Braman points out that US constitutional law 'forbids the use of language in laws or regulations that is vague (reasonable adults may not agree on its meaning) or over broad (covering far more activity and types of communication than is the intended target of a particular law or regulation).' Contract law does not. See 'Series Editor's Introduction', in Rikke Frank Jørgensen

(ed.), *Human Rights in the Age of Platforms*, MIT Press, Cambridge, MA, 2019, vii.

11 Scott R. Peppet, 'Regulating the Internet of Things: First Steps Toward Managing Discrimination, Privacy, Security, and Consent', *Texas Law Review*, Vol. 93, No. 83 (2014), 85–176.

12 Zuboff, *Surveillance Capitalism*, p. 237.

13 Nicolas P. Suzor, *Lawless: The Secret Rules that Govern Our Digital Lives*, Cambridge University Press, Cambridge, 2019, p. 11.

14 See Neil Richards and Woodrow Hartzog, 'Taking Trust Seriously in Privacy Law,' *Stanford Technology Law Review*, Vol. 19 (2016), 431-472; 'Privacy's Trust Gap: A Review', *Yale Law Journal*, Vol. 126, No. 4 (2017), 1180-1224; 'Trusting Big Data Research', *DePaul Law Review*, Vol. 66, No. 2 (2017), 579-590; 'Pathologies'; 'Privacy's Constitutional Moment and the Limits of Data Protection', *Boston College Law Review*, Vol. 61, No. 5 (2020), 1687–1761

15 Kate Cox, 'Unredacted suit shows Google's own engineers confused by privacy settings', *Ars Technica*, 25 August 2020 <https://arst echnica.com/tech-policy/2020/08/unredacted-suit-shows-goog les-own-engineers-confused-by-privacy-settings/> (accessed 20 August 2021).

16 Richards and Hartzog, 'Pathologies', p. 1489; Arielle Pardes, 'How Facebook and Other Sites Manipulate Your Privacy Choices', *Wired*, 12 August 2020 <https://www.wired.com/story/faceb ook-social-media-privacy-dark-patterns/> (accessed 20 August 2021).

17 Katherine J. Strandburg, 'Monitoring, Datafication, and Consent: Legal Approaches to Privacy in the Big Data Context', in Julia Lane, Victoria Stodden, Stefan Bender and Helen Nissenbaum (eds), *Privacy, Big Data, and the Public Good: Frameworks for Engagement*, Cambridge University Press, New York, 2014, p. 31; Barocas and Nissenbaum, 'Big Data's End Run', pp. 58–59.

18 Joshua A.T. Fairfield and Christoph Engel, 'Privacy as a Public Good', *Duke Law Journal*, Vol. 65, No. 3 (December 2015), 385–457.

19 Fairfield and Engel, 'Privacy as a Public Good'.

20 Karen Yeung and Martin Lodge, *Algorithmic Regulation*, Oxford University Press, Oxford, 2019, p. 36.

21 Fairfield and Engel, 'Privacy as a Public Good'.

22 Richards and Hartzog, 'Pathologies', p. 1479.

23 The GDPR should, in theory, go some way to mitigating this problem. Recital 34 provides that consent will not provide a legal basis for

data processing, where there is a significant imbalance between the position of the data subject and the controller and this imbalance makes it unlikely that consent was given freely. In addition, 'utmost account' must be given to whether the performance of a contract is made conditional on consent to data processing that is not necessary to perform the contacts (Article 7). However, these provisions do not appear to have significantly dented the consent trap in practice.

24 Fairfield and Engel, 'Privacy as a Public Good'; Stuart A. Thompson and Charlie Warzel, 'How to Track President Trump', The New York Times, 20 December 2019 <https://www.nytimes.com/interactive/2019/12/20/opinion/location-data-national-security.html> (accessed 20 August 2021).

25 Martijn van Otterlo, 'A machine learning view on profiling', in Mireille Hildebrandt and Katja de Vries (eds), Privacy, Due Process and the Computational Turn: The Philosophy of Law Meets the Philosophy of Technology, Routledge, Abingdon, 2013, p. 42. See also Zeynep Tufekci, 'The Latest Data Privacy Debacle', The New York Times, 30 January 2018 <https://www.nytimes.com/2018/01/30/opinion/strava-privacy.html> (accessed 20 August 2021).

26 Billings Learned Hand, The Spirit of Liberty, Vintage, New York, 1959, p. 144.

27 See also Julie E. Cohen, Configuring the Networked Self: Law, Code, and the Play of Everyday Practice, Yale University Press, New Haven, 2012, p. 149; Neil Richards, Intellectual Privacy: Rethinking Civil Liberties in the Digital Age, Oxford University Press, Oxford, 2017.

28 See Quentin Skinner, Liberty before Liberalism, Cambridge University Press, Cambridge, 1998, pp. 23–24.

29 Fairfield and Engel, 'Privacy as a Public Good'. See also Solove, 'Introduction: Privacy Self-Management'; Cohen, Between Truth and Power.

30 Stephen L. Elkin, Reconstructing the Commercial Republic: Constitutional Design after Madison, University of Chicago Press, Chicago, 2006, p. 4.

CHAPTER FIFTEEN

1 F. A. Hayek, The Constitution of Liberty, Routledge, Abingdon, 2006, p. 143.

2 Ibid., p. 399.

3   David Levi-Faur, 'From "Big Government" to "Big Governance"?', in David Levi-Faur (ed.), The Oxford Handbook of Governance, Oxford University Press, Oxford, 2014, p. 5.

4   See Roger Brownsword and Morag Goodwin, Law and the Technologies of the Twenty-First Century: Texts and Materials, Cambridge University Press, Cambridge, 2012, pp. 24–25. For clarity, I do not adopt the broader definition of 'regulation' which includes 'all forms of social or economic influence – where all mechanisms affecting behaviour – whether these be state-based or from other sources (e.g. markets) are deemed regulatory': Robert Baldwin, Martin Cave and Martin Lodge, Understanding Regulation: Theory, Strategy, and Practice, Oxford University Press, Oxford, 2012, p. 3.

5   Jonathan Sumption, Trials of the State: Law and the Decline of Politics, Profile Books, London, 2019, p. 4.

6   Martin Loughlin, Foundations of Public Law, Oxford University Press, Oxford, 2014, p. 101.

7   Ron Chernow, Grant, Head of Zeus, London, 2017, p. 644.

8   Paul Tucker, Unelected Power: The Quest for Legitimacy in Central Banking and the Regulatory State, Princeton University Press, Princeton, 2018, p. 28.

9   Michael Moran, 'The Rise of the Regulatory State', in Davin Coen, Wyn Grant and Graham Wilson (eds), The Oxford Handbook of Business and Government, Oxford University Press, Oxford, 2011, p. 387.

10   K. Sabeel Rahman, 'Regulating Informational Infrastructure: Internet Platforms as the New Public Utilities', Georgetown Law and Technology Review, Vol. 2, No. 2 (2018), 234–251.

11   Ibid.

12   Baldwin et al., Understanding Regulation, p. 4.

13   Stephen Sedley, Lions Under the Throne: Essays on the History of English Public Law, Cambridge University Press, Cambridge, 2015, p. 56.

14   Baldwin et al., Understanding Regulation, p. 4.

15   Moran, 'Rise', p. 387; Tucker, Unelected Power, p. 29.

16   Karen Yeung, 'The Regulatory State', in Robert Baldwin, Martin Cave and Martin Lodge (eds), The Oxford Handbook of Regulation, Oxford University Press, Oxford, 2013, p. 72.

17   See Nicholas Timmins, The Five Giants: A Biography of the Welfare State, William Collins, London, 2017.

18   Yeung, 'Regulatory State', p. 66.

19   Cass Sunstein, After the Rights Revolution: Reconceiving the Regulatory State, Harvard University Press, Harvard, 1993, pp. v, 22.

20   Yeung, 'Regulatory State', p. 66.

21   Ibid.
22   Ibid., p. 67.

CHAPTER SIXTEEN

1   For a powerful conceptual overview of this field, see the work of Julie E. Cohen, esp. *Configuring the Networked Self: Law, Code, and the Play of Everyday Practice*, Yale University Press, New Haven, 2012, and *Between Truth and Power: The Legal Constructions of Informational Capitalism*, Oxford University Press, Oxford, 2019.

2   James Ball, *The System: Who Owns the Internet, and How it Owns Us*, Bloomsbury, London, 2020, pp. 66–67.

3   Matthias C. Kettemann, *The Normative Order of the Internet*, Oxford University Press, Oxford, 2020, p. 46.

4   These acronyms stand for: the Food and Drug Administration (FDA), the Securities and Exchange Commission (SEC), the National Highway Traffic and Safety Administration (NHTSA), the Fair Credit Reporting Act (FRCA), the Gramm-Leach-Bliley Act (GLBA), the Health Insurance Portability and Accountability Act (HIPAA), the Family Educational Rights and Privacy Act (FERPA), the Children's Online Privacy Protection Rule (COPPA), the Electronic Communications Privacy Act (ECPA), and the Computer Fraud and Abuse Act (CFAA).

5   Section 5 of the Federal Trade Commission Act 1914.

6   John D. McKinnon and James V. Grimaldi, 'Justice Department, FTC Skirmish Over Antitrust Turf', *The Wall Street Journal*, 5 August 2019   <https://www.wsj.com/articles/justice-department-ftc-skirmish-over-antitrust-turf-11564997402>   (accessed   20 August 2021).

7   See, for a useful overview, Mitchell Noordyke, 'US state comprehensive privacy law comparison', *International Association of Privacy Professionals*, 18 April 2019 <https://iapp. org/news/a/us-state-comprehensive-privacy-law-comparison/> (accessed 20 August 2021).

8   Scott R. Peppet, 'Regulating the Internet of Things: First Steps Toward Managing Discrimination, Privacy, Security, and Consent', *Texas Law Review*, Vol. 93, No. 83 (2014), 85–176.

9   Neil Richards and Woodrow Hartzog, 'The Pathologies of Digital Consent', *Washington University Law Review*, Vol. 96 (2019), 1461–1503.

10   Katherine J. Strandburg, 'Monitoring, Datafication, and Consent: Legal Approaches to Privacy in the Big Data Context', in

Julia Lane, Victoria Stodden, Stefan Bender and Helen Nissenbaum (eds), *Privacy, Big Data, and the Public Good: Frameworks for Engagement*, Cambridge University Press, New York, 2014, p. 21.

11   Apart from the 1974 Privacy Act, which does not apply to the private sector.

12   Chris Jay Hoofnagle, Woodrow Hartzog and Daniel J. Solove, 'The FTC can rise to the privacy challenge, but not without help from Congress', *The Brookings Institution*, 8 August 2019 <https://www.brooki ngs.edu/blog/techtank/2019/08/08/the-ftc-can-rise-to-the-priv acy-challenge-but-not-without-help-from-congress/> (accessed 20 August 2021).

13   Nilay Patel, 'Facebook's $5 billion FTC fine is an embarrassing joke', *The Verge*, 12 July 2019 <https://www.theverge. com/2019/7/12/20692524/facebook-five-billion-ftc-fine-embarrassing-joke> (accessed 20 August 2021). Cf. Daniel J. Solove and Woodrow Hartzog, 'The FTC and the New Common Law of Privacy', *Columbia Law Review*, Vol. 114:583 (2014).

14   Lilian Edwards, *Law, Policy and the Internet*, Hart Publishing, Oxford, 2019, p. 75.

15   See Danielle Keats Citron and Benjamin Wittes, 'The Internet Will Not Break: Denying Bad Samaritans §230 Immunity', *Fordham Law Review*, Vol. 86, No. 2 (2017), 401–423; Olivier Sylvain, 'Discriminatory Designs on User Data', in David E. Pozen (ed.), *The Perilous Public Square: Structural Threats to Free Expression Today*, Columbia University Press, New York, 2020, pp. 181–184.

16   Citron and Wittes, 'The Internet Will Not Break'.

17   Kiran Jeevanjee et al., 'All the Ways Congress Wants to Change Section 230', *Slate*, 23 March 2021 <https://slate.com/technol ogy/2021/03/section-230-reform-legislative-tracker.html> (accessed 3 October 2021).

18   Jeff Kosseff, *The Twenty-Six Words that Created the Internet*, Cornell University Press, Ithaca, 2019, p. 3.

19   Ball, *The System*, p. 34.

20   See Tarleton Gillespie, *Custodians of the Internet: Platforms, Content Moderation, and the Hidden Decisions that Shape Social Media*, Yale University Press, New Haven, 2018; Citron and Wittes, 'The Internet Will Not Break'.

21   See Cohen, *Between Truth and Power*, p. 99: 'As James Grimmelmann has painstakingly demonstrated, search engines have become adept at insisting on their neutrality for purposes of section 230 even while claiming that their search results are their own constitutionally

protected speech. For the most part, courts have uncritically accepted both sets of arguments, concluding both that algorithmic intermediation doesn't make an intermediary a publisher of other people's speech and that the same processes of intermediation are speech-like in their own right.'

22 Citron and Wittes, 'The Internet Will Not Break'.

23 Sylvain, 'Discriminatory Designs', pp. 185–188.

24 Mary Anne Franks, 'How The Internet Unmakes Law', *The Ohio Technology Law Journal*, Vol. 16, No. 1 (2020), 10–24.

25 Citron and Wittes, 'The Internet Will Not Break'.

26 Cited in Danielle Keats Citron, 'Section 230's Challenge to Civil Rights and Civil Liberties', in Pozen (ed.), *Perilous*, p. 202.

27 For a good introduction, see Mireille Hildebrandt, *Law for Computer Scientists and Other Folk*, Oxford University Press, Oxford, 2020, p. 135 *et seq.*

28 Article 83.

29 Roger Brownsword, *Law, Technology and Society: Re-Imagining the Regulatory Environment*, Routledge, Abingdon, 2019, p. 99; Gillespie, *Custodians*, p. 33; Mark Leiser and Andrew Murray, 'The role of non-state actors and institutions in the governance of new and emerging digital technologies', in Roger Brownsword, Eloise Scotford and Karen Yeung (eds), *The Oxford Handbook of Law, Regulation, and Technology*, Oxford University Press, Oxford, 2017, p. 678; Agnès Callamard, 'The Human Rights Obligations of non-State Actors', in Rikke Frank Jørgensen (ed.), *Human Rights in the Age of Platforms*, MIT Press, Cambridge, MA, 2019, p. 205.

Note that the E-Commerce Directive 'harmonises the conditions under which intermediaries can be held liable for third-party infringement throughout the EU. It also employs a notice-and-takedown scheme for infringing content similar to that of the DCMA. However, in contrast to the DCMA, the E-Commerce Directive applies to a range of activities including copyright and trademark infringement, as well as defamation.' (Natasha Tusikov, *Chokepoints: Global Private Regulation on the Internet*, University of California Press, Oakland, 2017, p. 54). Moreover, Article 40 of the E-Commerce Directive calls upon member states . . . to develop 'rapid and reliable procedures for removing and disabling access to illegal information' and it allows that these could be developed 'on the basis of voluntary agreements between all parties concerned'. In 2004, the European Commission called upon industry to take an active role in anti-infringement efforts and promoted the development of non-legally binding codes of conduct as 'a supplementary means of bolstering the regulatory framework' (Tusikov, *Chokepoints*, p. 61). See also Lilian Edwards, *Law, Policy and the Internet*, Hart Publishing, Oxford, 2019, p. 268.

30   Olaf Storbeck, Madhumita Murgia and Rochelle Toplensky, 'Germany blocks Facebook from pooling user data without consent', *The Financial Times*, 7 February 2019 <https://www.ft.com/content/3a0351b6-2ab9-11e9-88a4-c32129756dd8> (accessed 20 August 2021); Adam Satariano, 'Europe Is Toughest on Big Tech, Yet Big Tech Still Reigns', *The New York Times*, 11 November 2019 <https://www.nytimes.com/2019/11/11/business/europe-technology-antitrust-regulation.html> (accessed 20 August 2021); Diane Coyle, *Markets, State, and People: Economics for Public Policy*, Princeton University Press, Princeton, 2020, p. 94; Cohen, *Between Truth and Power*, p. 177.

31   Whether or not related, the EU accounts for less than 4 per cent of the market capitalisation of the world's seventy largest platforms (America has 73 per cent): *The Economist*, 'The EU wants to set the rules for the world of technology', 20 February 2020 <https://www.economist.com/business/2020/02/20/the-eu-wants-to-set-the-rules-for-the-world-of-technology> (accessed 20 August 2021).

32   Charlie Taylor, 'Data Protection Commission criticised as WhatsApp decision nears', *Irish Times*, 15 January 2020 <https://www.irishtimes.com/business/technology/data-protection-commission-criticised-as-whatsapp-decision-nears-1.4139804> (accessed 20 August 2021); Nicole Kobie, 'Germany says GDPR could collapse as Ireland dallies on big fines', *Wired*, 27 April 2020 <https://www.wired.co.uk/article/gdpr-fines-google-facebook> (accessed 20 August 2021).

33   Johnny Ryan and Alan Toner, 'Europe's enforcement paralysis: ICCL's 2021 report on the enforcement capacity of data protection authorities', Irish Council for Civil Liberties, 2021 <https://www.iccl.ie/digital-data/2021-gdpr-report/> (accessed 3 October 2021).

34   Adam Satariano, 'Europe's Privacy Law Hasn't Shown Its Teeth, Frustrating Advocates', *The New York Times*, 27 April 2020 <https://www.nytimes.com/2020/04/27/technology/GDPR-priv

acy-law-europe.html> (accessed 20 August 2021); Johnny Ryan and Alan Toner, 'Europe's governments are failing the GDPR', Brave, 2020 <https://brave.com/wp-content/uploads/2020/04/Brave-2020-DPA-Report.pdf> (accessed 20 August 2021).

35 Midas Nouwens et al., 'Dark Patterns after the GDPR: Scraping Consent Pop-ups and Demonstrating their Influence', arXiv: 2001.02479 (2020).

36 Albeit that Article 5 GDPR provides extra protection.

37 Article 5(1)(b).

38 Article 6(4); see Waltraut Kotschy, 'Article 6. Lawfulness of processing', in Christopher Kuner, Lee A. Bygrave and Christopher Docksey (eds), The EU Data Protection Regulation (GDPR): A Commentary, Oxford University Press, Oxford, 2020.

39 Article 22.

40 Article 22(c).

41 'free given, specific, informed, and unambiguous': Article 4(11); Recital 32. In some circumstances the requirement is higher still.

CHAPTER SEVENTEEN

1 Cited in Shoshana Zuboff, The Age of Surveillance Capitalism: The Fight for a Human Future at the New Frontier of Power, Profile Books, London, 2019, p. 105.

2 The law could be very different. As Katharina Pistor observes, when craftsmen were first granted legal protection for 'new and ingenious devices' in fifteenth-century Venice, they could file cases against other merchants who used their inventions but, in exchange, had to allow the city itself to use them 'for its own use and needs'. The law's protection came with a requirement that inventions could be used for the public good. See The Code of Capital: How the Law Creates Wealth and Inequality, Princeton University Press, Princeton, 2019, p. 119.

3 Cited in James Boyle, The Public Domain: Enclosing the Commons of the Mind, Yale University Press, New Haven, 2008, p. 19.

4 James S. Coleman, Power and the Structure of Society, W. W. Norton & Company, New York, 1974, p. 23; Philip Pettit, 'Two Fallacies about Corporations', in Subramanian Rangan (ed.), Performance & Progress: Essays on Capitalism, Business, and Society, Oxford University Press, Oxford, 2017, p. 386.

5 Pettit, 'Two Fallacies', p. 390.

6   David Ciepley, 'Beyond Public and Private: Toward a Political Theory of the Corporation', *American Political Science Review*, Vol. 107, No. 1 (February 2013), 139–158.

7   Pistor, *Code of Capital*, p. 55.

8   Ibid., pp. 7, 71.

9   Ciepley, 'Beyond Public and Private'.

10 Pistor, *Code of Capital*, p. 14.

11 Pettit, 'Two Fallacies', p. 389.

12 See, most famously, Citizens United v. Federal Election Commission, 558 U.S. 310 (2010).

13 For a sceptical view, see Hugh Collins, *Regulating Contracts*, Oxford University Press, Oxford, 2002.

14 Collins, Regulating Contracts, p. 56.

15 See e.g. the writings of Julie E. Cohen; and Lyca Belli and Jamila Venturini, 'Private ordering and the rise of terms of service as cyber-regulation', *Internet Policy Review*, Vol. 5, No. 4 (2016).

16 Aileen Kavanagh, 'The Constitutional Separation of Powers', in David Dyzenhaus and Malcolm Thorburn (eds), *Philosophical Foundations of Constitutional Law*, Oxford University Press, Oxford, 2019, p. 230.

17 Margaret Jane Radin, *Boilerplate: The Fine Print, Vanishing Rights, and the Rule of Law*, Princeton University Press, Princeton, 2013, p. 33.

18 Cass Sunstein, *Democracy and the Problem of Free Speech*, Free Press, New York, 1995, p. 37.

19 Ira C. Magaziner, 'Creating a Framework for Global Electronic Commerce', The Progress & Freedom Foundation, *Future Insight*, July 1999 <http://www.pff.org/issues-pubs/futureinsights/fi6.1globaleconomiccommerce.html> (accessed 3 October 2021).

CHAPTER EIGHTEEN

1   Philip Pettit, *Just Freedom: A Moral Compass for a Complex World*, W. W. Norton and Company, New York, 2014, p. 77; Frank Lovett, *A General Theory of Domination and Justice*, Oxford University Press, Oxford, 2012, p112.

2   See Nicolas P. Suzor, *Lawless: The Secret Rules that Govern Our Digital Lives*, Cambridge University Press, Cambridge, 2019, p. 30.

3   See Rick Swedloff, 'The New Regulatory Imperative for Insurance', *Boston College Law Review*, Vol. 61, No. 6 (2020), 2031–2084.

4   Shirin Ghaffary and Jason Del Rey, 'The real cost of Amazon', *Vox*, 29 June 2020 <https://www.vox.com/recode/2020/6/29/21303643/

amazon-coronavirus-warehouse-workers-protest-jeff-bezos-chris-smalls-boycott-pandemic> (accessed 20 August 2021).

5 Sarah Kessler, 'Companies Are Using Employee Survey Data to Predict — and Squash — Union Organizing', *OneZero*, 30 July 2020 <https://onezero.medium.com/companies-are-using-employee-survey-data-to-predict-and-squash-union-organizing-a7e28a8c2158> (accessed 20 August 2021).

6 Lee Fang, 'Facebook Pitched New Tool Allowing Employers to Suppress Words Like "Unionize" in Workplace Chat Product', *The Intercept*, 11 June 2021 <https://theintercept.com/2020/06/11/facebook-workplace-unionize/> (accessed 20 August 2021).

7 Quite separately, a game of football was ruined for spectators at home when the 'AI camera', trained to follow the ball around the pitch, instead followed the bald, shiny head of one of the match officials. James Felton, 'AI Camera Ruins Soccer Game For Fans After Mistaking Referee's Bald Head For Ball', *IFLScience*, 29 October 2020 <https://www.iflscience.com/technology/ai-camera-ruins-soccar-game-for-fans-after-mistaking-referees-bald-head-for-ball/> (accessed 3 October 2021).

8 Julie E. Cohen, *Between Truth and Power: The Legal Constructions of Informational Capitalism*, Oxford University Press, Oxford, 2019, p. 179; Monique Mann and Tobias Matzner, 'Challenging algorithmic profiling: The limits of data protection and anti-discrimination in responding to emergent discrimination', *Big Data & Society*, Vol. 6, No. 2 (2019), 1–11; Sandra Wachter, Brent Mittelstadt and Chris Russell, 'Why fairness cannot be automated: Bridging the gap between EU non-discrimination law and AI', *Computer Law & Security Review*, Vol. 41 (2021), 105567–105597; Nizan Geslevich Packin and Yafit Lev-Aretz, 'Learning algorithms and discrimination', in Woodrow Barfield and Ugo Pagallo (eds), *Research Handbook on the Law of Artificial Intelligence*, Edward Elgar Publishing, Cheltenham, 2018.

9 Perhaps by reversing the burden of proof so that instead of employees having to show that they were discriminated against, employers have to show that they have not discriminated: Jason D. Lohr, Winston J. Maxwell and Peter Watts, 'Legal Practitioners' Approach to Regulating AI Risks', in Karen Yeung and Martin Lodge, *Algorithmic Regulation*, Oxford University Press, Oxford, 2019, p. 241.

10 For the inspiration for this principle, see John Braithwaite and Philip Pettit, Not Just Desserts: A Republican Theory of Criminal Justice, Oxford, Oxford University Press, 1992.

11 Friedrich Nietzsche, Thus Spoke Zarathustra, Penguin, London, 2003, p. 75.

12 See Jamie Susskind, Future Politics: Living Together in a World Transformed by Tech, Oxford University Press, Oxford, 2018.

13 See generally Bernard E. Harcourt, Exposed: Desire and Disobedience in the Digital Age, Harvard University Press, Cambridge, MA, 2015; David Kaye, 'The surveillance industry is assisting state suppression. It must be stopped', The Guardian, 26 November 2019 <https://www.theguardian.com/commentisfree/2019/nov/26/surveillance-industry-suppression-spyware> (accessed 20 August 2021).

14 Max Seddon and Madhumita Murgia, 'Apple and Google drop Navalny app after Kremlin piles on pressure', Financial Times, 17 September 2021 <https://www.ft.com/content/faaada81-73d6-428c-8d74-88d273adbad3> (accessed 3 October 2021).

15 Rory Van Loo, 'The New Gatekeepers: Private Firms as Public Enforcers', Virginia Law Review, Vol. 106, No. 2 (2020), 467–522; Molly K. Land, 'Regulating Private Harms Online: Content Regulation under Human Rights Law', in Rikke Frank Jørgensen (ed.), Human Rights in the Age of Platforms, MIT Press, Cambridge, MA, 2019, pp. 297–301.

16 Jack Goldsmith and Andrew Keane Woods, 'Internet Speech Will Never Go Back to Normal', The Atlantic, 25 April 2020 <https://www.theatlantic.com/ideas/archive/2020/04/what-covid-revealed-about-internet/610549/> (accessed 20 August 2021).

17 Natasha Tusikov, Chokepoints: Global Private Regulation on the Internet, University of California Press, Oakland, 2017.

18 Suzor, Lawless, p. 152.

19 Jack Goldsmith, 'The Failure of Internet Freedom', in David E. Pozen (ed.), The Perilous Public Square: Structural Threats to Free Expression Today, Columbia University Press, New York, 2020, p. 247.

20 Katherine J. Strandburg, 'Monitoring, Datafication, and Consent: Legal Approaches to Privacy in the Big Data Context', in Julia Lane, Victoria Stodden, Stefan Bender and Helen Nissenbaum (eds), Privacy, Big Data, and the Public Good: Frameworks for Engagement, Cambridge University Press, New York, 2014, p. 33.

21 Alexander Hamilton, James Madison and John Jay, The Federalist Papers, Penguin, New York, 2012, p. 94.

CHAPTER NINETEEN

1   Nicolas P. Suzor, *Lawless: The Secret Rules that Govern Our Digital Lives*, Cambridge University Press, Cambridge, 2019, p. 10.

2   Tarleton Gillespie, *Custodians of the Internet: Platforms, Content Moderation, and the Hidden Decisions that Shape Social Media*, Yale University Press, New Haven, 2018, p. 209.

3   Suzor, *Lawless*, p. 10.

4   Ibid.

5   Gillespie, *Custodians*.

6   Mariana Mazzucato, *The Entrepreneurial State: Debunking Public v Private Sector Myths*, Penguin, London, 2018.

7   While republicans and social democrats share some sympathies, republicans tend to be more anxious about excessive state power: see the *parsimony principle*.

8   Robert Baldwin, Martin Cave and Martin Lodge, *Understanding Regulation: Theory, Strategy, and Practice*, Oxford University Press, Oxford, 2012, p. 22; Bronwen Morgan and Karen Yeung, *An Introduction to Law and Regulation: Texts and Materials*, Cambridge University Press, Cambridge, 2007, p. 36.

9   Wendy Wagner, 'Regulating by the Stars', in Cary Coglianese (ed.), *Achieving Regulatory Excellence*, Brookings Institution Press, Washington DC, 2017, p. 40.

10 Jamie Susskind, *Future Politics: Living Together in a World Transformed by Tech*, Oxford University Press, Oxford, 2018, p. 117.

11 See Hod Lipson and Melba Kurman, *Driverless: Intelligent Cars and the Road Ahead*, MIT Press, Cambridge, MA, 2016, pp. 249–253.

12 These examples are taken from Antje von Ungern-Sternberg's fine article, 'Autonomous driving: regulatory challenges raised by artificial decision-making and tragic choices', in Woodrow Barfield and Ugo Pagallo (eds), *Research Handbook on the Law of Artificial Intelligence*, Edward Elgar Publishing, Cheltenham, 2018.

13 Ibid.

14 See Iason Gabriel, 'Artificial Intelligence, Values, and Alignment', *Minds and Machines*, Vol. 30, No. 3 (2020), 411-437.

15 Bruce Bimber, *The Politics of Expertise in Congress: the Rise and Fall of the Office of Technology Assessment*, State University of New York Press, Albany, NY, 1996, x.

16 Ibid., x; pp. 12–14.

17 Ibid., p. 66.

18 Friends of Andrew Yang, 'Revive the Office of Technology Assessment', 2020 <https://2020.yang2020.com/policies/revive ota/> (accessed 20 August 2021).

19 Karen Yeung, 'The Regulatory State', in Robert Baldwin, Martin Cave, and Martin Lodge (eds), The Oxford Handbook of Regulation, Oxford University Press, Oxford, 2013, p. 76; David Levi-Faur, 'Regulatory Excellence via Multiple Forms of Expertise', in Coglianese, Achieving, p. 227.

20 Peter Cane, Administrative Law, 5th edn, Oxford University Press, Oxford, 2011, p. 111.

21 See Jason D. Lohr, Winston J. Maxwell and Peter Watts, 'Legal Practitioners' Approach to Regulating AI Risks', in Karen Yeung and Martin Lodge, Algorithmic Regulation, Oxford University Press, Oxford, 2019, pp. 243–244.

22 Bruce Schneier, Click Here to Kill Everybody: Security and Survival in a Hyper-Connected World, W. W. Norton & Company, New York, 2018, pp. 147–148.

23 Catherine Miller, Jacob Ohrvik-Stott and Rachel Coldicutt, 'Regulating for Responsible Technology: Capacity, Evidence and Redress: a new system for a fairer future, Doteveryone, London, 2018; Geoff Mulgan, 'Anticipatory Regulation: 10 ways governments can better keep up with fast-changing industries', Nesta, 15 May 2017 <https://www.nesta.org.uk/ blog/anticipatory-regulation-10-ways-governments-can-bet ter-keep-up-with-fast-changing-industries/> (accessed 3 October 2021).

24 Cane, Administrative Law, pp. 111-112.

25 Ibid., p. 111; Paul Tucker, Unelected Power: The Quest for Legitimacy in Central Banking and the Regulatory State, Princeton University Press, Princeton, 2018, p. 11.

26 Frank Vibert, The Rise of the Unelected: Democracy and the New Separation of Powers, Cambridge University Press, Cambridge, 2007, p. 15.

27 See Susskind, Future Politics, p. 252 and citations therein.

CHAPTER TWENTY

1 Tony Blair, A Journey, Cornerstone, London, 2010.

2 James S. Fishkin, Democracy When the People are Thinking: Revitalising Our Politics Through Public Deliberation, Oxford University Press, Oxford, 2018, p. 1.

3 See generally the work of James S. Fishkin and Hélène Landemore.

4   See Cass Sunstein, *Democracy and the Problem of Free Speech*, Free Press, New York, 1995.

5   Ibid., p. 242.

6   See Richard Bellamy, 'The Republic of Reasons: Public Reasoning, Depoliticization, and Non-Domination', in Samantha Besson and José Luis Martí (eds), *Legal Republicanism: National and International Perspectives*, Oxford University Press, Oxford, 2009, pp. 102–105.

7   Fishkin, *Democracy*, pp. 52–53.

8   Josiah Ober, *Demopolis: Democracy Before Liberalism in Theory and Practice*, Cambridge University Press, Cambridge, 2017, pp. 19–20; Fishkin, *Democracy*, p. 70.

9   John S. Dryzek et al., 'The crisis of democracy and the science of deliberation', *Science*, Vol. 363, No. 6432 (2019), 1144-1146. The OECD identifies twelve different models of mini-public: Ieva Česnulaitytė, 'Models of representative deliberative processes', in *OECD* (ed.), *Innovative Citizen Participation and New Democratic Institutions*, OECD Publishing, Paris, 2020.

10 Česnulaitytė, 'Models of representative deliberative processes'.

11 Ibid.

12 Ibid.

13 See Simon Joss and John Durant (eds), *Public participation in science: The role of consensus conferences in Europe*, Antony Roe, Chippenham, 1995.

14 Ibid.

15 Ibid.

16 Claudia Chwalisz, 'Introduction: Deliberation and new forms of governance', in OECD, *Innovative Citizen Participation*.

17 Dryzek et al., 'The crisis of democracy'; Frank Fischer, 'Participatory Governance: from Theory to Practice', in David Levi-Faur (ed.), *The Oxford Handbook of Governance*, Oxford University Press, Oxford, 2014, p. 459.

18 Gordon Pennycook and David G. Rand, 'Lazy, not biased: Susceptibility to partisan fake news is better explained by lack of reasoning than by motivated reasoning', *Cognition*, Vol. 188 (2019), 39-50; Bence Bago, David G. Rand, and Gordon Pennycook, 'Fake News, Fast and Slow: Deliberation Reduces Belief in False (but Not True) News Headlines', *Journal of Experimental Psychology: General*, Vol. 149, No. 8 (2020), 1608-1613.

19 Vesa Koskimaa and Lauri Rapeli, 'Fit to govern? Comparing citizen and policymaker perceptions of deliberative democratic innovations', *Policy & Politics*, Vol. 48, No. 4 (2020), 637-652(16).

20 Nesta, 'vTaiwan', 2020 <https://www.nesta.org.uk/feature/six-pioneers-digital-democracy/vtaiwan/> (accessed 20 August 2021).

21 Carl Miller, 'Taiwan is making democracy work again. It's time we paid attention', *Wired*, 26 November 2019 <https://www.wired.co.uk/article/taiwan-democracy-social-media> (accessed 20 August 2021).

22 Ibid.

23 Ibid.

24 See OECD, *Innovative Citizen Participation*.

25 Ada Lovelace Institute, *The Citizens' Biometrics Council*, March 2021 <https://www.adalovelaceinstitute.org/report/citizens-bio metrics-council> (accessed 3 October 2021). I am grateful to Carly Kind for bringing the Council to my attention.

26 Canadian Citizens' Assembly on Democratic Express, Canadian Citizens' Assembly on Democratic Express: Recommendations to strengthen Canada's response to new digital technologies and reduce the harm caused by their misuse, Public Policy Forum, Ottawa, 2021.

27 For an example of arm's length commissioning, see Sciencewise, 'Supporting socially informed policy making', 2021 <https://scie ncewise.org.uk> (accessed 20 August 2021).

28 Fishkin, *Democracy*, p. 167.

29 OECD, *Innovative Citizen Participation*.

30 I am grateful to Tom Upchurch for his assistance with this point.

31 Melissa Lane, *Greek and Roman Political Ideas*, Penguin Books, London, 2014, p. 200.

32 Marcus Tullius Cicero, *The Republic and The Laws*, Oxford University Press, Oxford, 1998, p. 49.

CHAPTER TWENTY-ONE

1 See e.g. Andrew Kersely, 'Couriers say Uber's "racist" facial identification tech got them fired', *Wired*, 3 January 2021 <https://www.wired.co.uk/article/uber-eats-couriers-facial-recognition> (accessed 20 August 2021).

2 See Charles R. Beitz, *The Idea of Human Rights*, Oxford University Press, Oxford, 2013, p. 14.

3 See e.g. Committee of Experts on Internet Intermediaries, *Algorithms and Human Rights: Study on the human rights dimensions of automated data processing techniques and possible regulatory implications*, DGI(2017)12, Strasbourg,

Council of Europe, 2017; Expert Committee on Human Rights Dimensions of Automated Data Processing and Different Forms of Artificial Intelligence, *Responsibility and AI*, DGI(2019)05, Strasbourg, Council of Europe, 2019.

4 Philip Alston, 'Landmark ruling by Dutch court stops government attempts to spy on the poor – UN expert', *Office of the United Nations High Commissioner for Human Rights*, 5 February 2020 <https://www.ohchr.org/EN/NewsEvents/Pages/DisplayNews.aspx?NewsID=25522&LangID=E> (accessed 20 August 2021).

5 Agnès Callamard, 'The Human Rights Obligations of Non-State Actors', in Rikke Frank Jørgensen (ed.), *Human Rights in the Age of Platforms*, MIT Press, Cambridge, MA, 2019, p. 198. Some have horizontal effect.

6 Consent and contract are not strictly the same thing, but they overlap in principle and practice. Many of the same considerations apply to both.

7 See Article 4(11) of the GDPR.

8 John Eggerton, 'Hill Briefing: DETOUR Act on Right Road', *Multichannel News*, 25 June 2019 <https://www.nexttv.com/news/hill-briefing-detour-act-on-right-road> (accessed 20 August 2021). On 'dark patterns' and the GDPR, see Midas Nouwens et al., 'Dark Patterns after the GDPR: Scraping Consent Pop-ups and Demonstrating their Influence', *arXiv: 2001.02479* (2020).

9 Ryan Calo, 'Against Notice Skepticism in Privacy (and Elsewhere)', *Notre Dame Law Review*, Vol. 87, No. 3 (2012), 1027–1072; Daniel J. Solove, 'Introduction: Privacy Self-Management and the Consent Dilemma', *Harvard Law Review*, Vol. 126, No. 7 (2013), 1880–1903; Neil Richards and Woodrow Hartzog, 'The Pathologies of Digital Consent', *Washington University Law Review*, Vol. 96 (2019), 1461–1503.

10 See generally Martin Fries, 'Law and Autonomous Systems Series: Smart consumer contracts - The end of civil procedure?' *Oxford Business Law Blog*, 29 March 2018 <https://www.law.ox.ac.uk/business-law-blog/blog/2018/03/smart-consumer-contracts-end-civil-procedure> (accessed 20 August 2021).

11 John C. P. Goldberg and Benjamin C. Zipursky, 'Tort Law and Responsibility', in John Oberdiek (ed.), *Philosophical Foundations of the Law of Torts*, Oxford University Press, Oxford, 2018, p. 27.

12 Frank Pasquale, 'Data-Informed Duties in AI Development', *Columbia Law Review*, Vol. 119, No. 7 (2019), 1917–1940.

13 Ryan Abbott, 'The Reasonable Computer: Disrupting the Paradigm of Tort Liability', *George Washington Law Review*, Vol. 86, No. 1 (2018), 1–45.

14 Andrew S. Gold and Paul B. Miller, 'Introduction', in Andrew S. Gold and Paul B. Miller (eds), *Philosophical Foundations of Fiduciary Law*, Oxford University Press, Oxford, 2016, p. 1.

15 James Edelman, 'The Role of Status in the Law of Obligations: Common Callings, Implied Terms, and Lessons for Fiduciary Duties', in Gold and Miller (eds), *Philosophical Foundations of Fiduciary Law*, p. 23.

16 See Paul B. Miller, 'The Fiduciary Relationship', in Gold and Miller (eds), *Philosophical Foundations of Fiduciary Law*, p. 69.

17 Carissa Véliz, *Privacy is Power: Why and How you Should Take Back Control of Your Data*, Transworld Publishers, London, 2020, pp. 136–137; Jack M. Balkin, 'Information Fiduciaries in the Digital Age', *Balkinization*, 5 March 2014 <https://balkin.blogspot.com/2014/03/informat ion-fiduciaries-in-digital-age.html> (accessed 20 August 2021); Jack M. Balkin and Jonathan Zittrain, 'A Grand Bargain to Make Tech Companies Trustworthy', *The Atlantic*, 3 October 2016) <https:// www.theatlantic.com/technology/archive/2016/10/informat ion-fiduciary/502346/> (accessed 20 August 2021); Neil Richards and Woodrow Hartzog, 'Taking Trust Seriously in Privacy Law,' *Stanford Technology Law Review*, Vol. 19 (2016), 431–472; 'Trusting Big Data Research', *DePaul Law Review*, Vol. 66, No. 2 (2017), 579– 590; 'Privacy's Trust Gap: A Review', *Yale Law Journal*, Vol. 126, No. 4 (2017), 1180–1224. See also, in the US, the proposed Data Care Act of 2018.

18 Jonathan Zittrain, 'How to Exercise the Power You Didn't Ask For', *Harvard Business Review*, 19 September 2018 <https://hbr. org/2018/09/how-to-exercise-the-power-you-didnt-ask-for> (accessed 20 August 2021).

19 Rebecca Crootof, 'Accountability for the Internet of Torts', *The Law and Political Economy Project*, 17 July 2018 <https://lpeproject.org/ blog/accountability-for-the-internet-of-torts/> (accessed 20 August 2021).

CHAPTER TWENTY-TWO

1 See Julie E. Cohen, *Between Truth and Power: The Legal Constructions of Informational Capitalism*, Oxford University Press, Oxford, 2019, p. 151. This partly explains why privacy claims have struggled in American federal courts in recent years. It is hard to show the individual

damage required to establish standing. See generally Neil Richards and Woodrow Hartzog, 'Taking Trust Seriously in Privacy Law', *Stanford Technology Law* Review, Vol. 19 (2016), 431–472; 'Privacy's Trust Gap: A Review', *Yale Law Journal*, Vol. 126, No. 4 (2017), 1180–1224; 'Trusting Big Data Research', *DePaul Law Review*, Vol. 66, No. 2 (2017), 579–590; 'The Pathologies of Digital Consent', *Washington University Law Review*, Vol. 96 (2019), 1461–1503; 'Privacy's Constitutional Moment and the Limits of Data Protection', *Boston College Law Review*, Vol. 61, No. 5 (2020), 1687–1761.

2   Expert Committee on Human Rights Dimensions of Automated Data Processing and Different Forms of Artificial Intelligence, *Responsibility and AI*, DGI(2019)05, Strasbourg, Council of Europe, 2019.

3   See the illuminating discussion of this topic in Jacob Turner, *Robot Rules: Regulating Artificial Intelligence*, Palgrave Macmillan, London, 2019, pp. 222–225.

4   Jonathan Zittrain, 'A Jury of Random People Can Do Wonders for Facebook', *The Atlantic*, 14 November 2019 <https://www.theatlantic.com/ideas/archive/2019/11/let-juries-review-facebook-ads/601996/> (accessed 20 August 2021).

5   Lawrence Busch, *Standards: Recipes for Reality*, MIT Press, Cambridge, MA, 2013, p. 17.

6   Tim Büthe and Walter Mattli, 'International Standards and Standard-Setting Bodies', in Davin Coen, Wyn Grant and Graham Wilson (eds), *The Oxford Handbook of Business and Government*, Oxford University Press, Oxford, 2011, p. 440.

7   Colin Scott, 'Standard-Setting in Regulatory Regimes', in Robert Baldwin, Martin Cave and Martin Lodge (eds), *The Oxford Handbook of Regulation*, Oxford University Press, Oxford, 2013, p. 114.

8   Cary Coglianese and Even Mendelson, 'Meta-Regulation and Self-Regulation', in Baldwin et al., *Oxford Handbook of Regulation*, p. 148.

9   Scott, 'Standard-Setting', p. 110.

10   Büthe and Mattli, 'International Standards', p. 442.

CHAPTER TWENTY-THREE

1   I do not claim originality for the term 'counterpower', even if my usage is different from others'. 'Counterpower' is one of a constellation of terms that have been used to denote the idea of balancing forces in society. Perhaps the closest usage is by Manuel Castells – a fact of which I was unaware when I first used it (see e.g. Manuel Castells, 'Communication, Power and Counter-power

354

in the Network Society', *International Journal of Communication*, Vol. 1 (2007), 238–266). Philip Pettit writes of 'antipower' ('Freedom as Antipower', *Ethics*, Vol. 106, No. 3 (April 1996), 576–604.) J. K. Galbraith writes of 'countervailing power' (*American Capitalism: The Concept of Countervailing Power*, Martino Fine Books, Eastford, CT, 2012). Montesquieu and Karl Polanyi explored similar ideas.

2   See Richard Bellamy, 'The Republic of Reasons: Public Reasoning, Depoliticization, and Non-Domination', in Samantha Besson and José Luis Martí (eds), *Legal Republicanism: National and International Perspectives*, Oxford University Press, Oxford, 2009.

3   See Philip Pettit, *Republicanism: a Theory of Freedom and Government*, Oxford University Press, Oxford, pp. 184–187. See also John P. McCormick, *Machiavellian Democracy*, Cambridge University Press, Cambridge, 2013, viii: 'Machiavelli was arguably the only major intellectual advocate of republics in which the people vigorously contest and constrain the behaviour of political and economic elites by extra-electoral means.'

4   This chapter draws heavily on the work of Richard Susskind, who has been writing about the potential of online justice for decades.

5   Richard Susskind, *Online Courts and the Future of Justice*, Oxford University Press, Oxford, 2019, p. 27.

6   Ibid.

7   See the concept of 'outcome thinking' in Susskind, *Online Courts*.

8   Susskind, *Online Courts*, p. 53.

9   Article 17.

10 Sandra Wachter and Brent Mittelstadt, 'A Right to Reasonable Inferences: Re-Thinking Data Protection Law in the Age of Big Data and AI', *Columbia Business Law Review*, Vol. 2019, No. 2 (2019), 494–620.

11 Montesquieu, *The Spirit of the Laws*, Cambridge University Press, Cambridge, 2018, p. 157; Pettit, *Republicanism*, p. 179.

12 Facebook, 'Oversight Board Charter', September 2019 <https://about.fb.com/wp-content/uploads/2019/09/oversight_board_charter.pdf> (accessed 20 August 2021).

13 Ibid.

14 Ibid.

15 Oversight Board, 'Announcing the Oversight Board's first case decisions', January 2021 <https://oversightboard.com/news/165523235084273-announcing-the-oversight-board-s-first-case-decisions/> (accessed 20 August 2021).

16   Susskind, *Online Courts*, p. 169.
17   Ibid., p. 170.
18   Ibid.
19   Ibid., p. 168.
20   Richard Susskind, 'The Future of Courts', *The Practice*, Vol. 6, Issue 5, July/August 2020.
21   Sean McDonald, 'The Fiduciary Supply Chain', *Centre for International Governance Innovation Online*, 28 October 2019 <https://www.cigionline.org/articles/fiduciary-supply-chain> (accessed 20 August 2021).
22   Article 18.

CHAPTER TWENTY-FOUR

1   The National Labor Relations Act of 1935 and Occupational Safety and Health Act of 1970.
2   Sean Farhang, *The Litigation State: Public Regulation and Private Lawsuits in the U.S.*, Princeton University Press, Princeton, 2010, p. 3.
3   Ibid.
4   Stephen B. Burbank, Sean Farhang and Herbert Kritzer, 'Private Enforcement' (2013) University of Pennsylvania Law School Public Law Research Paper No. 13-22, 662; Farhang, *Litigation State*, p. 9.
5   Joseph Jerome, 'Private right of action shouldn't be a yes-no proposition in federal US privacy legislation', *International Association of Privacy Professionals*, 3 October 2019 <https://iapp. org/news/a/private-right-of-action-shouldnt-be-a-yes-no-proposition-in-federal-privacy-legislation/> (accessed 20 August 2021).
6   Reality Check Team, 'Social media: How do other governments regulate it?' BBC, 12 February 2020 <https://www.bbc.co.uk/news/technology-47135058> (accessed 20 August 2021).
7   Neil Gunningham, 'Enforcement and Compliance Strategies', in Robert Baldwin, Martin Cave and Martin Lodge (eds), *The Oxford Handbook of Regulation*, Oxford University Press, Oxford, 2013, p. 121.
8   Robert Baldwin, Martin Cave and Martin Lodge, *Understanding Regulation: Theory, Strategy, and Practice*, Oxford University Press, Oxford, 2012, p. 239.
9   See e.g. Mark Rumold, 'Regulating Surveillance through Litigation: Some Thoughts from the Trenches', in David Gray and Stephen E. Henderson (eds), *The Cambridge Handbook of Surveillance Law*, Cambridge University Press, New York, 2019.
10   Farhang, *Litigation State*, p. 10.

11 Ibid., p. 13.

12 Burbank et al., 'Private Enforcement', p. 646.

13 Ibid., p. 713.

14 See generally Julie E. Cohen, Between Truth and Power: The Legal Constructions of Informational Capitalism, Oxford University Press, Oxford, 2019, p. 147.

15 Roger Brownsword, Law, Technology and Society: Re-Imagining the Regulatory Environment, Routledge, Abingdon, 2019, p. 205.

16 Richard Dagger, 'Republicanism and the Foundations of Criminal Law', in R. A. Duff and Stuart P. Green (eds), Philosophical Foundations of Criminal Law, Oxford University Press, Oxford, 2013, p. 48.

17 Criminal Code Amendment (Sharing of Abhorrent Violent Material) Act 2019.

18 See Philip Pettit, Just Freedom: A Moral Compass for a Complex World, W. W. Norton and Company, New York, 2014, p. 95; Philip Pettit, On the People's Terms: A Republican Theory and Model of Democracy, Cambridge University Press, Cambridge, 2014, p. 119; Philip Pettit, Republicanism: a Theory of Freedom and Government, Oxford University Press, Oxford, p. 154.

CHAPTER TWENTY-FIVE

1 See Lawrence Busch, Standards: Recipes for Reality, MIT Press, Cambridge, MA, 2013.

2 Ibid., p. 201.

3 Ibid., p. 211.

4 Joint Committee on Human Rights, Oral Evidence: The Right to Privacy (Article 8) and the Digital Revolution (HC 1810).

5 A. Michael Froomkin, 'Regulating Mass Surveillance as Privacy Pollution: Learning from Environmental Impact Statements', University of Illinois Law Review, Vol. 2015, No. 5 (2015), 1713–1790; Michael Rovatsos, Brent Mittelstadt and Ansgar Koene, 'Landscape Summary: Bias in Algorithmic Decision-Making', Centre for Data Ethics and Innovation, 19 July 2019 <https://assets.publishing.service. gov.uk/government/uploads/system/uploads/attachment_data/ file/819055/Landscape_Summary_-_Bias_in_Algorithmic_Decis ion-Making.pdf> (accessed 20 August 2021); Dillon Reisman et al., 'Algorithmic Impact Assessments: A Practical Framework for Public Agency Accountability', AI Now Institute, April 2018 <https://ain owinstitute.org/aiareport2018.pdf> (accessed 3 October 2021).

6 Committee of Experts on Internet Intermediaries, *Algorithms and Human Rights: Study on the human rights dimensions of automated data processing techniques and possible regulatory implications*, DGI (2017) 12, Strasbourg, Council of Europe, 2017.

7 Expert Committee on Human Rights Dimensions of Automated Data Processing and Different Forms of Artificial Intelligence, *Responsibility and AI*, DGI(2019)05, Council of Europe, Strasbourg, 2019. See Aurelia Tamò-Larrieux, *Designing for Privacy and its Legal Framework: Data Protection by Design and Default for the Internet of Things*, Springer, Cham, 2018.

8 Julie E. Cohen, *Between Truth and Power: The Legal Constructions of Informational Capitalism*, Oxford University Press, Oxford, 2019, p. 92.

9 Mike Feintuck, 'Regulatory Rationales beyond the Economic: in search of the Public Interest', in Robert Baldwin, Martin Cave and Martin Lodge (eds), *The Oxford Handbook of Regulation*, Oxford University Press, Oxford, 2013, pp. 46–47.

10 Karen Yeung, 'Why Worry about Decision-Making by Machine?' in Karen Yeung and Martin Lodge, *Algorithmic Regulation*, Oxford University Press, Oxford, 2019, p. 40.

11 Expert Committee on Human Rights Dimensions of Automated Data Processing and Different Forms of Artificial Intelligence, *Responsibility and AI*.

12 Busch, *Standards*, p. 45.

13 Margaret Jane Radin, *Boilerplate: The Fine Print, Vanishing Rights, and the Rule of Law*, Princeton University Press, Princeton, 2013, p. 189.

14 GDPR, Recital 84; Andrew D. Selbst and Solon Barocas, 'The Intuitive Appeal of Explainable Machines', *Fordham Law Review*, Vol. 87, No. 3 (2018), 1085-1140.

15 See e.g. GDPR Articles 25, 42 and 43, and recitals 81 and 100. Article 25 is arguably advisory, however, and not a mandatory prerequisite to certification, which is itself voluntary: Article 42(3). See Lilian Edwards, *Law, Policy and the Internet*, Hart Publishing, Oxford, 2019, p. 111.

16 Promisingly, the EU's draft Artificial Intelligence Act proposes new forms of certification.

17 See Margot E. Kaminski and Andrew D. Selbst, 'The Legislation That Targets the Racist Impacts of Tech', *The New York Times*, 7 May 2019 <https://www.nytimes.com/2019/05/07/opinion/tech-racism-algorithms.html> (accessed 20 August 2021).

18   Gillian K. Hadfield, *Rules for a Flat World: Why Humans Invented Law and How to Reinvent it for a Complex Global Economy*, Oxford University Press, Oxford, 2017, p. 268.

19   Nicolas P. Suzor, *Lawless: The Secret Rules that Govern Our Digital Lives*, Cambridge University Press, Cambridge, 2019, p. 98.

20   Hadfield, *Rules*, p. 271.

21   Busch, *Standards*, p. 220.

CHAPTER TWENTY-SIX

1   Brent Mittelstadt, 'Principles alone cannot guarantee ethical AI', *Nature Machine Intelligence*, Vol. 1 (2019), 501-507.

2   Mary Wollstonecraft, *A Vindication of the Rights of Women*, Penguin, London, 2004, p. 26.

3   Richard Susskind and Daniel Susskind, *The Future of the Professions: How Technology will Transform the Work of Human Experts*, Oxford University Press, Oxford, 2015, pp. 10–15.

4   Ibid., p. 10.

5   For a thoughtful discussion of this issue, see chapter seven of Jacob Turner, *Robot Rules: Regulating Artificial Intelligence*, Palgrave Macmillan, London, 2019.

6   See Mittelstadt, 'Principles'.

7   Instead of suing the owners or manufacturers of self-driving cars, accident victims could be compensated by a no-fault insurance pool funded by taxes on vehicle manufacturers (similar to the system already used in New Zealand for road traffic accidents): see David Enoch, 'Tort Liability and Taking Responsibility', in John Oberdiek (ed.), *Philosophical Foundations of the Law of Torts*, Oxford University Press, Oxford, 2018, pp. 252–253. Several scholars have suggested that a similar model of insurance could be mandatory for all those who design powerful digital systems: see Jason D. Lohr, Winston J. Maxwell and Peter Watts, 'Legal Practitioners' Approach to Regulating AI Risks', in Karen Yeung and Martin Lodge, *Algorithmic Regulation*, Oxford University Press, Oxford, 2019, p. 245; Turner, *Robot Rules*, p. 317; Roger Brownsword, *Law, Technology and Society: Re-Imagining the Regulatory Environment*, Routledge, Abingdon, 2019, p. 188. See also Martin Eling, 'How insurance can mitigate AI risks', *The Brookings Institution*, 7 November 2019 <https://www.brookings.edu/research/how-insurance-can-mitigate-ai-risks/> (accessed 20 August 2021).

8   Financial Conduct Authority, 'Factsheet: Becoming an approved person', No. 029, 2020 <https://www.fca.org.uk/publication/other/fs029-becoming-an-approved-person.pdf> (accessed 20 August 2021).

9   Stuart Russell, *Human Compatible: AI and the Problem of Control*, Allen Lane, London, 2019, p. 250.

10   B. Hecht et al., 'It's Time to Do Something: Mitigating the Negative Impacts of Computing Through a Change to the Peer Review Process', *ACM Future of Computer Blog*, 29 March 2018 <https://acm-fca.org/2018/03/29/negativeimpacts/> (accessed 20 August 2021).

11   Mittelstadt, 'Principles'.

12   K. Sabeel Rahman, 'Regulating Informational Infrastructure: Internet Platforms as the New Public Utilities', *Georgetown Law and Technology Review*, Vol. 2, No. 2 (2018), 234–251.

CHAPTER TWENTY-SEVEN

1   Laura Denardis, *The Internet in Everything: Freedom and Security in a World with No Off Switch*, Yale University Press, New Haven, 2020, p. 200.

2   Ibid.

3   Lilian Edwards, *Law, Policy and the Internet*, Hart Publishing, Oxford, 2019, p. 78.

4   Robert Baldwin, Martin Cave and Martin Lodge, *Understanding Regulation: Theory, Strategy, and Practice*, Oxford University Press, Oxford, 2012, pp. 357–358.

5   Stuart Russell, *Human Compatible: AI and the Problem of Control*, Allen Lane, London, 2019, p. 182.

6   Will Knight, 'Washington Must Bet Big on AI or Lose Its Global Clout', *Wired*, 17 December 2019 <https://www.wired.com/story/washington-bet-big-ai-or-lose-global-clout/> (accessed 20 August 2021).

7   Steven Feldstein, 'How Should Democracies Confront China's Digital Rise? Weighing the Merits of a T-10 Alliance', *Council on Foreign Relations*, 30 November 2020 <https://www.cfr.org/blog/how-should-democracies-confront-chinas-digital-rise-weighing-merits-t-10-alliance> (accessed 20 August 2021); Jared Cohen and Richard Fontaine, 'Uniting the Techno-Democracies', *Foreign Affairs*, November/December 2020 <https://www.foreignaffairs.com/articles/united-states/2020-10-13/uniting-techno-democracies> (accessed 20 August 2021).

8   Monika Bickert, 'Defining the Boundaries of Free Speech on Social Media', in Lee C. Bollinger and Geoffrey R. Stone, *The Free Speech Century*, Oxford University Press, New York, 2019, p. 262.

9   Tim Wu, 'A TikTok Ban Is Overdue', *The New York Times*, 18 August 2020 <https://www.nytimes.com/2020/08/18/opinion/tiktok-wechat-ban-trump. html> (accessed 20 August 2021).

10 See e.g. Jennifer C. Daskal, 'Borders and Bits', *Vanderbilt Law Review*, Vol. 71, No. 1 (2018), 179–240.

11 Mathias Koenig-Archibugi, 'Global Regulation', in Robert Baldwin, Martin Cave and Martin Lodge (eds), *The Oxford Handbook of Regulation*, Oxford University Press, Oxford, 2013, p. 415.

12 Kieron O'Hara and Wendy Hall, 'Four Internets: The Geopolitics of Digital Governance' (December 2018) Centre for International Governance Innovation Paper No. 206. See also Nils Gilman and Henry Farrell, 'Three Moral Economies of Data', *The American Interest*, 7 November 2018 <https://www.the-american-interest. com/2018/11/07/three-moral-economies-of-data/> (accessed 20 August 2021).

13 Bickert, 'Defining', p. 262.

14 Natasha Tusikov, *Chokepoints: Global Private Regulation on the Internet*, University of California Press, Oakland, 2017, p. 163.

15 Koenig-Archibugi, 'Global Regulation', pp. 413–414.

16 Michèle Finck, *Blockchain Regulation and Governance in Europe*, Cambridge University Press, Cambridge, 2019, pp. 38–39.

17 Roger Brownsword, *Law, Technology and Society: Re-Imagining the Regulatory Environment*, Routledge, Abingdon, 2019, p. 333.

18 Brownsword, *Law, Technology and Society*, p. 33.

19 Nicolas P. Suzor, *Lawless: The Secret Rules that Govern Our Digital Lives*, Cambridge University Press, Cambridge, 2019, p. 50.

20 Melissa Lane, *Greek and Roman Political Ideas*, Penguin Books, London, 2014, p. 317.

CHAPTER TWENTY-EIGHT

1   Rory Van Loo, 'The Missing Regulatory State: Monitoring Businesses in an Age of Surveillance', *Vanderbilt Law Review*, Vol. 72, No. 5 (2019), 1563–1631.

2   See Article 28 of the draft Digital Services Act.

3   Michel Foucault, *Power/Knowledge*, Vintage Books, New York, 1980, p. 152.

4   Deven R. Desai and Joshua A. Kroll, 'Trust But Verify: A Guide to Algorithms and the Law', *Harvard Journal of Law & Technology*, Vol. 31, No. 1 (2017), 1–64.

5   Joshua A. Kroll et al., 'Accountable Algorithms', *University of Pennsylvania Law Review*, Vol. 165, No. 3 (2017), 633–705

6   Desai and Kroll, 'Trust But Verify', citing Kroll et al., 'Accountable Algorithms'.

7   Miles Brundage et al., 'Toward Trustworthy AI Development: Mechanisms for Supporting Verifiable Claims', *arXiv*: 2004.07213v2 (2020).

8   Jonathan Zittrain, 'How to Exercise the Power You Didn't Ask For', *Harvard Business Review*, 19 September 2018 <https://hbr.org/2018/09/how-to-exercise-the-power-you-didnt-ask-for> (accessed 20 August 2021).

9   Richard Susskind, *The Future of Law: Facing the Challenges of Legal Technology*, Oxford University Press, Oxford, 1996, p. 24.

10 See the discussion in Jacob Turner, *Robot Rules: Regulating Artificial Intelligence*, Palgrave Macmillan, London, 2019, p. 311.

11 Expert Committee on Human Rights Dimensions of Automated Data Processing and Different Forms of Artificial Intelligence, *Responsibility and AI*, DGI(2019)05, Strasbourg, Council of Europe, 2019.

CHAPTER TWENTY-NINE

1   Wikipedia, 'Draco (lawgiver)', 30 June 2021 <https://en.wikipedia.org/wiki/Draco_(lawgiver)> (accessed 20 August 2021).

2   Michèle Finck, *Blockchain Regulation and Governance in Europe*, Cambridge University Press, Cambridge, 2019, p. 45.

3   Lon Fuller, *The Morality of Law*, Yale University Press, New Haven and London, 1969.

4   Cass Sunstein, *Free Markets and Social Justice*, Oxford University Press, New York, 1997, p. 326.

5   Sunstein, *Free Markets*, p. 327.

6   Alex Engler, 'The case for AI transparency requirements', *The Brookings Institution*, 22 January 2020 <https://www.brookings.edu/research/the-case-for-ai-transparency-requirements/> (accessed 20 August 2021).

7   See Tarleton Gillespie, *Custodians of the Internet: Platforms, Content Moderation, and the Hidden Decisions that Shape Social Media*, Yale University Press, New Haven, 2018, p. 169.

8   Nicolas P. Suzor et al., 'What Do We Mean When We Talk About Transparency? Toward Meaningful Transparency in Commercial Content Moderation', *International Journal of Communication*, Vol. 13 (2019), 1526–1543.

9   See, seminally, Frank Pasquale, *The Black Box Society: The Secret Algorithms that Control Money and Information*, Harvard University Press, Cambridge, MA, 2015.

10 Amy Webb, *The Big Nine: How the Tech Titans & Their Thinking Machines Could Warp Humanity*, Hachette Book Group, New York, 2019, p. 111.

11 Ibid., p. 113.

CHAPTER THIRTY

1   Umang Bhatt et al., 'Explainable Machine Learning in Deployment', arXiv: 1909.06342 (2020).

2   Cynthia Rudin, 'Stop explaining black box machine learning models for high stakes decisions and use interpretable models instead', *Nature Machine Intelligence*, Vol. 1 (May 2019), 206–215.

3   Andrew D. Selbst and Solon Barocas, 'The Intuitive Appeal of Explainable Machines', *Fordham Law Review*, Vol. 87, No. 3 (2018), 1085–1140.

4   Selbst and Barocas, 'Intuitive Appeal'; cf. Rudin, 'Stop explaining'.

5   Rudin, 'Stop explaining'; Teresa Scantamburlo, Andrew Charlesworth and Nello Cristianini, 'Machine Decisions and Human Consequences', in Karen Yeung and Martin Lodge, *Algorithmic Regulation*, Oxford University Press, Oxford, 2019.

6   Nicole Rigillo, 'AI Must Explain Itself', *Noēma Magazine*, 16 June 2020 <https://www.noemamag.com/ai-must-explain-itself/> (accessed 20 August 2021).

7   Selbst and Barocas, 'Intuitive Appeal'.

8   Ibid.

9   Joshua A. Kroll et al., 'Accountable Algorithms', *University of Pennsylvania Law Review*, Vol. 165, No. 3 (2017), 633–705.

10 Finale Doshi-Velez et al., 'Accountability of AI Under the Law: The Role of Explanation', arXiv: 1711.01134 (2017).

11 Sandra Wachter, Brent Mittelstadt and Chris Russell, 'Counterfactual Explanations Without Opening the Black Box: Automated Decisions and the GDPR', *Harvard Journal of Law & Technology*, Vol. 31, No. 2 (2018), 841–888; Eliza Mik, 'Persuasive Technologies: From Loss of Privacy to Loss of Autonomy', in Kit Barker, Karen Fairweather and Ross

Grantham (eds), *Private Law in the 21st Century*, Hart Publishing, Oxford, 2017, p. 377.

12 Robert Baldwin, Martin Cave and Martin Lodge, *Understanding Regulation: Theory, Strategy, and Practice*, Oxford University Press, Oxford, 2012, p. 120; Cass Sunstein, *Free Markets and Social Justice*, Oxford University Press, New York, 1997, p. 338. On the possible risks of transparency generally, see Adrian Weller, 'Challenges for Transparency', *arXiv*: 1708.01870v1 (2017).

13 Ranking Digital Rights, '2019 Ranking Digital Rights Corporate Accountability Index', May 2019 <https://rankingdigitalrights.org/index2019/assets/static/download/RDRindex2019report.pdf> (accessed 20 August 2021).

14 Julie E. Cohen, *Between Truth and Power: The Legal Constructions of Informational Capitalism*, Oxford University Press, Oxford, 2019, p. 135.

15 Natasha Tusikov, *Chokepoints: Global Private Regulation on the Internet*, University of California Press, Oakland, 2017, p. 233.

16 Nicolas P. Suzor, *Lawless: The Secret Rules that Govern Our Digital Lives*, Cambridge University Press, Cambridge, 2019, pp. 136–137.

17 Cohen, *Between Truth and Power*, p. 136.

18 Ibid., p. 45.

19 Tarleton Gillespie, *Custodians of the Internet: Platforms, Content Moderation, and the Hidden Decisions that Shape Social Media*, Yale University Press, New Haven, 2018, p. 193.

20 Ibid., p. 199.

21 Cohen, *Between Truth and Power*, p. 189.

22 See e.g. GDPR Articles 12–15.

23 GDPR Article 22.

24 Sandra Wachter and Luciano Floridi, 'Why a right to explanation of automated decision-making does not exist in the General Data Protection Regulation', *International Data Privacy Law*, Vol. 7, No. 2 (2017), 76–99; cf. Andrew D. Selbst and Julia Powles, 'Meaningful information and the right to explanation', *International Data Privacy Law*, Vol. 7, No. 4 (2017), 233–242.

25 Lilian Edwards and Michael Veale, 'Slave to the Algorithm? Why a "Right to an Explanation" Is Probably Not the Remedy You Are Looking For', *Duke Law & Technology Review*, Vol. 16, No. 1 (2017), 18–84.

CHAPTER THIRTY-ONE

1 For a fine explanation of the importance of contestability, see Talia B. Gillis and Josh Simons, 'Explanation < Justification: GDPR and

the Perils of Privacy', *Pennsylvania Journal of Law and Innovation*, Vol. 2 (2019), 71–99.

2   See Gillis and Simons, 'Explanation < Justification'; Miles Brundage et al., 'Toward Trustworthy AI Development: Mechanisms for Supporting Verifiable Claims', arXiv: 2004.07213v2 (2020). Compare the proposed duty in Article 13 of the EU's draft Artificial Intelligence Act: 'High-risk AI systems shall be designed and developed in such a way to ensure that their operation is sufficiently transparent to enable users to interpret the system's output and use it appropriately.'

3   See Article 15 of the EU's draft Digital Services Act.

4   Finale Doshi-Velez et al., 'Accountability of AI Under the Law: The Role of Explanation', arXiv: 1711.01134 (2017).

5   Ibid.

6   Ibid; Sandra Wachter, Brent Mittelstadt, and Chris Russell, 'Counterfactual Explanations Without Opening the Black Box: Automated Decisions and the GDPR', *Harvard Journal of Law & Technology*, Vol. 31, No. 2 (2018), 841-888.

7   Lilian Edwards and Michael Veale, 'Slave to the Algorithm? Why a "Right to an Explanation" Is Probably Not the Remedy You Are Looking For', *Duke Law & Technology Review*, Vol. 16, No. 1 (2017), 18–84.

8   Doshi-Velez et al., 'Accountability'.

9   Ibid.

10 See Edwards and Veale, 'Slave'. See also Gillis and Simons, 'Explanation < Justification'.

11 In *Future Politics*, I referred to this as the rough and ready test of algorithmic justice: Does [the system, process, feature or function] deliver results that are consistent or inconsistent with a relevant principle of justice? Jamie Susskind, *Future Politics: Living Together in a World Transformed by Tech*, Oxford University Press, Oxford, 2018, p. 280.

12 Doshi-Velez et al., 'Accountability'.

13 Cliff Kuang and Robert Fabricant, *User Friendly: How the hidden rules of design are changing the way we live, work, and play*, WH Allen, London, 2019, p. 210.

14 Toby Walsh, *Android Dreams: The Past, Present and Future of Artificial Intelligence*, C. Hurst & Co (Publishers), London, 2017, p. 111.

15 See Article 52(1).

16 See e.g. GDPR Articles 24(1) and 82(3).

CHAPTER THIRTY-TWO

1   Tess Townsend, 'Keith Ellison and the New "Antitrust Caucus" Want to Know Exactly How Bad Mergers Have Been for the American Public', *Intelligencer*, 4 December 2017 <https://nymag.com/intelligencer/2017/12/antitrust-bill-from-keith-ellison-seek-info-on-mergers.html> (accessed 20 August 2021).

2   Senate Democrats, 'A Better Deal: Cracking Down on Corporate Monopolies', June 2017 <https://www.democrats.senate.gov/imo/media/doc/2017/07/A-Better-Deal-on-Competition-and-Costs-1.pdf> (accessed 20 August 2021).

3   Elizabeth Warren, 'Here's how we can break up Big Tech', *Team Warren*, 8 March 2019 <https://medium.com/@teamwarren/heres-how-we-can-break-up-big-tech-9ad9e0da324c> (accessed 20 August 2021).

4   Daniel A. Crane, 'Antitrust's Unconventional Politics' (2018) University of Michigan Law & Economics Working Paper No. 153.

5   Steve Lohr, 'The Week in Tech: How Is Antitrust Enforcement Changing?' *The New York Times*, 22 December 2019 <https://www.nytimes.com/2019/12/22/technology/the-week-in-tech-how-is-antitrust-enforcement-changing.html> (accessed 20 August 2021).

6   Clare Duffy, 'Marc Benioff says it's time to break up Facebook', CNN, 17 October 2019 <https://edition.cnn.com/2019/10/16/tech/salesforce-marc-benioff-break-up-facebook-boss-files/index.html> (accessed 20 August 2021).

7   Lina Khan, 'The New Brandeis Movement: America's Antimonopoly Debate', *Journal of European Competition Law & Practice*, Vol. 9, No. 3 (2018), 131–132.

8   Tim Wu, *The Curse of Bigness: Antitrust in the New Gilded Age*, Columbia Global Reports, New York, 2018.

9   Ibid., pp. 123–124.

10 Adam Satariano, 'Google Fined $1.7 Billion by E.U. for Unfair Advertising Rules', *The New York Times*, 20 March 2019 <https://www.nytimes.com/2019/03/20/business/google-fine-advertising.html> (accessed 20 August 2021).

11 Samuel Stolton, 'Vestager distances Commission from option of "Big Tech Breakups"', *Euractiv*, 27 October 2020 <https://www.euractiv.com/section/digital/news/vestager-distances-commission-from-option-of-big-tech-breakups/> (accessed 20 August 2021).

12   Martin Moore and Damian Tambini, 'Introduction', in Martin Moore and Damian Tambini (eds), *Digital Dominance: The Power of Google, Amazon, Facebook, and Apple*, Oxford University Press, Oxford, 2018.

13   Ibid., p. 5.

14   Alexi Mostrous and Peter Hoskin, 'Welcome to Apple: A one-party state', *Tortoise Media*, 6 January 2020 <https://www.tortoi semedia.com/2020/01/06/day-1-apple-state-of-the-nation-2/> (accessed 20 August 2021).

15   See Julie E. Cohen, *Between Truth and Power: The Legal Constructions of Informational Capitalism*, Oxford University Press, Oxford, 2019, p. 236.

16   Diane Coyle, *Markets, State, and People: Economics for Public Policy*, Princeton University Press, Princeton, 2020, p. 94.

17   Anthony Cuthbertson, 'Who Controls the Internet? Facebook and Google Dominance Could Cause the "Death of the Web"', *Newsweek*, 11 February 2017 <https://www.newsweek.com/facebook-goo gle-internet-traffic-net-neutrality-monopoly-699286> (accessed 20 August 2021).

18   Moore and Tambini, 'Introduction', p. 5.

19   Binyamin Applebaum, *The Economists' Hour: False Prophets, Free Markets, and the Fracture of Society*, Little, Brown and Company, New York, 2019, p. 156.

20   See David Dayen, 'The Final Battle in Big Tech's War to Dominate Your World', *The New Republic*, 8 April 2019 <https://newrepublic. com/article/153515/final-battle-big-techs-war-dominate-world> (accessed 20 August 2021).

21   Ibid.

22   See the discussion of Metcalfe's Law in Jamie Susskind, *Future Politics: Living Together in a World Transformed by Tech*, Oxford University Press, Oxford, p. 320.

23   Zephyr Teachout, *Break 'Em Up: Recovering our Freedom from Big Ag, Big Tech, and Big Money*, All Points Books, New York, 2020, p. 50.

24   Paul Arnold, 'How enforcement of antitrust law can enliven American innovation', *San Francisco Chronicle*, 14 August 2020 <https://www. sfchronicle.com/opinion/openforum/article/How-enforcem ent-of-antitrust-law-can-enliven-15482922.php> (accessed 20 August 2021).

25   Jacques Crémer, Yves-Alexandre de Montjoye and Heike Schweitzer, 'Competition policy for the digital era: Final Report', European Commission, Luxembourg, 2019.

26   Susskind, *Future Politics*, p. 322.

27 See Gerald F. Davis, 'Corporate Power in the Twenty-First Century', in Subramanian Rangan (ed.), *Performance & Progress: Essays on Capitalism, Business, and Society*, Oxford University Press, Oxford, 2017.

28 Jane Chung, 'Big Tech, Big Cash: Washington's New Power Players', Public Citizen, March 2021 <https://www.citizen.org/article/big-tech-lobbying-update/> (accessed 20 August 2021). I am grateful to Meredith Broussard for bringing this research to my attention.

29 Chung, 'Big Tech'.

30 Rana Faroohar, *Don't be Evil: The Case Against Big Tech*, Allen Lane, London, 2019, xviii, p. 216; Chung, 'Big Tech'.

31 Lawrence Lessig, *Republic, Lost: How Money Corrupts Congress – and a Plan to Stop It*, Hachette Book Group, New York, 2011; Alexi Mostrous and Peter Hoskin, 'Domestic policy: facing both ways', *Tortoise Media*, 8 January 2020 <https://www.tortoisemedia.com/2020/01/08/tech-states-apple-domestic-policy/> (accessed 20 August 2021).

32 Faroohar, *Don't be Evil*, p. 230.

33 See e.g. David Dayen, 'Fiona, Apple, and Amazon: How Big Tech Pays to Win the Battle of Ideas', *The American Prospect*, 20 July 2020 <https://prospect.org/power/fiona-apple-amazon-how-big-tech-pays-to-win-battle-ideas/> (accessed 20 August 2021).

34 Tony Romm, 'Amazon, Facebook and Google turn to deep network of political allies to battle back antitrust probes', *Washington* Post, 10 June 2020 <https://www.washingtonpost.com/technology/2020/06/10/amazon-facebook-google-political-allies-antitrust/> (accessed 20 August 2021); Alex Kantrowitz, 'Inside Big Tech's Years-Long Manipulation Of American Op-Ed Pages', *Big Technology*, 16 July 2020 <https://bigtechnology.substack.com/p/inside-big-techs-years-long-manipulation> (accessed 20 August 2021).

35 Melissa Lane, *Greek and Roman Political Ideas*, Penguin Books, London, 2014, pp. 21–22; Philip Pettit, *Republicanism: a Theory of Freedom and Government*, Oxford University Press, Oxford, p. 179.

36 John W. Maynor, *Republicanism in the Modern World*, Polity Press, Cambridge, 2003, p. 21.

37 Bernard Bailyn, *The Ideological Origins of the American Revolution*, Harvard University Press, Cambridge, MA, 1992, p. 76; Hannah Arendt, *On Revolution*, Penguin, New York, 2006, pp. 142 *et seq*.

38   Judith N. Shklar, 'Montesquieu and the new republicanism', in Gisela Bock, Quentin Skinner and Maurizio Viroli (eds), *Machiavelli and Republicanism*, Cambridge University Press, Cambridge, 1993, p. 269.

39   Melvin Richter, *The Political Theory of Montesquieu*, Cambridge University Press, New York, 1977, p. 3.

40   Bailyn, *Ideological Origins*, p. 76.

41   Alexander Hamilton, James Madison and John Jay, *The Federalist Papers*, Penguin, New York, 2012, p. 17.

42   Madison, *Federalist Papers*, p. 47.

43   Prateek Raj, '"Antimonopoly Is as Old as the Republic"', *Promarket*, 22 May 2017 <https://promarket.org/2017/05/22/antimonop oly-old-republic/> (accessed 20 August 2021).

44   See Teachout, *Break 'Em Up*, xi; Wu, *Curse*.

45   Cited in K. Sabeel Rahman, 'Monopoly Men', *Boston Review*, 11 October 2017 <https://bostonreview.net/class-inequality/k-sab eel-rahman-monopoly-men> (accessed 3 October 2021).

46   Crémer et al., 'Competition policy', Makan Delrahim, '"I'm Free": Platforms and Antitrust Enforcement in the Zero-Price Economy', Department of Justice, Boulder, CO, 2019.

47   See Stigler Committee on Digital Platforms, 'Final Report', Stigler Centre for the Study of the Economy and the State, 2019 <https:// www.chicagobooth.edu/-/media/research/stigler/pdfs/digital- platforms---committee-report---stigler-center.pdf> (accessed 20 August 2021).

48   Wu, *Curse*, pp. 128–129.

49   Ibid., p. 133.

50   Rory Cellan-Jones, 'Parler social network sues Amazon for pulling support', BBC, 11 January 2021 <https://www.bbc.co.uk/news/ technology-55615214> (accessed 20 August 2021).

51   Teachout, *Break 'Em Up*, p. 216. See also K. Sabeel Rahman, 'Regulating Informational Infrastructure: Internet Platforms as the New Public Utilities', *Georgetown Law and Technology Review*, Vol. 2, No. 2 (2018), 234–251.

52   Wu, *Curse*, p. 17. See also Daniel A. Crane, 'The Tempting of Antitrust: Robert Bork and the Goals of Antitrust Policy', *Antitrust Law Journal*, Vol. 79, No. 3 (2014), 835–853.

53   These are the dates of the Sherman Act and the Clayton Act, the main federal statutes concerning antitrust. See Lina M. Khan, 'Amazon's

Antitrust Paradox', *The Yale Law Journal*, Vol. 126, No. 3 (2017), 710–805 (note).

54   Wu, *Curse*, p. 74.

55   Megan Slack, 'From the Archives: President Teddy Roosevelt's New Nationalism Speech', *The White House: President Barack Obama*, 6 December 2011 <https://obamawhitehouse.archives.gov/blog/2011/12/06/archives-president-teddy-roosevelts-new-nationalism-speech> (accessed 20 August 2021).

CHAPTER THIRTY-THREE

1   Megan Slack, 'From the Archives: President Teddy Roosevelt's New Nationalism Speech', *The White House: President Barack Obama*, 6 December 2011 <https://obamawhitehouse.archives.gov/blog/2011/12/06/archives-president-teddy-roosevelts-new-nationalism-speech> (accessed 20 August 2021).

2   Paul De Hert et al., 'The right to data portability in the GDPR: Towards user-centric interoperability of digital services', *Computer Law & Security Review*, Vol. 34, No. 2 (2018), 193–203.

3   Inge Graef, Martin Husovec and Nadezhda Purtova, 'Data Portability and Data Control: Lessons for an Emerging Concept in EU Law', *German Law Journal*, Vol. 19, No. 6 (2018), 1359–1398.

4   De Hert et al., 'Right to data portability'.

5   Taylor Lyles, 'Facebook's new tool makes it easy to transfer photos and videos to Google Photos', *The Verge*, 30 April 2020 <https://www.theverge.com/2020/4/30/21241093/facebook-google-photos-transfer-tool-data> (accessed 20 August 2021).

6   As Graef et al. observe in 'Data portability and Data Control', 'the impact of the RtDP will likely be limited because the right can be invoked only—following Article 20(1) GDPR—with regard to personal data processed based on consent or on a contract. This caveat effectively excludes an obligation for the controller to provide a copy of the data processed under all other grounds, including legitimate interest. This raises the question whether controllers will be able to preclude data subjects from relying on the RtDP by invoking a legitimate interest as a ground for processing personal data instead of consent or a contract.'

7   Lilian Edwards, *Law, Policy and the Internet*, Hart Publishing, Oxford, 2019, p. 109.

8   See Digital Competition Expert Panel, 'Unlocking digital competition', Crown, 2019 <https://assets.publishing.service.

gov.uk/government/uploads/system/uploads/attachment_data/file/785547/unlocking_digital_competition_furman_review_web.pdf> (accessed 20 August 2021).

9   Tom Wilson and Kate Starbird, 'Cross-platform disinformation campaigns: Lessons learned and next steps', The Harvard Kennedy School Misinformation Review, Vol. 1, No. 1 (2020).

10 Shin-Shin Hua and Haydn Belfield, AI & Antitrust: Reconciling Tensions Between Competition Law and Cooperative AI Development, 23 Yale J.L. & Tech. 360 (2021) (forthcoming).

11 J. Matthew Hoye and Jeffrey Monaghan, 'Surveillance, freedom and the republic', European Journal of Political Theory, Vol. 17, No. 3 (2018), 343–363.

12 This reflects current European Commission policy: Natasha Lomas, 'Don't break up big tech – regulate data access, says EU antitrust chief', TechCrunch, 11 March 2019 <https://techcrunch.com/2019/03/11/dont-break-up-big-tech-regulate-data-access-says-eu-antitrust-chief/> (accessed 20 August 2021).

13 Tim Wu, The Curse of Bigness: Antitrust in the New Gilded Age, Columbia Global Reports, New York, 2018, p. 54.

14 Jamie Susskind, Future Politics: Living Together in a World Transformed by Tech, Oxford University Press, Oxford, 2018, chapter 19.

15 Rana Faroohar, Don't be Evil: The Case Against Big Tech, London, Allen Lane, 2019, p. 190.

16 Lina Khan, 'The Separation of Platforms and Commerce', Columbia Law Review, Vol. 119, No. 4 (2019), 973–1098; Martin Moore and Damian Tambini, 'Introduction', in Martin Moore and Damian Tambini (eds), Digital Dominance: The Power of Google, Amazon, Facebook, and Apple, Oxford University Press, Oxford, 2018, p. 3.

17 Daniel A. Crane, 'A Premature Postmortem on the Chicago School of Antitrust', Business History Review, Vol. 93, No. 4: New Perspectives in Regulatory History (2019), 759–776.

CHAPTER THIRTY-FOUR

1   Neil Richards and Woodrow Hartzog, 'Privacy's Constitutional Moment and the Limits of Data Protection', Boston College Law Review, Vol. 61, No. 5 (2020), 1687–1761.

2   Ibid.

3   Katherine J. Strandburg, 'Monitoring, Datafication, and Consent: Legal Approaches to Privacy in the Big Data Context', in Julia Lane, Victoria Stodden, Stefan Bender and Helen Nissenbaum

(eds), *Privacy, Big Data, and the Public Good: Frameworks for Engagement*, Cambridge University Press, New York, 2014, p. 6; Abraham L. Newman, 'The Governance of Privacy', in David Levi-Faur (ed.), *The Oxford Handbook of Governance*, Oxford University Press, Oxford, 2014, p. 601.

4   Strandburg, 'Monitoring', pp. 7–8.

5   Richards and Hartzog, 'Privacy's Constitutional Moment'.

6   Ibid.

7   Ibid.

8   See, seminally, Viktor Mayer-Schönberger and Kenneth Cukier, *Big Data: A Revolution That Will Transform How We Live, Work and Think*, John Murray, London, 2013.

9   Data Dividend Project, 2021 <https://www.datadividendproject. com> (accessed 20 August 2021).

10   See Joris van Hoboken, 'The Privacy Disconnect', in Rikke Frank Jørgensen (ed.), *Human Rights in the Age of Platforms*, MIT Press, Cambridge, MA, 2019, p. 257.

11   Carissa Véliz, *Privacy is Power: Why and How you Should Take Back Control of Your Data*, Transworld Publishers, London, 2020, pp. 126–129. Véliz argues that this should include even anonymous data. Given that data can be reidentified or deanonymised rather easily, this makes sense.

12   Véliz, *Privacy is Power*, p. 136.

13   Darrell M. West, '10 actions that will protect people from facial recognition software', *The Brookings Institution*, 31 October 2019 <https://www.brookings.edu/research/10-acti ons-that-will-protect-people-from-facial-recognition-software/> (accessed 20 August 2021); Véliz, *Privacy is Power*, p. 147.

14   Lilian Edwards and Michael Veale, 'Slave to the Algorithm? Why a "Right to an Explanation" Is Probably Not the Remedy You Are Looking For', *Duke Law & Technology Review*, Vol. 16, No. 1 (2017), 18–84.

15   Anouk Ruhaak, 'When One Affects Many: The Case For Collective Consent', *Mozilla*, 13 February 2020 <https://foundation.mozilla. org/en/blog/when-one-affects-many-case-collective-consent/> (accessed 20 August 2021); 'Data trusts: what are they and how do they work?', *The RSA*, 11 June 2020 <https://www.thersa.org/ blog/2020/06/data-trusts-protection> (accessed 20 August 2021); Matt Prewitt, 'A View Of The Future Of Our Data', *Noēma Magazine*, 23 February 2021 <https://www.noemamag.com/a-view-of-the-fut ure-of-our-data/> (accessed 20 August 2021).

16   Ruhaak, 'Data trusts'.

17   See in part Anna Artyushina, 'The EU is launching a market for personal data. Here's what that means for privacy', MIT Technology Review, 11 August 2020 <https://www.technologyreview.com/2020/08/11/1006555/eu-data-trust-trusts-project-privacy-policy-opinion/> (accessed 20 August 2021).

18   Lilian Edwards, Law, Policy and the Internet, Oxford, Hart Publishing, 2019, pp. 162–163.

19   Tom Sorell and John Guelke, 'Liberal Democratic Regulation and Technological Advance', in Roger Brownsword, Eloise Scotford and Karen Yeung (eds), The Oxford Handbook of Law, Regulation, and Technology, Oxford University Press, Oxford, 2017, p. 108.

20   Melissa Lane, Greek and Roman Political Ideas, Penguin Books, London, 2014, xx.

21   See Christopher Kuner, Lee A. Bygrave, and Christopher Docksey (eds), The EU Data Protection Regulation (GDPR): A Commentary, Oxford, Oxford University Press, 2020, p. 330.

22   Article 6(4): 'Where the processing for a purpose other than that for which the personal data have been collected is not based on the data subject's consent.'. 'The fact that consent is, despite non-applicability of Article 6(1) and despite lack of any reference to consent in Article 23, explicitly named as a legal basis for compatible further processing in Article 6(4), must be accepted as acknowledgement of the legal possibility for data subjects to waive a fundamental right': Waltraut Kotschy, 'Article 6. Lawfulness of processing', in Christopher Kuner, Lee A. Bygrave and Christopher Docksey (eds), The EU Data Protection Regulation (GDPR): A Commentary, Oxford University Press, Oxford, 2020, p. 343.

23   Véliz, Privacy is Power, p. 131. See European Parliament, Commission evaluation report on the implementation of the General Data Protection Regulation two years after its application <https://www.europarl.europa.eu/doceo/document/TA-9-2021-0111_EN.pdf> (accessed 4 October 2021): '"Legitimate interest" is very often abusively mentioned as a legal ground for processing; points out that controllers continue to rely on legitimate interest without conducting the required test of the balance of interests, which includes a fundamental rights assessment; is particularly concerned by the fact that some Member States are adopting national legislation to determine conditions for processing based on legitimate interest by providing for the balancing of the respective interests of the

controller and of the individuals concerned, while the GDPR obliges each and every controller to undertake this balancing test individually, and to avail themselves of that legal ground'.

24 Ruhaak, 'When One Affects Many'.

25 'At present this remains, ambiguously, in Recital 43' of the GDPR: Edwards, *Law, Policy, and the Internet*, p. 98.

26 Sandra Wachter and Brent Mittelstadt, 'A Right to Reasonable Inferences: Re-Thinking Data Protection Law in the Age of Big Data and AI', *Columbia Business Law Review*, Vol. 2019, No. 2 (2019), 494–620.

27 Van Hoboken, 'The Privacy Disconnect', p. 264.

28 Helen Nissenbaum has written extensively, and seminally, on the importance of contextual privacy.

CHAPTER THIRTY-FIVE

1 Katie Deighton, 'Tech Firms Train Voice Assistants to Understand Atypical Speech', *The Wall Street Journal*, 24 February 2021 <https://www.wsj.com/articles/tech-firms-train-voice-assistants-to-understand-atypical-speech-11614186019> (accessed 20 August 2021).

2 Abubakar Abid, Maheen Farooqi and James Zou, 'Persistent Anti-Muslim Bias in Large Language Models', *arXiv*: 2101.05783v1 (2021).

3 Thanks to Sam Gilbert for help with this metaphor.

4 Sandra Wachter and Brent Mittelstadt, 'A Right to Reasonable Inferences: Re-Thinking Data Protection Law in the Age of Big Data and AI', *Columbia Business Law Review*, Vol. 2019, No. 2 (2019), 494–620.

5 Solon Barocas and Andrew D. Selbst, 'Big Data's Disparate Impact', *California Law Review*, Vol. 104, No. 3 (2016), 671–732; Charles A. Sullivan, 'Employing AI', *Villanova Law Review*, Vol. 63, No. 3 (2018), 395–430.

6 Tyler Vigen, *Spurious Correlations*, Hachette Books, New York, 2015, p. 5.

7 James Grimmelmann and Daniel Westreich, 'Incomprehensible Discrimination', *California Law Review*, Vol. 7 (2017), 164–177.

8 See generally Wachter and Mittelstadt, 'A Right to Reasonable Inferences'.

9 This recognition is related to anti-discrimination law, which prohibits unfavourable decisions made directly or indirectly on the basis of protected characteristics like age or race. And it would be

a natural extension of the administrative law principle that forbids government decision-makers from taking into account irrelevant considerations (like refusing to grant a restaurant licence because the official doesn't like Chinese food). A policy of 'forbidden variables' would mark a timely evolution of the law, not a departure from it: Rick Swedloff, 'The New Regulatory Imperative for Insurance', *Boston College Law Review*, Vol. 61, No. 6 (2020), 2031–2084.

10   Frederick Schauer, *Profiles, Probabilities, and Stereotypes*, Harvard University Press, Cambridge, MA, 2003, p. 46.

11   Philip Sales, 'Algorithms, Artificial Intelligence and the Law', *Sir Henry Brooke Lecture for BAILII*, Freshfields Bruckhaus Deringer LLP, London, 12 November 2019; Schauer, *Profiles*, p46.

12   Schauer, *Profiles*, pp. 48–49.

13   Ibid.

14   See, for a recent example of a system confusing satire with sincerity, Mike Isaac, 'For Political Cartoonists, the Irony Was That Facebook Didn't Recognize Irony', *The New York Times*, 10 June 2021 <https://www.nytimes.com/2021/03/19/technology/political-cartooni sts-facebook-satire-irony.html> (accessed 3 October 2021).

15   David Weinberger, 'How Machine Learning Pushes Us to Define Fairness', *Harvard Business Review*, 6 November 2019 <https://hbr.org/2019/11/how-machine-learning-pushes-us-to-define-fairn ess> (accessed 20 August 2021).

16   Adrian Weller, 'Challenges for Transparency', arXiv: 1708.01870v1 (2017); Karen Yeung, 'Why Worry about Decision-Making by Machine?' in Karen Yeung and Martin Lodge, *Algorithmic Regulation*, Oxford University Press, Oxford, 2019, p. 29.

CHAPTER THIRTY-SIX

1   Facebook's approach is summarised by Monika Bickert, Head of Global Policy: 'The goal of our Community Standards is to create a place for expression and give people voice. Building community and bringing the world closer together depends on people's ability to share diverse views, experiences, ideas and information. We want people to be able to talk openly about the issues that matter to them, even if some may disagree or find them objectionable. In some cases, we allow content which would otherwise go against our Community Standards—if it is newsworthy and in the public interest. We do this only after weighing the public interest value

against the risk of harm, and we look to international human rights standards to make these judgments.'

2   Cass Sunstein, *Democracy and the Problem of Free Speech*, Free Press, New York, 1995, p. 122.

3   Ibid., p. 3.

4   Irini Katsirea, '"Fake News": reconsidering the value of untruthful expression in the face of regulatory uncertainty', *Journal of Media Law*, Vol. 10, No. 2 (2018), 159–188, p. 173; Frederick Schauer, 'Every Possible Use of Language?' in Lee C. Bollinger and Geoffrey R. Stone, *The Free Speech Century*, Oxford University Press, New York, 2019.

5   Neil Chilson and Casey Mattox, '[The] Breakup Speech: Can Antitrust Fix the Relationship Between Platforms and Free Speech Values?' *Knight First Amendment Institute*, 5 March 2020 <https://knightcolum bia.org/content/the-breakup-speech-can-antitrust-fix-the-relat ionship-between-platforms-and-free-speech-values> (accessed 20 August 2021).

6   Monika Bickert, 'Defining the Boundaries of Free Speech on Social Media', in Bollinger and Stone, *Free Speech Century*, p. 268; Tarleton Gillespie, *Custodians of the Internet: Platforms, Content Moderation, and the Hidden Decisions that Shape Social Media*, Yale University Press, New Haven, 2018, p. 88.

7   Io Dodds and Mike Wright, 'Instagram removed nearly 10,000 suicide and self-harm images per day after the Molly Russell scandal', *The Telegraph*, 13 November 2019 <https://www.telegraph.co.uk/technology/2019/11/13/instagram-removed-nearly-10000-suic ide-self-harm-images-day/> (accessed 20 August 2021).

8   Gillespie, *Custodians*, p. 77, 142.

9   Bickert, 'Defining', p. 269.

10 Julie E. Cohen, *Between Truth and Power: The Legal Constructions of Informational Capitalism*, Oxford University Press, Oxford, 2019, p. 95.

11 See e.g. the work of Siva Vaidhyanathan, esp. *Antisocial Media: How Facebook Disconnects Us And Undermines Democracy*, Oxford University Press, Oxford, 2018.

12 Nathaniel Persily, *The Internet's Challenge to Democracy: Framing the Problem and Assessing Reforms*, Kofi Annan Foundation, Stanford, CA, 2019.

13 See Gillespie, *Custodians*, p. 19.

14 Shoshana Zuboff, *The Age of Surveillance Capitalism: The Fight for a Human Future at the New Frontier of Power*, Profile Books, London, 2019.

15 Uri Bram and Martin Schmalz, *The Business of Big Data: How to Create Lasting Value in the Age of AI*, Amazon, UK, 2019, p. 99. See also Zuboff,

Surveillance Capitalism, p. 93; Evgeny Morozov, 'Capitalism's New Clothes', *The Baffler*, 4 February 2019 <https://thebaffler.com/lat est/capitalisms-new-clothes-morozov> (accessed 19 August 2021).

16 Ian Bogost and Alexis C. Madrigal, 'How Facebook Works For Trump', *The Atlantic*, 17 April 2020 <https://www.theatlantic. com/technology/archive/2020/04/how-facebooks-ad-technol ogy-helps-trump-win/606403/> (accessed 20 August 2021).

17 See James Ball, *The System: Who Owns the Internet, and How it Owns Us*, Bloomsbury, London, 2020, p. 122 et seq.

18 Anthony Nadler, Matthew Crain and Joan Donovan, 'Weaponizing the Digital Influence Machine: The Political Perils of Online Ad Tech', Data & Society Research Institute, 2018 <https://datasoci ety.net/library/weaponizing-the-digital-influence-machine/> (accessed 20 August 2021).

19 Ryan Mac and Craig Silverman, 'Facebook Has Been Showing Military Gear Ads Next To Insurrection Posts', *BuzzFeed News*, 13 January 2021 <https://www.buzzfeednews.com/article/ryan mac/facebook-profits-military-gear-ads-capitol-riot> (accessed 20 August 2021).

20 Zuboff, *Surveillance Capitalism*, pp. 509–510.

21 Joe Tidy, 'Twitter apologises for letting ads target neo-Nazis and bigots', BBC, 16 January 2020 <https://www.bbc.co.uk/news/tec hnology-51112238> (accessed 20 August 2021).

22 See e.g. Stuart Russell, *Human Compatible: AI and the Problem of Control*, Allen Lane, London, 2019, p. 105.

23 Shanti Das and Geoff White, 'Instagram sends paedophiles to accounts of children as young as 11', *The Sunday Times*, 1 December 2019 <https://www.thetimes.co.uk/article/instagram-sends- predators-to-accounts-of-children-as-young-as-11-j2gn5hq83> (accessed 20 August 2021).

24 Mark Townsend, 'Facebook algorithm found to 'actively promote' Holocaust denial', *The Guardian*, 16 August 2020 <https://www.theg uardian.com/world/2020/aug/16/facebook-algorithm-found-to- actively-promote-holocaust-denial> (accessed 20 August 2021); Jakob Guhl and Jacob Davey, 'Hosting the 'Holohoax': A Snapshot of Holocaust Denial Across Social Media', Institute for Strategic Dialogue, London, 10 August 2020 <https://www.isdglobal.org/ isd-publications/hosting-the-holohoax-a-snapshot-of-holocaust- denial-across-social-media/> (accessed 20 August 2021).

25 Cliff Kuang and Robert Fabricant, *User Friendly: How the hidden rules of design are changing the way we live, work, and play*, WH Allen, London, 2019, p. 292; James Williams, *Stand Out of Our Light: Freedom and Resistance in the Attention Economy*, Cambridge University Press, Cambridge, 2018, p. 33; Rana Faroohar, *Don't be Evil: The Case Against Big Tech*, Allen Lane, London, 2019, xvi; Mark Bergen, 'YouTube Executives Ignored Warnings, Letting Toxic Videos Run Rampant', Bloomberg, 2 April 2019 <https://www.bloomberg.com/news/features/2019-04-02/youtube-executives-ignored-warnings-letting-toxic-videos-run-rampant> (accessed 20 August 2021).

26 See esp. Williams, *Stand Out*.

27 Persily, *Internet's Challenge*; Center for Humane Technology, 'Ledger of Harms', June 2021 <https://ledger.humanetech.com> (accessed 20 August 2021).

28 Davey Alba, 'On Facebook, Misinformation Is More Popular Now Than in 2016', *The New York Times*, 12 October 2020 <https://www.nytimes.com/2020/10/12/technology/on-facebook-misinformation-is-more-popular-now-than-in-2016.html> (accessed 3 October 2021).

29 Joey D'Urso and Alex Wickham, 'YouTube Is Letting Millions Of People Watch Videos Promoting Misinformation About The Coronavirus', *BuzzFeed News*, 19 March 2020 <https://www.buzzfeed.com/joeydurso/youtube-coronavirus-misinformation> (accessed 20 August 2021); Ella Hollowood and Alexi Mostrous, 'The Infodemic: Fake news in the time of C-19', *Tortoise Media*, 23 March 2020 <https://www.tortoisemedia.com/2020/03/23/the-infodemic-fake-news-coronavirus/> (accessed 20 August 2021).

30 James Callery and Jacqui Goddard, 'Most-clicked link on Facebook spread doubt about Covid vaccine', *The Times*, 23 August 2021 <https://www.thetimes.co.uk/article/most-clicked-link-on-facebook-spread-doubt-about-covid-vaccine-flknpp9n5> (accessed 3 October 2021).

31 Center for Humane Technology, 'Ledger'.

32 Anjana Susarla, 'Biases in algorithms hurt those looking for information on health', *The Conversation*, 14 July 2020 <https://theconversation.com/biases-in-algorithms-hurt-those-looking-for-information-on-health-140616> (accessed 20 August 2021).

33 Craig Silverman and Jane Lytvynenko, 'Amazon Is Pushing Readers Down A "Rabbit Hole" Of Conspiracy Theories About The Coronavirus', *BuzzFeed News*, 15 March 2021 <https://www.

buzzfeednews.com/article/craigsilverman/amazon-covid-conspir
acy-books> (accessed 3 October 2021).

34 Gilad Edelman, 'Stop Saying Facebook Is "Too Big to Moderate"',
    Wired, 28 July 2020 <https://www.wired.com/story/stop-saying-
    facebook-too-big-to-moderate/> (accessed 20 August 2021).

35 Alexander Hamilton, James Madison and John Jay, *The Federalist Papers*,
    Penguin, New York, 2012, p. 4.

36 Marchamont Nedham [1656], cited in Frank Lovett, 'Algernon
    Sidney, Republican Stability, and the Politics of Virtue', *Journal of
    Political Science*, Vol. 48, No. 1, Article 3 (2020), 59–83. See Axel
    Bruns, *Are Filter Bubbles Real?*, Cambridge, Polity Press, 2019.

37 Yochai Benkler, Robert Faris and Hal Roberts, *Network
    Propaganda: Manipulation, Disinformation, and Radicalization in American Politics*,
    New York, Oxford University Press, 2018.

38 Benkler et al., *Network Propaganda*, p. 13.

39 Elizabeth Dubois and Grant Blank, 'The echo chamber is
    overstated: the moderating effect of political interest and
    diverse media', *Information, Communication & Society*, Vol. 21, No.
    5: Communication, Information Technologies, and Media
    Sociology (CITAMS) Special Issue (2018), 729–745; Judith Möller
    et al., 'Do Not Blame It on the Algorithm: An Empirical Assessment
    of Multiple Recommender Systems and Their Impact on Content
    Diversity', *Information, Communication & Society*, Vol. 21, No. 7 (2018),
    959–977.

40 Clive Thompson, *Coders: Who They Are, What They Think And How They Are
    Changing Our World*, Picador, London, 2019, p. 318. See the work of
    Joan Donavan, Renee DiResta and Alice Marwick.

41 Williams, *Stand Out*, p. 34

42 Center for Humane Technology, 'Ledger'.

43 Alexandra A. Siegel, 'Online Hate Speech', in Nathaniel Persily
    and Joshua A. Tucker (eds), *Social Media and Democracy: The State of the
    Field and Prospects for Reform*, Cambridge University Press, New York,
    2020, p. 63.

44 Kevin Roose, 'The Making of a YouTube Radical', *The New York Times*, 8
    June 2019 <https://www.nytimes.com/interactive/2019/06/08/
    technology/youtube-radical.html> (accessed 20 August 2021).

45 Persily, *Internet's Challenge*.

46 Thompson, *Coders*, p. 317.

47 Max Fisher and Amanda Taub, 'How YouTube Radicalized Brazil',
    The New York Times, 11 August 2019 <https://www.nytimes.

com/2019/08/11/world/americas/youtube-brazil.html>
(accessed 20 August 2021); cf. Bergen, 'YouTube Executives Ignored
Warnings'.

48   Fisher and Taub, 'How YouTube Radicalized Brazil'.

49   Muhammad Ali et al., 'Ad Delivery Algorithms: The Hidden Arbiters
of Political Messaging', arXiv: 1912.04255v3 (2019).

50   See John Rawls, 'The Idea of Public Reason Revisited', esp. p. 445
et seq, in John Rawls, Political Liberalism, Columbia University Press,
New York, 2015.

51   Private companies do it too: Jeremy B. Merrill, 'How Facebook's
Ad System Lets Companies Talk Out of Both Sides of Their
Mouths', The Markup, 13 April 2021 <https://themarkup.
org/news/2021/04/13/how-facebooks-ad-system-lets-
companies-talk-out-of-both-sides-of-their-mouths> (accessed 3
October 2021).

52   Sue Halpern, 'The Problem of Political Advertising on Social Media',
The New Yorker, 24 October 2019 <https://www.newyorker.com/
tech/annals-of-technology/the-problem-of-political-advertis
ing-on-social-media> (accessed 20 August 2021).

53   Faroohar, Don't be Evil, p. 233.

54   Niall Ferguson, 'What Is To Be Done? Safeguarding Democratic
Governance In The Age Of Network Platforms', Hoover Institution,
Governance in an Emerging World, Fall Series, No. 318 (2018).

55   BBC, 'Russian disinformation "ongoing problem" says FBI chief',
6 February 2020 <https://www.bbc.co.uk/news/technology-
51399568> (accessed 20 August 2021).

56   Robin Emmott, 'Russia deploying coronavirus disinformation to sow
panic in West, EU document says', Reuters, 18 March 2020 <https://
www.reuters.com/article/us-health-coronavirus-disinformation-
idUSKBN21518F> (accessed 20 August 2021).

57   Andrew M. Guess and Benjamin A. Lyons, 'Misinformation,
Disinformation, and Online Propaganda', in Persily and Tucker
(eds), Social Media and Democracy, p. 14.

58   Donie O'Sullivan, 'The biggest Black Lives Matter page on Facebook is
fake', CNN, 9 April 2018 <https://money.cnn.com/2018/04/09/
technology/fake-black-lives-matter-facebook-page/index.html>
(accessed 20 August 2021).

59   Daphne Keller, 'Dolphins in the Net: Internet Content Filters and
the Advocate General's Glawischnig-Piesczek v. Facebook Ireland Opinion',
Stanford Center for Internet and Society, 4 September 2019 <https://

cyberlaw.stanford.edu/files/Dolphins-in-the-Net-AG-Analysis.
pdf> (accessed 20 August 2021).

60 Matthew Arnold (1822–1888), 'Dover Beach' <https://www.
poetryfoundation.org/poems/43588/dover-beach> (accessed 25
September 2021).

61 Peter Pomerantsev, *This Is Not Propaganda: Adventures in the War Against
Reality*, Faber & Faber, London, 2019, p. 166.

62 John Milton, *Areopagitica and Other Writings*, Penguin, London, 2014.

63 J. S. Mill, *On Liberty and other writings*, Cambridge University Press,
Cambridge, 2008, p. 49.

64 Persily, *Internet's Challenge*.

<h3 style="text-align:center">CHAPTER THIRTY-SEVEN</h3>

1 Louis Edward Ingelhart, *Press and Speech Freedoms in the World, from Antiquity
until 1998: A Chronology*, Greenwood Press, Westport, 1998, pp. 4-5.

2 Ibid., p. 15.

3 Ibid., p. 31.

4 Ibid., p. 39.

5 James S. Coleman, *Power and the Structure of Society*, W. W. Norton &
Company, New York, 1974, p. 8.

6 Ibid., ix.

7 Robert Walker, 'The First Amendment and Article Ten: Sisters Under
The Skin?' *Holdsworth Club Presidential Address*, Birmingham Law School,
Birmingham, 24 October 2008; Monika Bickert, 'European Court
Ruling Raises Questions About Policing Speech', *Facebook*, 14 October
2019   <https://about.fb.com/news/2019/10/european-court-
ruling-raises-questions-about-policing-speech/> (accessed 20
August 2021).

8 Priscilla M. Regan, 'Reviving the Public Trustee Concept and
Applying it to Information Privacy Policy', *Maryland Law Review*, Vol.
76, No. 4 (2017), 1025–1043, p. 1025. See the Radio Act 1927
and the Communications Act 1934; Mike Feintuck, 'Regulatory
Rationales beyond the Economic: in search of the Public Interest',
in Robert Baldwin, Martin Cave and Martin Lodge (eds), *The Oxford
Handbook of Regulation*, Oxford University Press, Oxford, 2013, p. 52.

9 Cass Sunstein, *Democracy and the Problem of Free Speech*, Free Press,
New York, 1995, p. 54.

10 Robert Britt Horwitz, *The Irony of Regulatory Reform: The Deregulation of
American Telecommunications*, Oxford University Press, New York,
1989, p. 13.

11 Feintuck, 'Regulatory Rationales', p. 53.

12 Sunstein, *Democracy*, p. 49.

13 By contrast, since 1996, the FCC has required broadband internet providers to operate in the public interest: Regan, 'Reviving', pp. 1032–1033.

14 See Victor Pickard, 'The Strange Life and Death of the Fairness Doctrine: Tracing the Decline of Positive Freedoms in American Policy Discourse', *International Journal of Communication*, Vol. 12 (2018), 3434–3453.

15 Jacob Rowbottom, *Media Law*, Oxford, Hart Publishing, 2018, p. 257.

16 Ibid., p. 280.

17 Ibid., p. 259.

18 Office of Communications, The Ofcom Broadcasting Code (with the Cross-promotion Code and the On Demand Programme Service Rules), 2020.

19 Communications Act 2003, s.321(2).

20 Rowbottom, *Media Law*, p. 281.

21 Ibid., p. 279.

22 Kevin Roose, 'The Making of a YouTube Radical', *The New York Times*, 8 June 2019 <https://www.nytimes.com/interactive/2019/06/08/technology/youtube-radical.html> (accessed 20 August 2021).

23 Yochai Benkler, Robert Faris and Hal Roberts, *Network Propaganda: Manipulation, Disinformation, and Radicalization in American Politics*, Oxford University Press, New York, 2018, p. 6.

24 Francis Fukuyama and Andrew Grotto, 'Comparative Media Regulation in the United States and Europe', in Nathaniel Persily and Joshua A. Tucker (eds), *Social Media and Democracy: The State of the Field and Prospects for Reform*, Cambridge University Press, New York, 2020, p. 201.

25 Rowbottom, *Media Law*, p. 286.

CHAPTER THIRTY-EIGHT

1 Redmond-Bate v Director of Public Prosecutions (1999) 7 BHRC 375, per Sedley LJ at [20].

2 Lee C. Bollinger, 'Dialogue', in Lee C. Bollinger and Geoffrey R. Stone, *The Free Speech Century*, Oxford University Press, New York, 2019, p. 1.

3 Ibid., p. 4.

4 Ibid., p. 5.

5 See especially the work of Zeynep Tufekci, Peter Pomerantsev and Tim Wu.

6   Tim Wu, 'Is the First Amendment Obsolete?', in Bollinger and Stone (eds), *Free Speech Century*, p. 272; see also Peter Pomerantsev, *This Is Not Propaganda: Adventures in the War Against Reality*, Faber & Faber, London, 2019.

7   Milan Kundera, *The Unbearable Lightness of Being*, Harper & Row Publishers, 1984, p. 99.

8   Damian Tambini, Danilo Leonardi and Chris Marsden, 'The privatisation of censorship: self regulation and freedom of expression' in Damian Tambini, Danilo Leonardi, and Chris Marsden (eds), *Codifying cyberspace: communications self-regulation in the age of internet convergence*, Routledge, Abingdon, 2008; Daphne Keller, 'Six Constitutional Hurdles For Platform Speech Regulation', *Stanford Center for Internet and Society Blog*, 22 January 2021 <http://cyberlaw. stanford.edu/blog/2021/01/six-constitutional-hurdles-platform-speech-regulation-0> (accessed 20 August 2021).

9   Virginia v. Black 538 U.S. 343.

10 Emily B. Laidlaw, *Regulating Speech in Cyberspace: Gatekeepers, Human Rights and Corporate Responsibility*, Cambridge University Press, Cambridge, 2017.

11 European Court of Human Rights, *Research Report: Positive obligations on member States under Article 10 to protect journalists and prevent impunity*, Council of Europe and the European Court of Human Rights, December 2011.

12 Cass Sunstein, *Democracy and the Problem of Free Speech*, Free Press, New York, 1995, xiv.

13 Laura Weinrib, 'Rethinking the Myth of the Modern First Amendment', in Bollinger and Stone (eds), *Free Speech Century*, p. 53.

14 Red Lion Broadcasting Co. v FCC 395 US 367 (1969).

15 Tim Wu, 'Is the First Amendment Obsolete?'

16 See the discussion of *Packingham v. North Carolina* in Heather Whitney, 'Search Engines, Social Media, and the Editorial Analogy', in David E. Pozen (ed.), *The Perilous Public Square: Structural Threats to Free Expression Today*, Columbia University Press, New York, 2020, p. 117.

17 See Alexander Meiklejohn, *Free Speech and its Relation to Self-Government*, The Lawbook Exchange, Clark, NJ, 2014.

18 Sunstein, *Democracy*, p. 19.

19 Meiklejohn, *Free Speech*, p. 16.

20 Ibid., pp. 25–26.

21 R v *Shayler* [2003] 1 AC 247, per Lord Bingham at [21].

22 See e.g. Daphne Keller, 'Amplification and its Discontents', Knight First Amendment Institute, 8 June 2021 <https://knightcolumbia.org/

content/amplification-and-its-discontents> (accessed 12 November 2012); 'Six Constitutional Hurdles'.

CHAPTER THIRTY-NINE

1   John Milton, *Areopagitica and Other Writings*, Penguin, London, 2014, p. 136.

2   Jillian C. York and Ethan Zuckerman, 'Moderating the Public Sphere', in Rikke Frank Jørgensen (ed.), *Human Rights in the Age of Platforms*, MIT Press, Cambridge, MA, 2019, p. 145. See also Philip M. Napoli, *Social Media and the Public Interest: Media Regulation in the Disinformation Age*, Columbia University Press, New York, 2019, pp. 197–198.

3   See Eric Goldman and Jess Miers, 'Regulating Internet Services by Size' (May 2021), *CPI Antitrust Chronicle*, Santa Clara University Legal Studies Research Paper.

4   Draft Digital Services Act.

5   Draft Online Safety Bill.

6   See e.g. Department for Digital, Culture, Media and Sport, *Online Harms White Paper* (White Paper, Cm 57, 2019); William Perrin, 'Regulation, misinformation and COVID19', *Carnegie UK*, 30 April 2020 <https://www.carnegieuktrust.org.uk/blog-posts/reg ulation-misinformation-and-covid19/> (accessed 20 August 2021); Monika Bickert; *Online Content Regulation: Charting a Way Forward*, Facebook, 2020; draft Digital Services Act Articles 26 and 27. See generally the work of William Perrin and Lorna Woods.

7   Richard Allan, 'Harm Reduction Plans', *regulate.tech*, 29 December 2020 <https://www.regulate.tech/harm-reduction-plans-29th-dec-2020/> (accessed 20 August 2021).

8   See e.g. Nathaniel Persily and Joshua A. Tucker (eds), *Social Media and Democracy: The State of the Field and Prospects for Reform*, Cambridge University Press, New York, 2020.

9   I believe that the idea originates with the American journalist Ezra Klein.

10 A considerable amount of content moderation is already done by external providers: Jem Bartholomew, 'The rise of the truth industry', *New Humanist*, 28 June 2021 <https://newhumanist.org. uk/articles/5818/the-rise-of-the-truth-industry> (accessed 3 October 2021); Adam Satariano and Mike Isaac, 'The Silent Partner Cleaning Up Facebook for $500 Million a Year', *The New York Times*, 31 August 2021 <https://www.nytimes.com/2021/08/31/technol ogy/facebook-accenture-content-moderation.html> (accessed 3

October 2021). It would not be a bad thing for the market for such services to continue to grow.

11   See Monika Bickert, 'European Court Ruling Raises Questions About Policing Speech', *Facebook*, 14 October 2019 <https://about.fb.com/news/2019/10/european-court-ruling-raises-questions-about-policing-speech/> (accessed 20 August 2021); Perrin, 'Regulation'.

12   Neil Gunningham, 'Enforcement and Compliance Strategies', in Robert Baldwin, Martin Cave and Martin Lodge (eds), *The Oxford Handbook of Regulation*, Oxford University Press, Oxford, 2013, p. 136.

13   See the EU's draft Digital Services Act, Articles 12, 13, 23, 24, 29, 30, 31, 33.

14   Daphne Keller and Paddy Leerssen, 'Facts and Where to Find Them: Empirical Research on Internet Platforms and Content Moderation', in Persily and Tucker (eds), *Social Media and Democracy*, pp. 232–233.

15   As Daphne Keller has observed, requiring platforms to repost content which they erroneously removed may be problematic under the First Amendment, as it arguably amounts to forced speech: Daphne Keller, 'Who Do You Sue? State and Platform Hybrid Power Over Online Speech', *Aegis Series Paper*, Hoover Institution, No. 1902 (2019). Platforms would, in any event, have to be immune from liabilities arising out of content they were forced to host.

16   On conditional liability, see generally the contributions in David E. Pozen (ed.), *The Perilous Public Square: Structural Threats to Free Expression Today*, Columbia University Press, New York, 2020, especially Danielle Keats Citron, 'Section 230's Challenge to Civil Rights and Civil Liberties'; Danielle Keats Citron and Benjamin Wittes, 'The Internet Will Not Break: Denying Bad Samaritans §230 Immunity', *Fordham Law Review*, Vol. 86, No. 2 (2017), 401–423; Niall Ferguson, 'What Is To Be Done? Safeguarding Democratic Governance In The Age Of Network Platforms', Hoover Institution, *Governance in an Emerging World*, Fall Series, No. 318 (2018); Karen Kornbluh and Ellen P. Goodman, 'Safeguarding Digital Democracy: Digital Innovation and Democracy Initiative Roadmap', Washington DC, The German Marshall Fund of the United States, 2020.

17   See generally the work of Tim Wu.

CONCLUSION: THE DIGITAL REPUBLIC

1   Stephen L. Elkin, *Reconstructing the Commercial Republic: Constitutional Design after Madison*, Chicago, University of Chicago Press, 2006. He was actually speaking of a 'constitution'.

2   See e.g. Yong Suk Lee et al., 'AI regulation and firm behaviour', *VoxEU. org*, 14 December 2019 <https://voxeu.org/article/ai-regulation-and-firm-behaviour> (accessed 20 August 2021).

3   Tim Büthe and Walter Mattli, 'International Standards and Standard-Setting Bodies', in Davin Coen, Wyn Grant and Graham Wilson (eds), *The Oxford Handbook of Business and Government*, Oxford University Press, Oxford, 2011, p. 442.

4   Diane Coyle, 'Three Cheers for Regulation', *Project Syndicate*, 17 July 2018 <https://www.project-syndicate.org/commentary/positive-effects-market-regulation-by-diane-coyle-2018-07> (accessed 20 August 2021).

5   Büthe and Mattli, 'International Standards', p. 442.

6   Coyle, 'Three Cheers for Regulation'.

7   Mireille Hildebrandt, *Law for Computer Scientists and Other Folk*, Oxford University Press, Oxford, 2020, p. 136.

8   Coyle, 'Three Cheers for Regulation'.

9   Julie E. Cohen, *Between Truth and Power: The Legal Constructions of Informational Capitalism*, Oxford University Press, Oxford, 2019, p. 92.

10 Tarleton Gillespie, *Custodians of the Internet: Platforms, Content Moderation, and the Hidden Decisions that Shape Social Media*, Yale University Press, New Haven, 2018, p. 108; Cohen, *Between Truth and Power*, p. 190.

Bibliography

Abbott, Ryan, 'The Reasonable Computer: Disrupting the Paradigm of Tort Liability', *George Washington Law Review*, Vol. 86, No. 1 (2018), 1–45.

Abid, Abubakar, Maheen Farooqi and James Zou, 'Persistent Anti-Muslim Bias in Large Language Models', *arXiv*: 2101.05783v1 (2021).

Ada Lovelace Institute, *The Citizens' Biometrics Council*, March 2021 <https://www.adalovelaceinstitute.org/report/citizens-biometrics-council> (accessed 3 October 2021).

Addis, Donna Rose, Alana T. Wong and Daniel L. Schacter, 'Remembering the Past and Imagining the Future: Common and Distinct Neural Substrates During Event Construction and Elaboration', *Neuropsychologia* 45, No. 7 (2007), 1363–1367.

Ajunwa, Ifeoma, 'Algorithms at Work: Productivity Monitoring Applications and Wearable Technology as the New Data-Centric Research Agenda for Employment and Labor Law', *Saint Louis University Law Journal*, Vol. 63, No. 1, Article 4 (2018), 21–54.

Alba, Davey, 'On Facebook, Misinformation Is More Popular Now Than in 2016', *The New York Times*, 12 October 2020 <https://www.nytimes.com/2020/10/12/technology/on-facebook-misinformation-is-more-popular-now-than-in-2016.html> (accessed 3 October 2021).

—, 'Facial Recognition Moves Into a New Front: Schools', *The New York Times*, 6 February 2020 <https://www.nytimes.com/2020/02/06/business/facial-recognition-schools.html> (accessed 20 August 2021).

—, 'Fake "Likes" Remain just a Few Dollars Away, Researchers Say', *The New York Times*, 6 December 2019 <https://www.nytimes.com/2019/12/06/technology/fake-social-media-manipulation.html> (accessed 20 August 2021).

Ali, Muhammad et al., 'Ad Delivery Algorithms: The Hidden Arbiters of Political Messaging', *arXiv*: 1912.04255v3 (2019).

Allan, Richard, 'Harm Reduction Plans', *regulate.tech*, 29 December 2020 <https://www.regulate.tech/harm-reduction-plans-29th-dec-2020/> (accessed 20 August 2021).

Allenby, Braden R. and Daniel Sarewitz, *The Techno-Human Condition*, MIT Press, Cambridge, MA, 2011.

Alpaydin, Ethem, *Machine Learning: the New AI*, MIT Press, Cambridge, MA, 2016.

Alston, Philip, 'Landmark ruling by Dutch court stops government attempts to spy on the poor – UN expert', *Office of the United Nations High Commissioner for Human Rights*, 5 February 2020 <https://www.ohchr.org/EN/NewsEvents/Pages/DisplayNews.aspx?NewsID=25522&LangID=E> (accessed 20 August 2021).

Amoore, Louise, 'Why "Ditch the algorithm" is the future of political protest', *The Guardian*, 19 August 2020 <https://www.theguardian.com/commentisfree/2020/aug/19/ditch-the-algorithm-generation-students-a-levels-politics> (accessed 20 August 2021).

Angwin, Julia, *Dragnet Nation: A Quest for Privacy, Security, and Freedom in a World of Relentless Surveillance*, St. Martin's Press, New York, 2014.

Applebaum, Binyamin, *The Economists' Hour: False Prophets, Free Markets, and the Fracture of Society*, Little, Brown and Company, New York, 2019.

Appleby, Joyce, *Capitalism and a New Social Order: The Republican Vision of the 1790s*, New York University Press, New York, 1984.

Arendt, Hannah, *The Origins of Totalitarianism*, Penguin Classics, New York, 2017.

—, *On Revolution*, Penguin Classics, New York, 2006.

—, *The Human Condition*, University of Chicago Press, Chicago, 1998.

Arnold, Matthew, 'Dover Beach' <https://www.poetryfoundation.org/poems/43588/dover-beach> (accessed 25 September 2021).

Arnold, Paul, 'How enforcement of antitrust law can enliven American innovation', *San Francisco Chronicle*, 14 August 2020 <https://www.sfchronicle.com/opinion/openforum/article/How-enforcement-of-antitrust-law-can-enliven-15482922.php> (accessed 20 August 2021).

Artyushina, Anna, 'The EU is launching a market for personal data. Here's what that means for privacy', *MIT Technology Review*, 11 August 2020 <https://www.technologyreview.com/2020/08/11/1006555/eu-data-trust-trusts-project-privacy-policy-opinion/> (accessed 20 August 2021).

Atiyah, P. S., *The Damages Lottery*, Hart Publishing, Oxford, 1997.

—, *The Rise and Fall of Freedom of Contract*, Oxford University Press, Oxford, 1979.

Azevedo, Mary Ann, 'Untapped Opportunity: Minority Founders Still Being Overlooked', *crunchbase*, 27 February 2019 <https://news.crunchbase.com/news/untapped-opportunity-minority-found ers-still-being-overlooked/> (accessed 20 August 2021).

Bago, Bence, David G. Rand and Gordon Pennycook, 'Fake News, Fast and Slow: Deliberation Reduces Belief in False (but Not True) News Headlines', *Journal of Experimental Psychology: General*, Vol. 149, No. 8 (2020), 1608–1613.

Bailyn, Bernard, *The Ideological Origins of the American Revolution*, Harvard University Press, Cambridge, MA, 1992.

Baker, Tom and Benedict G. C. Dellaert, 'Regulating Robo Advice Across the Financial Services Industry', *University of Pennsylvania Law School Faculty Scholarship*, 2018, 170.0.

Baldwin, Robert, Martin Cave and Martin Lodge (eds), *The Oxford Handbook of Regulation*, Oxford University Press, Oxford, 2013.

—, *Understanding Regulation:Theory, Strategy, and Practice*, Oxford University Press, Oxford, 2012.

Balkin, Jack M., 'Information Fiduciaries in the Digital Age', *Balkinization*, 5 March 2014 <https://balkin.blogspot.com/2014/03/information-fiduciaries-in-digital-age.html> (accessed 20 August 2021).

Balkin, Jack M. and Jonathan Zittrain, 'A Grand Bargain to Make Tech Companies Trustworthy', *The Atlantic*, 3 October 2016 <https://www.theatlantic.com/technology/archive/2016/10/information-fiduci ary/502346/> (accessed 20 August 2021).

Ball, James, *The System:Who Owns the Internet, and How it Owns Us*, Bloomsbury, London, 2020.

Ball, Kirstie and Lauren Snider (eds), *The Surveillance-Industrial Complex: A political economy of surveillance*, Routledge, Abingdon, 2013.

Barber, Gregory, 'Deepfakes Are Getting Better, But They're Still Easy to Spot', *Wired*, 26 May 2019 <https://www.wired.com/story/deepfa kes-getting-better-theyre-easy-spot/> (accessed 20 August 2021).

Barocas, Solon and Andrew D. Selbst, 'Big Data's Disparate Impact', *California Law Review*, Vol. 104, No. 3 (2016), 671–732.

Barfield, Woodrow and Ugo Pagallo (eds), *Research Handbook on the Law of Artificial Intelligence*, Edward Elgar Publishing, Cheltenham, 2018.

Barker, Kit, Karen Fairweather and Ross Grantham (eds), *Private Law in the 21st Century*, Hart Publishing, Oxford, 2017.

Bartholomew, Jem, 'The rise of the truth industry', *New Humanist*, 28 June 2021 <https://newhumanist.org.uk/articles/5818/the-rise-of-the-truth-industry> (accessed 3 October 2021).

Bartlett, Jamie, *The People Vs Tech: How the internet is killing democracy (and how we save it)*, Ebury Press, London, 2018.

Basu, Tanya, 'How a ban on pro-Trump patterns unraveled the online knitting world', MIT *Technology Review*, 6 March 2020 <https://www.technologyreview.com/2020/03/06/905472/ravelry-ban-on-pro-trump-patterns-unraveled-the-online-knitting-world-censors hip-free/> (accessed 20 August 2021).

BBC, 'Russian disinformation "ongoing problem" says FBI chief', 6 February 2020 <https://www.bbc.co.uk/news/technology-51399 568> (accessed 20 August 2021).

—, 'Apple "sorry" that workers listened to Siri voice recordings', 28 August 2019 <https://www.bbc.co.uk/news/technology-49502 292> (accessed 20 August 2021).

Beer, David, *The Quirks of Digital Culture*, Emerald Publishing, Bingley, 2019.

Beer, Samuel H., *To Make a Nation:The Rediscovery of American Federalism*, Harvard University Press, Cambridge, MA, 1993.

Beitz, Charles R., *The Idea of Human Rights*, Oxford University Press, Oxford, 2013.

Belli, Lyca and Jamila Venturini, 'Private ordering and the rise of terms of service as cyber-regulation', *Internet Policy Review*, Vol. 5, No. 4 (2016).

Belong.co <https://belong.co/hireplus/> (accessed 20 August 2021).

Benjamin, Ruha, *Race After Technology: Abolitionist Tools for the New Jim Code*, Policy Press, Cambridge, 2019.

Benaich, Nathan and Ian Hogarth, 'State of AI Report', 28 June 2019 <https://www.stateof.ai/2019> (accessed 20 August 2021).

Benkler, Yochai, Robert Faris and Hal Roberts, *Network Propaganda: Manipulation, Disinformation, and Radicalization in American Politics*, Oxford University Press, New York, 2018.

Berg, Tobias et al., 'On the Rise of FinTechs – Credit Scoring using Digital Footprints' (2018) National Bureau of Economic Research Working Paper No. 24551.

Bergen, Mark, 'YouTube Executives Ignored Warnings, Letting Toxic Videos Run Rampant', *Bloomberg*, 2 April 2019 <https://www.bloomberg.com/news/features/2019-04-02/youtube-executives-ignored-warnings-letting-toxic-videos-run-rampant> (accessed 20 August 2021).

Bernholz, Lucy, Hélène Landemore and Rob Reich (eds), *Digital Technology and Democratic Theory*, University of Chicago Press, London, 2021.

Besson, Samantha and John Tasioulas (eds), *The Philosophy of International Law*, Oxford University Press, Oxford, 2010.

Besson, Samantha and José Luis Martí (eds), *Legal Republicanism: National and International Perspectives*, Oxford University Press, Oxford, 2009.

Bhatt, Umang et al., 'Explainable Machine Learning in Deployment', *arXiv*: 1909.06342 (2020).

Bickert, Monika, *Online Content Regulation: Charting a Way Forward*, Facebook, 2020.

—, 'European Court Ruling Raises Questions About Policing Speech', *Facebook*, 14 October 2019 <https://about.fb.com/news/2019/10/european-court-ruling-raises-questions-about-policing-speech/> (accessed 20 August 2021).

Biddle, Sam, Paulo Victor Ribeiro and Tatiana Dias, 'Invisible Censorship: TikTok Told Moderators to Suppress Posts by "Ugly" People and the Poor to Attract New Users', *The Intercept*, 16 March 2020 <https://theintercept.com/2020/03/16/tiktok-app-moderators-users-discrimination/> (accessed 20 August 2021).

Bietti, Elettra, 'Consent as a Free Pass: Platform Power and the Limits of the Informational Turn', *Pace Law Review*, Vol. 40, No. 1 (2020), 307–397.

Bijker, Wiebe E., Thomas P. Hughes and Trevor Pinch (eds), *The Social Construction of Technological Systems: New Directions in the Sociology and History of Technology*, MIT Press, Cambridge, MA, 2012.

Bimber, Bruce, *The Politics of Expertise in Congress: the Rise and Fall of the Office of Technology Assessment*, State University of New York Press, Albany, NY, 1996.

Birks, Peter, *The Roman Law of Obligations*, Oxford University Press, Oxford, 2014.

Blair, Tony, *A Journey*, Cornerstone, London, 2010.

Blavatskyy, Pavlo, 'Obesity of politicians and corruption in post-Soviet countries', *Economics of Transition and Institutional Change*, Vol. 29, No. 2 (2020), 343–356.

Bloch-Wehba, Hannah, 'Automation in Moderation' (2020) Texas A&M University School of Law Legal Studies Research Paper Series, Research Paper No. 20-33.

Bock, Gisela, Quentin Skinner and Maurizio Viroli (eds), *Machiavelli and Republicanism*, Cambridge University Press, Cambridge, 1993.

Bogost, Ian and Alexis C. Madrigal, 'How Facebook Works For Trump', *The Atlantic*, 17 April 2020 <https://www.theatlantic.com/technology/archive/2020/04/how-facebooks-ad-technology-helps-trump-win/606403/> (accessed 20 August 2021).

Bohman, James and William Rehg (eds), *Deliberative Democracy: Essays on Reason and Politics*, MIT Press, Cambridge, MA, 2002.

Bollinger, Lee C. and Geoffrey R. Stone, *The Free Speech Century*, Oxford University Press, New York, 2019.

Bourke, Richard and Raymond Geuss (eds), *Political Judgement: Essays for John Dunn*, Cambridge University Press, Cambridge, 2009.

Bowles, Samuel, *Microeconomics: Behaviour, Institutions, and Evolution*, Princeton University Press, Princeton, 2004.

Bowles, Samuel and Herbert Gintis, *Democracy and Capitalism: Property, Community, and the Contradictions of Modern Social Thought*, Basic Books, London, 1986.

Boyle, James, *The Public Domain: Enclosing the Commons of the Mind*, Yale University Press, New Haven, 2008.

Bradner, Eric and Sarah Mucha, 'Biden campaign launches petition lambasting Facebook over refusal to remove political misinformation', CNN, 11 June 2020 <https://edition.cnn.com/2020/06/11/politics/joe-biden-facebook-open-letter/index.html> (accessed 19 August 2021).

Braithwaite, John and Philip Pettit, *Not Just Desserts: A Republican Theory of Criminal Justice*, Oxford University Press, Oxford, 1992.

Bram, Uri and Martin Schmalz, *The Business of Big Data: How to Create Lasting Value in the Age of AI*, Amazon, UK, 2019.

Broad, Ellen, *Made by Humans*, Melbourne University Press, Carlton, 2018.

Brockman, John (ed.), *Possible Minds: 25 Ways of Looking at AI*, 1st edn, Penguin Press, New York, 2019.

Broussard, Meredith, *Artificial Unintelligence: How Computers Misunderstand the World*, MIT Press, Cambridge, MA, 2019.

Brownsword, Roger, *Law, Technology and Society: Re-Imagining the Regulatory Environment*, Routledge, Abingdon, 2019.

Brownsword, Roger and Morag Goodwin, *Law and the Technologies of the Twenty-First Century: Texts and Materials*, Cambridge University Press, Cambridge, 2012.

Brownsword, Roger, Eloise Scotford and Karen Yeung (eds), *The Oxford Handbook of Law, Regulation, and Technology*, Oxford University Press, Oxford, 2017.

Brundage, Miles et al., 'Toward Trustworthy AI Development: Mechanisms for Supporting Verifiable Claims', arXiv: 2004.07213v2 (2020).

Bruns, Axel, *Are Filter Bubbles Real?* Polity Press, Cambridge, 2019.

Bucher, Taina, *If…Then: Algorithmic Power and Politics*, Oxford University Press, Oxford, 2018.

Burbank, Stephen B., Sean Farhang and Herbert Kritzer, 'Private Enforcement' (2013) University of Pennsylvania Law School Public Law Research Paper No. 13–22.

Busch, Lawrence, *Standards: Recipes for Reality*, MIT Press, Cambridge, MA, 2013.

Caliskan-Islam, Aylin, Joanna J. Bryson and Arvind Narayanan, 'Semantics derived automatically from language corpora necessarily contain human biases', arXiv: 1608.07187v2 (2016).

Callery, James and Jacqui Goddard, 'Most-clicked link on Facebook spread doubt about Covid vaccine', *The Times*, 23 August 2021 <https://www.thetimes.co.uk/article/most-clicked-link-on-facebook-spread-doubt-about-covid-vaccine-flknpp9n5> (accessed 3 October 2021).

Calo, Ryan, 'Against Notice Skepticism in Privacy (and Elsewhere)', *Notre Dame Law Review*, Vol. 87, No. 3 (2012), 1027–1072.

Canadian Citizens' Assembly on Democratic Express, *Canadian Citizens' Assembly on Democratic Express: Recommendations to strengthen Canada's response to new digital technologies and reduce the harm caused by their misuse*, Public Policy Forum, Ottawa, 2021.

Cane, Peter, *Administrative Law*, 5th edn, Oxford University Press, Oxford, 2011.

Carr, Nicholas, *The Glass Cage: Who Needs Humans Anyway?* Vintage, London, 2015.

Case, R. H., 'The Market for Goods and the Market for Ideas', *The American Economic Review*, Vol. 64, No. 2 (May 1974), 381–391.

Castells, Manuel, 'Communication, Power and Counter-power in the Network Society', *International Journal of Communication*, Vol. 1 (2007), 238–266.

Cellan-Jones, Rory, 'Parler social network sues Amazon for pulling support', BBC, 11 January 2021 <https://www.bbc.co.uk/news/technology-55615214> (accessed 20 August 2021).

Center for Humane Technology, 'Ledger of Harms', June 2021 <https://ledger.humanetech.com> (accessed 20 August 2021).

Centre for Data Ethics and Innovation, *Snapshot Paper – Deepfakes and Audiovisual Disinformation*, 12 September 2019 <https://www.gov.uk/

government/publications/cdei-publihes-its-first-series-of-three-snapshot-papers-ethical-issues-in-ai/snapshot-paper-deepfakes-and-audiovisual-disinformation> (accessed 20 August 2021).

Česnulaitytė, Ieva, 'Models of representative deliberative processes', in OECD (ed.), *Innovative Citizen Participation and New Democratic Institutions*, OECD Publishing, Paris, 2020.

Cheney-Lippold, John, *We Are Data: Algorithms and the Making of our Digital Selves*, New York University Press, New York, 2017.

Chernow, Ron, *Grant*, Head of Zeus, London, 2017.

Chesney, Robert and Danielle Keats Citron, 'Deep Fakes: A Looming Challenge for Privacy, Democracy, and National Security', 107 California Law Review 1753 (2019).

Chilson, Neil and Casey Mattox, '[The] Breakup Speech: Can Antitrust Fix the Relationship Between Platforms and Free Speech Values?' *Knight First Amendment Institute*, 5 March 2020 <https://knightcolumbia.org/content/the-breakup-speech-can-antitrust-fix-the-relationship-between-platforms-and-free-speech-values> (accessed 20 August 2021).

Chung, Jane, 'Big Tech, Big Cash: Washington's New Power Players', Public Citizen, March 2021 <https://www.citizen.org/article/big-tech-lobbying-update/> (accessed 20 August 2021).

Chwalisz, Claudia, 'Introduction: Deliberation and new forms of governance', in OECD (ed.), *Innovative Citizen Participation and New Democratic Institutions*, OECD Publishing, Paris, 2020.

Cicero, Marcus Tullius, *The Republic and The Laws*, Oxford University Press, Oxford, 1998.

Ciepley, David, 'Beyond Public and Private: Toward a Political Theory of the Corporation', *American Political Science Review*, Vol. 107, No. 1 (February 2013), 139–158.

Cihon, Peter, Matthijs M. Maas and Luke Kemp, 'Should Artificial Intelligence Governance be Centralised? Design Lessons from History', *Proceedings of the AAAI/ACM Conference on AI, Ethics and Society* (2020), 228–234.

Citron, Danielle Keats and Mary Anne Franks, 'The Internet as a Speech Machine and Other Myths Confounding Section 230 Reform', *University of Chicago Legal Forum*, Vol. 2020, Article 3 (2020), 45–75.

Citron, Danielle Keats and Benjamin Wittes, 'The Internet Will Not Break: Denying Bad Samaritans §230 Immunity', *Fordham Law Review*, Vol. 86, No. 2 (2017), 401–423.

Coeckelbergh, Mark, *AI Ethics*, MIT Press, Cambridge, MA, 2020.

Coen, David, Wyn Grant and Graham Wilson (eds), *The Oxford Handbook of Business and Government*, Oxford University Press, Oxford, 2011.

Coglianese, Cary (ed.), *Achieving Regulatory Excellence*, Brookings Institution Press, Washington DC, 2017.

—, *Regulatory Breakdown: The Crisis of Confidence in U.S. Regulation*, University of Pennsylvania Press, Philadelphia, 2012.

Cohen, Jared and Richard Fontaine, 'Uniting the Techno-Democracies', *Foreign Affairs*, November/December 2020 <https://www.foreignaffa irs.com/articles/united-states/2020-10-13/uniting-techno-demo cracies> (accessed 20 August 2021).

Cohen, Julie E., 'How (Not) to Write a Privacy Law', *Knight First Amendment Institute*, 23 March 2021 <https://knightcolumbia.org/content/ how-not-to-write-a-privacy-law> (accessed 20 August 2021).

—, *Between Truth and Power: The Legal Constructions of Informational Capitalism*, Oxford University Press, Oxford, 2019.

—, *Configuring the Networked Self: Law, Code, and the Play of Everyday Practice*, Yale University Press, New Haven, 2012.

Coleman, James S., *Power and the Structure of Society*, W. W. Norton & Company, New York, 1974.

Collins, Hugh, *Regulating Contracts*, Oxford University Press, Oxford, 2002.

Committee of Experts on Internet Intermediaries, *Algorithms and Human Rights: Study on the human rights dimensions of automated data processing techniques and possible regulatory implications*, DGI (2017)12, Council of Europe, Strasbourg, 2017.

Constant, Benjamin, *The Liberty of Ancients Compared with that of Moderns* [1816], Liberty Fund Inc E-Book.

Cook, Katy, *The Psychology of Silicon Valley: Ethical Threats and Emotional Unintelligence in the Tech Industry*, Palgrave Macmillan, London, 2020.

Copeland, David A., *The Idea of a Free Press: The Enlightenment and Its Unruly Legacy*, Northwestern University Press, Evanston, 2006.

Cox, Joseph, 'Zoom iOS App Sends Data to Facebook Even if You Don't Have a Facebook Account', *Vice*, 26 March 2020 <https://www. vice.com/en/article/k7e599/zoom-ios-app-sends-data-to-faceb ook-even-if-you-dont-have-a-facebook-account> (accessed 20 August 2021).

Cox, Kate, 'Unredacted suit shows Google's own engineers confused by privacy settings', *Ars Technica*, 25 August 2020 <https://arst echnica.com/tech-policy/2020/08/unredacted-suit-shows-goog les-own-engineers-confused-by-privacy-settings/> (accessed 20 August 2021).

Coyle, Diane, *Markets, State, and People: Economics for Public Policy*, Princeton University Press, Princeton, 2020.

—, 'Three Cheers for Regulation', *Project Syndicate*, 17 July 2018 <https://www.project-syndicate.org/commentary/positive-effects-market-regulation-by-diane-coyle-2018-07> (accessed 20 August 2021).

Crane, Daniel A., 'A Premature Postmortem on the Chicago School of Antitrust', *Business History Review*, Vol. 93, No. 4: New Perspectives in Regulatory History (2019), 759–776.

—, 'Antitrust's Unconventional Politics' (2018) University of Michigan Law & Economics Working Paper No. 153.

—, 'The Tempting of Antitrust: Robert Bork and the Goals of Antitrust Policy', *Antitrust Law Journal*, Vol. 79, No. 3 (2014), 835–853.

Crémer, Jacques, Yves-Alexandre de Montjoye and Heike Schweitzer, 'Competition policy for the digital era: Final Report', European Commission, Luxembourg, 2019.

Croft, Jane, 'Chatbots join the legal conservation', *Financial Times*, 7 June 2018 <https://www.ft.com/content/0eabcf44-4c83-11e8-97e4-13afc22d86d4> (accessed 20 August 2021).

Crootof, Rebecca, 'Accountability for the Internet of Torts', *The Law and Political Economy Project*, 17 July 2018 <https://lpeproject.org/blog/accountability-for-the-internet-of-torts/> (accessed 20 August 2021).

Cuthbertson, Anthony, 'Google Admits Workers Listen to Private Audio Recordings From Google Home Smart Speakers', *The Independent*, 11 July 2019 <https://www.independent.co.uk/life-style/gadgets-and-tech/news/google-home-smart-speaker-audio-recordings-privacy-voice-spy-a9000616.html> (accessed 20 August 2021).

—, 'Who Controls the Internet? Facebook and Google Dominance Could Cause the 'Death of the Web'', *Newsweek*, 11 February 2017 <https://www.newsweek.com/facebook-google-internet-traffic-net-neutrality-monopoly-699286> (accessed 20 August 2021).

Dagger, Richard, 'Neo-republicanism and the civic economy', *Politics, Philosophy & Economics*, Vol. 5, No 2. (2006), 151–173.

Das, Shanti and Geoff White, 'Instagram sends paedophiles to accounts of children as young as 11', *The Sunday Times*, 1 December 2019 <https://www.thetimes.co.uk/article/instagram-sends-predators-to-accounts-of-children-as-young-as-11-j2gn5hq83> (accessed 20 August 2021).

Daskal, Jennifer C., 'Borders and Bits', *Vanderbilt Law Review*, Vol. 71, No. 1 (2018), 179–240.

Data Dividend Project, 2021 <https://www.datadividendproject.com> (accessed 20 August 2021).

Dayen, David, 'Fiona, Apple, and Amazon: How Big Tech Pays to Win the Battle of Ideas', *The American Prospect*, 20 July 2020 <https://prospect.org/power/fiona-apple-amazon-how-big-tech-pays-to-win-battle-ideas/> (accessed 20 August 2021).

—, 'The Final Battle in Big Tech's War to Dominate Your World', *The New Republic*, 8 April 2019 <https://newrepublic.com/article/153515/final-battle-big-techs-war-dominate-world> (accessed 20 August 2021).

Dearden, Lizzie, 'Iran's Supreme Leader claims gender equality is "Zionist plot" aiming to corrupt role of women in society', *The Independent*, 21 March 2017 <https://www.independent.co.uk/news/world/middle-east/iran-supreme-leader-ayatollah-khamenei-gender-equality-women-zionist-plot-society-role-islamic-leader-theocracy-a7641041.html> (accessed 20 August 2021).

deepsense.ai <https://deepsense.ai> (accessed 20 August 2021).

De Hert, Paul et al., 'The right to data portability in the GDPR: Towards user-centric interoperability of digital services', *Computer Law & Security Review*, Vol. 34, No. 2 (2018), 193–203.

Deighton, Katie, 'Tech Firms Train Voice Assistants to Understand Atypical Speech', *The Wall Street Journal*, 24 February 2021 <https://www.wsj.com/articles/tech-firms-train-voice-assistants-to-understand-atypical-speech-11614186019> (accessed 20 August 2021).

Dekker, Sidney, *Drift into Failure: From Hunting Broken Components to Understanding Complex Systems*, CRC Press, Boca Raton, FL, 2011.

Delrahim, Makan, '"I'm Free": Platforms and Antitrust Enforcement in the Zero-Price Economy', Department of Justice, Boulder, CO, 2019.

Denardis, Laura, *The Internet in Everything: Freedom and Security in a World with No Off Switch*, Yale University Press, New Haven, 2020.

Department for Digital, Culture, Media and Sport, *Online Harms White Paper* (White Paper, Cm 57, 2019).

Desai, Deven R. and Joshua A. Kroll, 'Trust But Verify: A Guide to Algorithms and the Law', *Harvard Journal of Law & Technology*, Vol. 31, No. 1 (2017), 1–64.

Desmaris, Sacha, Pierre Dubreuil and Benoît Loutrel, 'Creating a French Framework to Make Social Media Platforms More Accountable: Acting in France with a European Vision', Final Mission Report Submitted to the French Secretary of State for Digital Affairs (May 2019).

Desrosières, Alain, *The Politics of Large Numbers: A History of Statistical Reasoning*, Harvard University Press, Cambridge, MA, 1998.

Develle, Yuji, 'Why we cannot trust Big Tech to be apolitical', *Wonk Bridge*, 4 May 2020 <https://medium.com/wonk-bridge/why-we-cannot-trust-big-tech-to-be-apolitical-f031af9386cf> (accessed 20 August 2021).

Dewan, Shaila and Serge F. Kovaleski, 'Thousands of Complaints Do Little to Change Police Ways', *The New York Times*, 8 June 2020 <https://www.nytimes.com/2020/05/30/us/derek-chauvin-george-floyd.html> (accessed 20 August 2021).

Dewey, John, *The Public and its Problems: An Essay in Political Inquiry*, Athens, Press, Athens, OH, 2016.

Dickson, Ben, 'Your Next Car Will Be Watching You More Than It's Watching the Road', *Gizmodo*, 28 November 2019 <https://gizmodo.com/your-next-car-will-be-watching-you-more-than-its-watchi-184 0055386> (accessed 20 August 2021).

Digital Competition Expert Panel, 'Unlocking digital competition', Crown, 2019 <https://assets.publishing.service.gov.uk/governm ent/uploads/system/uploads/attachment_data/file/785547/ unlocking_digital_competition_furman_review_web.pdf> (accessed 20 August 2021).

Dodds, Io and Mike Wright, 'Instagram removed nearly 10,000 suicide and self-harm images per day after the Molly Russell scandal', *The Telegraph*, 13 November 2019 <https://www.telegraph.co.uk/tec hnology/2019/11/13/instagram-removed-nearly-10000-suic ide-self-harm-images-day/> (accessed 20 August 2021).

Dormehl, Luke, *The Formula: How Algorithms Solve All our Problems...and Create More*, WH Allen, London, 2014.

Doshi-Velez, Finale et al., 'Accountability of AI Under the Law: The Role of Explanation', arXiv: 1711.01134 (2017).

Dryzek, John S. et al., 'The crisis of democracy and the science of deliberation', *Science*, Vol. 363, No. 6432 (2019), 1144–1146.

Dryzek, John S., *Democracy in Capitalist Times: Ideas, Limits, and Struggles*, Oxford University Press, New York, 1996.

Dubois, Elizabeth and Grant Blank, 'The echo chamber is overstated: the moderating effect of political interest and diverse media', *Information, Communication & Society*, Vol. 21, No. 5: Communication, Information Technologies, and Media Sociology (CITAMS) Special Issue (2018), 729–745.

Duff, R. A. and Stuart P. Green (eds), *Philosophical Foundations of Criminal Law*, Oxford University Press, Oxford, 2013.

Duffy, Clare, 'Marc Benioff says it's time to break up Facebook', CNN, 17 October 2019 <https://edition.cnn.com/2019/10/16/tech/salesforce-marc-benioff-break-up-facebook-boss-files/index.html> (accessed 20 August 2021).

D'Urso, Joey and Alex Wickham, 'YouTube Is Letting Millions Of People Watch Videos Promoting Misinformation About The Coronavirus', *BuzzFeed News*, 19 March 2020 <https://www.buzzfeed.com/joeydurso/youtube-coronavirus-misinformation> (accessed 20 August 2021).

Dyzenhaus, David and Malcolm Thorburn (eds), *Philosophical Foundations of Constitutional Law*, Oxford University Press, Oxford, 2019.

Dzhanova, Yelena, 'Facebook did not hire Black employees because they were not a "culture fit," report says', *Business Insider*, 6 April 2021 <https://www.businessinsider.com/facebook-workplace-hiring-eeoc-black-employees-culture-fit-2021-4> (accessed 3 October 2021).

The Economist, 'The EU wants to set the rules for the world of technology', 20 February 2020 <https://www.economist.com/business/2020/02/20/the-eu-wants-to-set-the-rules-for-the-world-of-technology> (accessed 20 August 2021).

Edelman, Gilad, 'Stop Saying Facebook Is "Too Big to Moderate"', *Wired*, 28 July 2020 <https://www.wired.com/story/stop-saying-facebook-too-big-to-moderate/> (accessed 20 August 2021).

Edwards, Lilian, *Law, Policy and the Internet*, Hart Publishing, Oxford, 2019.

Edwards, Lilian, Burkhard Schafer and Edina Harbinja (eds), *Future Law: Emerging Technology, Regulation and Ethics*, Edinburgh University Press, Edinburgh, 2020.

Edwards, Lilian and Michael Veale, 'Slave to the Algorithm? Why a "Right to an Explanation" Is Probably Not the Remedy You Are Looking For', *Duke Law & Technology Review*, Vol. 16, No. 1 (2017), 18–84.

Eggerton, John, 'Hill Briefing: DETOUR Act on Right Road', *Multichannel News*, 25 June 2019 <https://www.nexttv.com/news/hill-briefing-detour-act-on-right-road> (accessed 20 August 2021).

Eichstaedt, Johannes C. et al., 'Facebook Language Predicts Depression in Medical Records', *Proceedings of the National Academy of Sciences of the United States of America*, Vol. 115, No. 44 (2018), 11203–11208.

Elazar, Yiftah and Geneviève Rousselière (eds), *Republicanism and the Future of Democracy*, Cambridge University Press, Cambridge, 2019.

Eling, Martin, 'How insurance can mitigate AI risks', *The Brookings Institution*, 7 November 2019 <https://www.brookings.edu/research/how-insurance-can-mitigate-ai-risks/> (accessed 20 August 2021).

Elkin, Stephen L., *Reconstructing the Commercial Republic: Constitutional Design after Madison*, University of Chicago Press, Chicago, 2006.

Elster, Jon and Aanund Hylland, *Foundations of Social Choice Theory*, Cambridge University Press, Cambridge, 1989.

Emmott, Robin, 'Russia deploying coronavirus disinformation to sow panic in West, EU document says', *Reuters*, 18 March 2020 <https://www.reuters.com/article/us-health-coronavirus-disinformation-idUSKBN21518F> (accessed 20 August 2021).

Engler, Alex, 'The case for AI transparency requirements', *The Brookings Institution*, 22 January 2020 <https://www.brookings.edu/research/the-case-for-ai-transparency-requirements/> (accessed 20 August 2021).

Epstein, Robert and Robert E. Robertson, 'The search engine manipulation effect (SEME) and its possible impact on the outcomes of elections', *Proceedings of the National Academy of Sciences of the United States of America*, Vol. 112, No. 33 (2015), E4512–E4521.

Epstein, Robert, Roger Mohr Jr. and Jeremy Martinez, 'The Search Suggestion Effect (SSE): How Search Suggestions Can Be Used to Shift Opinions and Voting Preferences Dramatically and Without People's Awareness', *98th Annual Meeting of the Western Psychological Association*, Portland, OR, 26 April 2018.

Eubanks, Virginia, *Automating Inequality: How High-Tech Tools Profile, Police, and Punish the Poor*, Picador, New York, 2019.

European Commission, 'Liability for Emerging Digital Technologies', Commission Staff Working Document SWD (2018) 137 final.

European Court of Human Rights, *Research Report: Positive obligations on member States under Article 10 to protect journalists and prevent impunity*, Council of Europe and the European Court of Human Rights, December 2011.

European Parliament, *Commission evaluation report on the implementation of the General Data Protection Regulation two years after its application* <https://www.europarl.europa.eu/doceo/document/TA-9-2021-0111_EN.pdf> (accessed 4 October 2021).

Expert Committee on Human Rights Dimensions of Automated Data Processing and Different Forms of Artificial Intelligence, *Responsibility and AI*, DGI(2019)05, Council of Europe, Strasbourg, 2019.

Facebook, 'Oversight Board Charter', September 2019 <https://about. fb.com/wp-content/uploads/2019/09/oversight_board_charter. pdf> (accessed 20 August 2021).

Fairfield, Joshua A. T. and Christoph Engel, 'Privacy as a Public Good', *Duke Law Journal*, Vol. 65, No. 3 (December 2015), 385–457.

Fang, Lee, 'Facebook Pitched New Tool Allowing Employers to Suppress Words Like "Unionize" in Workplace Chat Product', *The Intercept*, 11 June 2021 <https://theintercept.com/2020/06/11/facebook-workplace-unionize/> (accessed 20 August 2021).

Farhang, Sean, *The Litigation State: Public Regulation and Private Lawsuits in the U.S.*, Princeton University Press, Princeton, 2010.

Farkas, Johan and Jannick Schou, *Post-Truth, Fake News and Democracy: Mapping the Politics of Falsehood*, Routledge, New York, 2020.

Faroohar, Rana, *Don't be Evil: The Case Against Big Tech*, Allen Lane, London, 2019.

Fassler, Ella, 'Here's How Easy It Is for Cops to Get Your Facebook Data', *OneZero*, 17 June 2020 <https://onezero.medium.com/cops-are-incre asingly-requesting-data-from-facebook-and-you-probably-wont-get-notified-if-they-5b7a2297df17> (accessed 20 August 2021).

Feiner, Lauren and Megan Graham, 'Pelosi says advertisers should use their "tremendous leverage" to force social media companies to stop spreading false and dangerous information', *CNBC*, 16 June 2020 <https://www.cnbc.com/2020/06/16/pelosi-says-adve rtisers-should-push-platforms-to-combat-disinformation.html> (accessed 19 August 2021).

Feldstein, Steven, 'How Should Democracies Confront China's Digital Rise? Weighing the Merits of a T-10 Alliance', *Council on Foreign Relations*, 30 November 2020 <https://www.cfr.org/blog/how-should-demo cracies-confront-chinas-digital-rise-weighing-merits-t-10-alliance> (accessed 20 August 2021).

Felton, James, 'AI Camera Ruins Soccer Game For Fans After Mistaking Referee's Bald Head For Ball', *IFLScience*, 29 October 2020 <https://www.iflscience.com/technology/ai-camera-ruins-soccar-game-for-fans-after-mistaking-referees-bald-head-for-ball/> (accessed 3 October 2021).

Ferguson, Adam, *An Essay on the History of Civil Society*, Cambridge University Press, Cambridge, 2007.

Ferguson, Niall, 'What Is To Be Done? Safeguarding Democratic Governance In The Age Of Network Platforms', *Hoover Institution, Governance in an Emerging World*, Fall Series, No. 318 (2018).

Fernandez, Sonia, 'WiFi System Identifies People Through Walls By Their Walk', Futurity, 1 October 2019 <https://www.futurity.org/wifi-video-identification-through-walls-2173442/> (accessed 20 August 2021).

Fiesler, Casey, 'What do we teach when we teach tech & AI ethics?' CUInfoScience, 17 January 2020 <https://medium.com/cuinfoscience/what-do-we-teach-when-we-teach-tech-ai-ethics-81059b710e11> (accessed 20 August 2021).

Financial Conduct Authority, 'Factsheet: Becoming an approved person', No. 029, 2020 <https://www.fca.org.uk/publication/other/fs029-becoming-an-approved-person.pdf> (accessed 20 August 2021).

Finck, Michèle, Blockchain Regulation and Governance in Europe, Cambridge University Press, Cambridge, 2019.

Fisher, Max and Amanda Taub, 'How YouTube Radicalized Brazil', The New York Times, 11 August 2019 <https://www.nytimes.com/2019/08/11/world/americas/youtube-brazil.html> (accessed 20 August 2021).

Fishkin, James S., Democracy When the People are Thinking: Revitalising Our Politics Through Public Deliberation, Oxford University Press, Oxford, 2018.

Fjeld, Jessica et al., 'Principled Artificial Intelligence: Mapping Consensus in Ethical and Rights-Based Approaches to Principles for AI' (2020) Berkman Klein Center Research Publication No. 2020-1.

Fleischman, Gary M. et al., 'Ethics Versus Outcomes: Managerial Responses to Incentive-Driven and Goal-Induced Employee Behavior', Journal of Business Ethics, Vol. 158, No. 4 (2019), 951–967.

Foucault, Michel, Power/Knowledge, Vintage Books, New York, 1980.

Fowler, Geoffrey A., 'You downloaded FaceApp. Here's what you've just done to your privacy', Washington Post, 17 July 2019 <https://www.washingtonpost.com/technology/2019/07/17/you-downloaded-faceapp-heres-what-youve-just-done-your-privacy/> (accessed 20 August 2021).

Franks, Mary Anne, 'How The Internet Unmakes Law', The Ohio Technology Law Journal, Vol. 16, No. 1 (2020), 10–24.

Frenkel, Sheera and Cecilia Kang, An Ugly Truth: Inside Facebook's Battle for Domination, The Bridge Street Press, London, 2021.

Friends of Andrew Yang, 'Revive the Office of Technology Assessment', 2020 <https://2020.yang2020.com/policies/reviveota/> (accessed 20 August 2021).

Frier, Bruce W., The Rise of the Roman Jurists: Studies in Cicero's "Pro Caecina", Princeton University Press, Princeton, 1985.

Fries, Martin, 'Law and Autonomous Systems Series: Smart consumer contracts - The end of civil procedure?' *Oxford Business Law Blog*, 29 March 2018 <https://www.law.ox.ac.uk/business-law-blog/blog/2018/03/smart-consumer-contracts-end-civil-procedure> (accessed 20 August 2021).

Froomkin, A. Michael, 'Regulating Mass Surveillance as Privacy Pollution: Learning from Environmental Impact Statements', *University of Illinois Law Review*, Vol. 2015, No. 5 (2015), 1713–1790.

Fukuyama, Francis, *Identity: Contemporary Identity Politics and the Struggle for Recognition*, Profile Books, London, 2018.

Fuller, Lon, *The Morality of Law*, Yale University Press, New Haven and London, 1969.

Gabriel, Iason, 'Artificial Intelligence, Values, and Alignment', *Minds and Machines*, Vol. 30, No. 3 (2020), 411–437.

Galbraith, J. K., *American Capitalism:The Concept of Countervailing Power*, Martino Fine Books, Eastford, CT, 2012.

Gardels, Nathan and Nicolas Berggruen, *Renovating Democracy: Governing in the Age of Globalisation and Digital Capitalism*, University of California Press, Oakland, 2019.

Garton Ash, Timothy, Robert Gorwa and Danaë Metaxa, 'Glasnost! Nine ways Facebook can make itself a better forum for free speech and democracy', *Oxford-Stanford Report*, Reuters Institute for the Study of Journalism, Oxford, 2019.

Gaus, Gerald F., 'Backwards into the Future: Neorepublicanism as a Postsocialist Critique of Market Society', *Social Philosophy and Policy*, Vol. 20, No. 1 (2003), 59–91.

Geistfeld, Mark A., 'Tort Law in the Age of Statutes', *Iowa Law Review*, Vol. 99, No. 3 (2014), 967–1020.

Gershgorn, Dave, 'Amazon's "holy grail" recruiting tool was actually just biased against women', *Quartz*, 10 October 2018 <https://qz.com/1419228/amazons-ai-powered-recruiting-tool-was-biased-against-women/> (accessed 20 August 2021).

Ghaffary, Shirin, 'Facebook is taking down some, but not all, quarantine protest event pages', *Vox*, 20 April 2020 <https://www.vox.com/recode/2020/4/20/21228224/facebook-coronavirus-covid-19-protests-taking-down-content-moderation-freedom-speech-debate> (accessed 20 August 2021).

Ghaffary, Shirin and Jason Del Rey, 'The real cost of Amazon', *Vox*, 29 June 2020 <https://www.vox.com/recode/2020/6/29/21303643/ama

zon-coronavirus-warehouse-workers-protest-jeff-bezos-chris-smalls-boycott-pandemic> (accessed 20 August 2021).

Ghosh, Dipayan, *Terms of Disservice: How Silicon Valley is Destructive by Design*, Brookings Institution Press, Washington, DC, 2020.

Gillespie, Tarleton, *Custodians of the Internet: Platforms, Content Moderation, and the Hidden Decisions that Shape Social Media*, Yale University Press, New Haven, 2018.

Gillis, Talia B. and Josh Simons, 'Explanation < Justification: GDPR and the Perils of Privacy', *Pennsylvania Journal of Law and Innovation*, Vol. 2 (2019), 71–99.

Gilman, Nils and Henry Farrell, 'Three Moral Economies of Data', *The American Interest*, 7 November 2018 <https://www.the-american-inter est.com/2018/11/07/three-moral-economies-of-data/> (accessed 20 August 2021).

Glaeser, Edward L. and Andre Shleifer, 'The Rise of the Regulatory State', *Journal of Economic Literature*, Vol. 41, No. 2 (2003), 401–425.

Glaser, April, 'Is a Tech Company Ever Neutral?' *Slate*, 11 October 2019 <https://slate.com/technology/2019/10/apple-chinese-gov ernment-microsoft-amazon-ice.html> (accessed 20 August 2021).

Gold, Andrew S. and Paul B. Miller (eds), *Philosophical Foundations of Fiduciary Law*, Oxford University Press, Oxford, 2016.

Goldman, Eric and Jess Miers, 'Regulating Internet Services by Size' (May 2021), *CPI Antitrust Chronicle*, Santa Clara University Legal Studies Research Paper.

Goldsmith, Jack and Andrew Keane Woods, 'Internet Speech Will Never Go Back to Normal', *The Atlantic*, 25 April 2020 <https://www.thea tlantic.com/ideas/archive/2020/04/what-covid-revealed-about-internet/610549/> (accessed 20 August 2021).

Gourevitch, Alex, *From Slavery to the Cooperative Commonwealth: Labor and Republican Liberty in the Nineteenth Century*, Cambridge University Press, Cambridge, 2015.

Graef, Inge, Martin Husovec and Nadezhda Purtova, 'Data Portability and Data Control: Lessons for an Emerging Concept in EU Law', *German Law Journal*, Vol. 19, No. 6 (2018), 1359–1398.

Gray, David and Stephen E. Henderson (eds), *The Cambridge Handbook of Surveillance Law*, Cambridge University Press, New York, 2019.

Gray, Tim, *Freedom*, Macmillan Education, Basingstoke, 1991.

Greene, Daniel, Anna Lauren Hoffman and Luke Stark, 'Better, Nicer, Clearer, Fairer: A Critical Assessment of the Movement for Ethical

Artificial Intelligence and Machine Learning', *Proceedings of the 52nd Hawaii International Conference on System Sciences* (2019), 2122–2131.

Greenfield, Adam, *Radical Technologies: The Design of Everyday life*, Verso, London, 2017.

Grimmelmann, James, 'Saving Facebook', *NYLS Legal Studies Research Paper*, No. 08/09-7 (2008).

Grimmelmann, James and Daniel Westreich, 'Incomprehensible Discrimination', *California Law Review*, Vol. 7 (2017), 164–177.

Grossman, Vasily, *Life and Fate*, Vintage, London, 2006.

Guhl, Jakob and Jacob Davey, 'Hosting the 'Holohoax': A Snapshot of Holocaust Denial Across Social Media', Institute for Strategic Dialogue, London, 10 August 2020 <https://www.isdglobal.org/isd-publicati ons/hosting-the-holohoax-a-snapshot-of-holocaust-denial-across-social-media/> (accessed 20 August 2021).

Gurley, Lauren Kaori and Joseph Cox, 'Inside Amazon's Secret Program to Spy On Workers' Private Facebook Groups', *Vice*, 1 September 2020 <https://www.vice.com/en/article/3azegw/amazon-is-spying-on-its-workers-in-closed-facebook-groups-internal-repo rts-show> (accessed 3 October 2021).

Haas, Leonard and Sebastian Gießler, with Veronika Thiel, 'In the realm of paper tigers – exploring the failings of AI ethics guidelines', *Algorithm Watch*, 28 April 2020 <https://algorithmwatch.org/en/ai-ethics-gui delines-inventory-upgrade-2020/> (accessed 20 August 2021).

Hadavas, Chloe, 'Why We Should Care That Facebook Accidentally Deplatformed Hundreds of Users', *Slate*, 12 June 2020 <https://slate. com/technology/2020/06/facebook-anti-racist-skinheads.html> (accessed 20 August 2021).

Hadfield, Gillian K., *Rules for a Flat World: Why Humans Invented Law and How to Reinvent it for a Complex Global Economy*, Oxford University Press, Oxford, 2017.

Halegoua, Germaine R., *Smart Cities*, MIT Press, Cambridge, MA, 2020.

Halpern, Sue, 'The Problem of Political Advertising on Social Media', *The New Yorker*, 24 October 2019 <https://www.newyorker.com/tech/ annals-of-technology/the-problem-of-political-advertising-on-soc ial-media> (accessed 20 August 2021).

Hamilton, Alexander, James Madison and John Jay, *The Federalist Papers*, Penguin Classics, New York, 2012.

Hao, Karen, 'We read the paper that forced Timnit Gebru out of Google. Here's what it says.' *MIT Technology Review*, 4 December 2020 <https:// www.technologyreview.com/2020/12/04/1013294/google-ai-eth

ics-research-paper-forced-out-timnit-gebru/> (accessed 3 October 2021).

Harcourt, Bernard E., *Exposed: Desire and Disobedience in the Digital Age*, Harvard University Press, Cambridge, MA, 2015.

—, *The Illusion of Free Markets: Punishment and the Myth of Natural Order*, Harvard University Press, Cambridge, MA, 2012.

Harrington, James, *The Commonwealth of Oceana and A System of Politics*, Cambridge University Press, Cambridge, 2008.

Harvard Business Review, *HBR's 10 must reads on AI, analytics, and the new machine age*, Harvard Business Publishing Corporation, Cambridge, MA, 2019.

Hart, H. L. A., *The Concept of Law*, Oxford University Press, Oxford, 1997.

Haugaard, Mark, *Power: A Reader*, Manchester University Press, Manchester, 2002.

Hayek, F. A., *The Constitution of Liberty*, Routledge, Abingdon, 2006.

Hazell, Will, 'A-level results 2020: 39% of teacher predicted grades downgraded by algorithm amid calls for U-turn', *inews*, 13 August 2020 <https://inews.co.uk/news/education/a-level-results-2020-grades-downgraded-algorithm-triple-lock-u-turn-result-day-578 194> (accessed 20 August 2021).

Heaven, Will Douglas, 'IBM's Debating AI Just Got a Lot Closer to Being a Useful Tool', *MIT Technology Review*, 21 January 2020 <https://www.technologyreview.com/2020/01/21/276156/ibms-debat ing-ai-just-got-a-lot-closer-to-being-a-useful-tool/> (accessed 20 August 2021).

Hecht, B. et al., 'It's Time to Do Something: Mitigating the Negative Impacts of Computing Through a Change to the Peer Review Process', *ACM Future of Computer Blog*, 29 March 2018 <https://acm-fca. org/2018/03/29/negativeimpacts/> (accessed 20 August 2021).

Helberger, Natali, 'The Political Power of Platforms: How Current Attempts to Regulate Misinformation Amplify Opinion Power', *Digital Journalism*, Vol. 8, No. 6 (2020), 842–854.

Held, David, *Models of Democracy*, Polity Press, Cambridge, 2006.

Hellman, Deborah and Sophia Moreau, *Philosophical Foundations of Discrimination Law*, Oxford University Press, Oxford, 2013.

Hern, Alex, 'TikTok's local moderation guidelines ban pro-LGBT content', *The Guardian*, 26 September 2019 <https://www.theguard ian.com/technology/2019/sep/26/tiktoks-local-moderation-gui delines-ban-pro-lgbt-content> (accessed 20 August 2021).

—, 'Revealed: how TikTok censors videos that do not please Beijing', *The Guardian*, 25 September 2019 <https://www.theguardian.

com/technology/2019/sep/25/revealed-how-tiktok-censors-vid eos-that-do-not-please-beijing> (accessed 20 August 2021).

Hildebrandt, Mireille, Law for Computer Scientists and Other Folk, Oxford University Press, Oxford, 2020.

Hildebrandt, Mireille and Katja de Vries (eds), Privacy, Due Process and the Computational Turn: The Philosophy of Law Meets the Philosophy of Technology, Routledge, Abingdon, 2013.

Hill, Kashmir, 'Wrongfully Accused by an Algorithm', The New York Times, 24 June 2020 <https://www.nytimes.com/2020/06/24/technol ogy/facial-recognition-arrest.html> (accessed 20 August 2021).

Hobbes, Thomas, Leviathan, Cambridge University Press, Cambridge, 1996.

Hodge, Lord Patrick, 'The Potential and Perils of Financial Technology: Can the Law adapt to cope?' Edinburgh FinTech Law Lecture, University of Edinburgh, Edinburgh, 14 March 2019.

Hoffman, Reid, Blitzscaling: The Lightning-Fast Path to Building Massively Valuable Companies, HarperCollins, London, 2018.

Hogarth, Ian, 'AI Nationalism', Ian Hogarth Blog, 13 June 2018 <https:// www.ianhogarth.com/blog/2018/6/13/ai-nationalism> (accessed 19 august 2021).

Hollowood, Ella and Alexi Mostrous, 'The Infodemic: Fake news in the time of C-19', Tortoise Media, 23 March 2020 <https://www.tortoi semedia.com/2020/03/23/the-infodemic-fake-news-coronavi rus/> (accessed 20 August 2021).

Honohan, Iseult, Civic Republicanism, Routledge, Abingdon, 2002.

Honohan, Iseult and Jeremy Jennings (eds), Republicanism in Theory and Practice, Routledge, Abingdon, 2006.

Hoofnagle, Chris Jay, Woodrow Hartzog and Daniel J. Solove, 'The FTC can rise to the privacy challenge, but not without help from Congress', The Brookings Institution, 8 August 2019 <https://www.brooki ngs.edu/blog/techtank/2019/08/08/the-ftc-can-rise-to-the-priv acy-challenge-but-not-without-help-from-congress/> (accessed 20 August 2021).

Horwitz, Jeff, 'The Facebook Files: Facebook Says Its Rules Apply to All. Company Documents Reveal a Secret Elite That's Exempt', The Wall Street Journal, 13 September 2021 <https://www.wsj.com/articles/ the-facebook-files-11631713039> (accessed 3 October 2021).

Horwitz, Morton J., The Transformation of American Law 1870-1960: The Crisis of Legal Orthodoxy, Oxford University Press, New York, 1992.

Horwitz, Robert Britt, The Irony of Regulatory Reform: The Deregulation of American Telecommunications, Oxford University Press, New York, 1989.

Hosonagar, Kartik, *A Human's Guide to Machine Intelligence: How Algorithms Are Shaping Our Lives and How We Can Stay in Control*, Penguin Books, New York, 2020.

Hoye, J. Matthew and Jeffrey Monaghan, 'Surveillance, freedom and the republic', *European Journal of Political Theory*, Vol. 17, No. 3 (2018), 343–363.

Hua, Shin-Shin and Haydn Belfield, *AI & Antitrust: Reconciling Tensions Between Competition Law and Cooperative AI Development*, 23 Yale J.L. & Tech. 360 (2021) (forthcoming).

Hurley, Mikella and Julius Adebayo, 'Credit Scoring in the Era of Big Data', *Yale Journal of Law and Technology*, Vol. 18, No. 1 (2016), 148–216.

Hyland, Paul and Neil Sammells (eds), *Writing & Censorship in Britain*, Routledge, London, 1992.

Hymas, Charles, 'AI used for first time in job interviews in UK to find best applicants', *The Telegraph*, 27 September 2019 <https://www.telegraph.co.uk/news/2019/09/27/ai-facial-recognition-used-first-time-job-interviews-uk-find/> (accessed 20 August 2021).

IBM AI Research, 'Project Debater', IBM, 2021 <https://www.research.ibm.com/artificial-intelligence/project-debater/> (accessed 20 August 2021).

Ingelhart, Louis Edward, *Press and Speech Freedoms in the World, from Antiquity until 1998: A Chronology*, Greenwood Press, Westport, 1998.

Instagram Inc, 'Community Guidelines', 2021 <https://help.instagram.com/477434105621119> (accessed 20 August 2021).

Isaac, Mike, 'For Political Cartoonists, the Irony Was That Facebook Didn't Recognize Irony', *The New York Times*, 10 June 2021 <https://www.nytimes.com/2021/03/19/technology/political-cartoonists-facebook-satire-irony.html> (accessed 3 October 2021).

Jackson, Lauren and Desiree Ibekwe, 'Jack Dorsey on Twitter's Mistakes', *The New York Times*, 7 August 2020 <https://www.nytimes.com/2020/08/07/podcasts/the-daily/Jack-dorsey-twitter-trump.html> (accessed 20 August 2021).

Jacobs, Julia, 'Will Instagram Ever "Free the Nipple"?' *The New York Times*, 22 November 2019 <https://www.nytimes.com/2019/11/22/arts/design/instagram-free-the-nipple.html> (accessed 20 August 2021).

Jeevanjee, Kiran et al., 'All the Ways Congress Wants to Change Section 230', *Slate*, 23 March 2021 <https://slate.com/technology/2021/03/section-230-reform-legislative-tracker.html> (accessed 3 October 2021).

Jerome, Joseph, 'Private right of action shouldn't be a yes-no proposition in federal US privacy legislation', *International Association of Privacy Professionals*,

3 October 2019 <https://iapp. org/news/a/private-right-of-action-shouldnt-be-a-yes-no-proposition-in-federal-privacy-legislation/> (accessed 20 August 2021).

Johnson, Khari, 'Google employee group urges Congress to strengthen whistleblower protections for AI researchers', *VentureBeat*, 8 March 2021 <https://venturebeat.com/2021/03/08/google-emplo yee-group-urges-congress-to-strengthen-whistleblower-protecti ons-for-ai-researchers/> (accessed 3 October 2021).

Joint Committee on Human Rights, *Oral Evidence: The Right to Privacy (Article 8) and the Digital Revolution* (HC 1810).

—, *The Right to Privacy (Article 8) and the Digital Revolution* (HC 122).

Jones, Lora, "I monitor my staff with software that takes screenshots", BBC, 29 September 2020 <https://www.bbc.co.uk/news/business-54289152> (accessed 3 October 2021).

Jørgensen, Rikke Frank (ed.), *Human Rights in the Age of Platforms*, MIT Press, Cambridge, MA, 2019.

Joss, Simon and John Durant (eds), *Public participation in science: The role of consensus conferences in Europe*, Antony Roe, Chippenham, 1995.

Kaminski, Margot E. and Andrew D. Selbst, 'The Legislation That Targets the Racist Impacts of Tech', *The New York Times*, 7 May 2019 <https://www.nytimes.com/2019/05/07/opinion/tech-racism-algorithms.html> (accessed 20 August 2021).

Kantrowitz, Alex, 'Inside Big Tech's Years-Long Manipulation Of American Op-Ed Pages', *Big Technology*, 16 July 2020 <https://bigtec hnology.substack.com/p/inside-big-techs-years-long-manipulat ion> (accessed 20 August 2021).

Karr, Timothy, 'Why Facebook Filtering Will Ultimately Fail', *Start It Up*, 15 November 2019 <https://medium.com/swlh/why-facebook-filter ing-will-ultimately-fail-90606ec98c11> (accessed 20 August 2021).

Katsirea, Irini, '"Fake News": reconsidering the value of untruthful expression in the face of regulatory uncertainty', *Journal of Media Law*, Vol. 10, No. 2 (2018), 159–188.

Kaye, David, 'The surveillance industry is assisting state suppression. It must be stopped', *The Guardian*, 26 November 2019 <https://www.theguardian.com/commentisfree/2019/nov/26/surveillance-indus try-suppression-spyware> (accessed 20 August 2021).

—, *Speech Police: The Global Struggle to Govern the Internet*, Columbia Global Reports, New York, 2019.

Keller, Daphne, 'Amplification and its Discontents', *Knight First Amendment Institute*, 8 June 2021 <https://knightcolumbia.org/content/amplification-and-its-discontents> (accessed 12 November 2012).

—, 'Six Constitutional Hurdles For Platform Speech Regulation', *Stanford Center for Internet and Society Blog*, 22 January 2021 <http://cyberlaw.stanford.edu/blog/2021/01/six-constitutional-hurdles-platform-speech-regulation-0> (accessed 20 August 2021).

—, 'Dolphins in the Net: Internet Content Filters and the Advocate General's *Glawischnig-Piesczek v. Facebook Ireland* Opinion', Stanford Center for Internet and Society, 4 September 2019 <https://cyberlaw.stanford.edu/files/Dolphins-in-the-Net-AG-Analysis.pdf> (accessed 20 August 2021).

—, 'Who Do You Sue? State and Platform Hybrid Power Over Online Speech' (2019) Hoover Institution, Aegis Series Paper No. 1902.

Kersely, Andrew, 'Couriers say Uber's 'racist' facial identification tech got them fired', *Wired*, 3 January 2021 <https://www.wired.co.uk/article/uber-eats-couriers-facial-recognition> (accessed 20 August 2021).

Kessler, Sarah, 'Companies Are Using Employee Survey Data to Predict — and Squash — Union Organizing', *OneZero*, 30 July 2020 <https://onezero.medium.com/companies-are-using-employee-survey-data-to-predict-and-squash-union-organizing-a7e28a8c2158> (accessed 20 August 2021).

Kettemann, Matthias C., *The Normative Order of the Internet*, Oxford University Press, Oxford, 2020.

Khan, Lina, 'The Separation of Platforms and Commerce', *Columbia Law Review*, Vol. 119, No. 4 (2019), 973–1098.

—, 'The New Brandeis Movement: America's Antimonopoly Debate', *Journal of European Competition Law & Practice*, Vol. 9, No. 3 (2018), 131–132.

—, 'Amazon's Antitrust Paradox', *The Yale Law Journal*, Vol. 126, No. 3 (2017), 710–805 (note).

Klass, Gregory, George Letsas and Prince Saprai (eds), *Philosophical Foundations of Contract Law*, Oxford University Press, Oxford, 2016.

Klonick, Kate, 'The New Governors: The People, Rules, and Processes Governing Online Speech', *Harvard Law Review*, Vol. 131, No. 6 (2018), 1598–1670.

Knight, Will, 'Washington Must Bet Big on AI or Lose Its Global Clout', *Wired*, 17 December 2019 <https://www.wired.com/story/washington-bet-big-ai-or-lose-global-clout/> (accessed 20 August 2021).

Kobie, Nicole, 'Germany says GDPR could collapse as Ireland dallies on big fines', *Wired*, 27 April 2020 <https://www.wired.co.uk/article/gdpr-fines-google-facebook> (accessed 20 August 2021).

Kornbluh, Karen and Ellen P. Goodman, 'Safeguarding Digital Democracy: Digital Innovation and Democracy Initiative Roadmap', The German Marshall Fund of the United States, Washington DC, 2020.

Kosinski, Michal, David Stillwell and Thore Graepel, 'Private traits and attributes are predictable from digital records of human behaviour', *Proceedings of the National Academy of Sciences of the United States of America*, Vol. 110, No. 15 (2013), 5802–5805.

Koskimaa, Vesa and Lauri Rapeli, 'Fit to govern? Comparing citizen and policymaker perceptions of deliberative democratic innovations', *Policy & Politics*, Vol. 48, No. 4 (2020), 637–652(16).

Kosseff, Jeff, *The Twenty-Six Words that Created the Internet*, Cornell University Press, Ithaca, 2019.

Kozlowska, Hanna, 'Each platform's approach to political ads in one table', *Quartz*, 13 December 2019 <https://qz.com/1767145/how-facebook-twitter-and-others-approach-political-advertising/> (accessed 20 August 2021).

Kroll, Joshua A. et al., 'Accountable Algorithms', *University of Pennsylvania Law Review*, Vol. 165, No. 3 (2017), 633–705.

Kuang, Cliff and Robert Fabricant, *User Friendly: How the hidden rules of design are changing the way we live, work, and play*, WH Allen, London, 2019.

Kundera, Milan, *The Unbearable Lightness of Being*, Harper & Row Publishers, 1984.

Kuner, Christopher, Lee A. Bygrave and Christopher Docksey (eds), *The EU Data Protection Regulation (GDPR): A Commentary*, Oxford University Press, Oxford, 2020.

Laborde, Cécile and John Maynor (eds), *Republicanism and Political Theory*, Blackwell Publishing, Oxford, 2008.

Laidlaw, Emily B., *Regulating Speech in Cyberspace: Gatekeepers, Human Rights and Corporate Responsibility*, Cambridge University Press, Cambridge, 2017.

LaJeunesse, Ross, 'I Was Google's Head of International Relations. Here's Why I Left', *Ross LaJeunesse*, 2 January 2020 <https://medium.com/@rossformaine/i-was-googles-head-of-international-relations-here-s-why-i-left-49313d23065> (accessed 20 August 2021).

Lakoff, George, *Moral Politics: How Liberals and Conservatives Think*, University of Chicago Press, Chicago, 2016.

Lande, Robert H., 'A Traditional and Textualist Analysis of the Goals of Antitrust: Efficiency, Preventing Theft from Consumers, and Consumer Choice' (2013) University of Baltimore Legal Studies Research Paper No. 2013-10.

Landemore, Hélène, *Open Democracy: Reinventing Popular Rule for the Twenty-First Century*, Princeton, Princeton University Press, 2020.

Lane, Julia, Victoria Stodden, Stefan Bender and Helen Nissenbaum (eds), *Privacy, Big Data, and the Public Good: Frameworks for Engagement*, Cambridge University Press, New York, 2014.

Lane, Melissa, *Greek and Roman Political Ideas*, Penguin Books, London, 2014.

Lanier, Jaron, *Dawn of the New Everything: A Journey Through Virtual Reality*, The Bodley Head, London, 2017.

Larson, Rob, *Bit Tyrants: The Political Economy of Silicon Valley*, Haymarket Books, Chicago, 2020.

Latham, Mark, Victor E. Schwartz and Christopher E. Appel, 'The Intersection of Tort and Environmental Law: Where the Twains Should Meet and Depart', *Fordham Law Review*, Vol. 80, No. 2 (2011), 737–773.

Lay, Paul, *Providence Lost: The Rise & Fall of Cromwell's Protectorate*, Head of Zeus, London, 2020.

Lazaro, Christophe and Daniel Le Métayer, 'Control over Personal Data: true Remedy or Fairy Tale?' *Scripted*, Vol. 12, No. 1 (June 2015), 3–34.

Learned Hand, Billings, *The Spirit of Liberty*, Vintage, New York, 1959.

Lecher, Colin, 'How Amazon Automatically Tracks and Fires Warehouse Workers for 'Productivity'', *The Verge*, 25 April 2019 <https://www.theverge.com/2019/4/25/18516004/amazon-warehouse-fulfillment-centers-productivity-firing-terminations> (accessed 20 August 2021).

Lee, Yong Suk et al., 'AI regulation and firm behaviour', *VoxEU.org*, 14 December 2019 <https://voxeu.org/article/ai-regulation-and-firm-behaviour> (accessed 20 August 2021).

LePan, Nicholas, 'Visualizing the Length of the Fine Print, for 14 Popular Apps', *Visual Capitalist*, 18 April 2020 <https://www.visualcapitalist.com/terms-of-service-visualizing-the-length-of-internet-agreements/> (accessed 20 August 2021).

Leprince-Ringuet, Daphne, 'Facial Recognition: This New AI Tool Can Spot When You Are Nervous or Confused', ZDNet, 21 October 2019 <https://www.zdnet.com/article/this-new-ai-tool-can-spot-if-you-are-nervous-or-confused/> (accessed 20 August 2021).

Lessig, Lawrence, *Republic, Lost: How Money Corrupts Congress — and a Plan to Stop It*, Hachette Book Group, New York, 2011.

—, *Code Version 2.0*, Basic Books, New York, 2006.

Levi-Faur, David (ed.), *The Oxford Handbook of Governance*, Oxford University Press, Oxford, 2014.

Lipson, Hod and Melba Kurman, *Driverless: Intelligent Cars and the Road Ahead*, MIT Press, Cambridge, MA, 2016.

Litman-Navarro, Kevin, 'We Read 150 Privacy Policies. They Were an Incomprehensible Disaster', *The New York Times*, 12 June 2019 <https://www.nytimes.com/interactive/2019/06/12/opinion/facebook-google-privacy-policies.html> (accessed 20 August 2021).

Liu, Hin-Yan, 'The power structure of artificial intelligence', *Law, Innovation and Technology*, Vol. 10, No. 2 (2018), 197–229.

Liu, Wendy, *Abolish Silicon Valley: How to Liberate Technology from Capitalism*, Repeater Books, London, 2020.

Lohr, Steve, 'The Week in Tech: How Is Antitrust Enforcement Changing?' *The New York Times*, 22 December 2019 <https://www.nytimes.com/2019/12/22/technology/the-week-in-tech-how-is-antitrust-enforcement-changing.html> (accessed 20 August 2021).

Lomas, Natasha, 'Don't break up big tech — regulate data access, says EU antitrust chief', *TechCrunch*, 11 March 2019 <https://techcrunch.com/2019/03/11/dont-break-up-big-tech-regulate-data-access-says-eu-antitrust-chief/> (accessed 20 August 2021).

Loughlin, Martin, *Foundations of Public Law*, Oxford University Press, Oxford, 2014.

Lovett, Frank, 'Algernon Sidney, Republican Stability, and the Politics of Virtue', *Journal of Political Science*, Vol. 48, No. 1, Article 3 (2020), 59–83.

—, *A General Theory of Domination and Justice*, Oxford University Press, Oxford, 2012.

Lukes, Steven, *Individualism*, ECPR Press, Colchester, 2006.

Lunt, Peter and Sonia Livingstone, *Media Regulation: Governance and the Interests of Citizens and Consumers*, Sage Publications, London, 2012.

Lyles, Taylor, 'Facebook's new tool makes it easy to transfer photos and videos to Google Photos', *The Verge*, 30 April 2020 <https://www.theverge.com/2020/4/30/21241093/facebook-google-photos-transfer-tool-data> (accessed 20 August 2021).

Lynn, Barry C., *Liberty from all Masters: the new American Autocracy vs. the Will of the People*, St. Martin's Press, New York, 2020.

Lyons, Kim, 'Twitter removes tweets by Brazil, Venezuela presidents for violating COVID-19 content rules', *The Verge*, 30 March 2020 <https://

www.theverge.com/2020/3/30/21199845/twitter-tweets-brazil-venezuela-presidents-covid-19-coronavirus-jair-bolsonaro-maduro> (accessed 20 August 2021).

Mac, Ryan, 'Facebook Apologizes After A.I. Puts "Primates" Label on Video of Black Men', The New York Times, 14 September 2021 <https://www.nytimes.com/2021/09/03/technology/facebook-ai-race-primates.html> (accessed 3 October 2021).

Mac, Ryan and Craig Silverman, 'Facebook Has Been Showing Military Gear Ads Next To Insurrection Posts', BuzzFeed News, 13 January 2021 <https://www.buzzfeednews.com/article/ryanmac/facebook-profits-military-gear-ads-capitol-riot> (accessed 20 August 2021).

Macaulay, Thomas, 'Someone let a GPT-3 bot loose on Reddit - it didn't end well', The Next Web, 7 October 2020 <https://thenextweb.com/news/someone-let-a-gpt-3-bot-loose-on-reddit-it-didnt-end-well> (accessed 3 October 2021).

Machiavelli, Niccolò, Discourses on Livy, Oxford University Press, Oxford, 2008.

Magaziner, Ira C., 'Creating a Framework for Global Electronic Commerce', The Progress & Freedom Foundation, Future Insight, July 1999 <http://www.pff.org/issues-pubs/futureinsights/fi6.1globale conomiccommerce.html> (accessed 3 October 2021).

Mann, Monique and Tobias Matzner, 'Challenging algorithmic profiling: The limits of data protection and anti-discrimination in responding to emergent discrimination', Big Data & Society, Vol. 6, No. 2 (2019), 1–11.

Marchant, Gary E., Kenneth W. Abbott and Braden Allenby (eds), Innovative Governance Models for Emerging Technologies, Edward Elgar Publishing, Cheltenham, 2013.

Marcus, Gary and Ernest Davis, Rebooting AI: Building Artificial Intelligence We Can Trust, Pantheon Books, New York, 2019.

'Mark Zuckerberg Testimony Transcript: Zuckerberg Testifies on Facebook Cryptocurrency Libra', Rev, 23 October 2019 <https://www.rev.com/blog/transcripts/mark-zuckerberg-testimony-transcript-zuc kerberg-testifies-on-facebook-cryptocurrency-libra> (accessed 20 August 2021).

Margetts, Helen, Peter John, Scott Hale and Taha Yasseri, Political Turbulence: How Social Media Shape Collective Action, Princeton University Press, Woodstock, 2016.

Marsh, Sarah, 'Councils let firms track visits to webpages on benefits and disability', The Guardian, 4 February 2020 <https://www.theguardian.com/technology/2020/feb/04/councils-let-firms-track-visits-to-webpages-on-benefits-and-disability> (accessed 20 August 2021).

Marx, Karl and Friedrich Engels, The Communist Manifesto, Penguin Classics, London, 2002.

Mayer-Schönberger, Viktor and Kenneth Cukier, Big Data: A Revolution That Will Transform How We Live, Work and Think, John Murray, London, 2013.

Mayer-Schönberger, Viktor and Thomas Ramge, Reinventing Capitalism in the Age of Big Data, John Murray, London, 2018.

Maynor, John W., Republicanism in the Modern World, Polity Press, Cambridge, 2003.

Mazzucato, Mariana, The Entrepreneurial State: Debunking Public v Private Sector Myths, Penguin Books, London, 2018.

McCormick, John P., Machiavellian Democracy, Cambridge University Press, Cambridge, 2013.

McDonald, Sean, 'The Fiduciary Supply Chain', Centre for International Governance Innovation Online, 28 October 2019 <https://www.cigionline.org/articles/fiduciary-supply-chain> (accessed 20 August 2021).

McGurk, Brendan, Data Profiling and Insurance Law, Hart Publishing, London, 2019.

McKinnon, John D. and James V. Grimaldi, 'Justice Department, FTC Skirmish Over Antitrust Turf', The Wall Street Journal, 5 August 2019 <https://www.wsj.com/articles/justice-department-ftc-skirmish-over-antitrust-turf-11564997402> (accessed 20 August 2021).

McNamee, Roger, Zucked: Waking Up to the Facebook Catastrophe, HarperCollins, London, 2019.

McStay, Andrew, Emotional AI: The Rise of Empathic Media, Sage Publications, London, 2018.

Meadows, Donella H., Thinking in Systems: A Primer, Chelsea Green Publishing, London, 2008.

Meiklejohn, Alexander, Free Speech and its Relation to Self-Government, The Lawbook Exchange, Clark, NJ, 2014.

Menke, Christoph, Critique of Rights, Polity Press, Cambridge, 2020.

Merrill, Jeremy B., 'How Facebook's Ad System Lets Companies Talk Out of Both Sides of Their Mouths', The Markup, 13 April 2021 <https://themarkup.org/news/2021/04/13/how-facebooks-ad-system-lets-companies-talk-out-of-both-sides-of-their-mouths> (accessed 3 October 2021).

Metcalf, Jacob, Emanuel Moss and danah boyd, 'Owning Ethics: Corporate Logics, Silicon Valley, and the Institutionalization of Ethics', *Social Research: An International Quarterly*, Vol. 82, No. 2 (2019), 449–476.

Mighty Recruiter <https://www.mightyrecruiter.com> (accessed 20 August 2021).

Mill, J. S., *On Liberty and other writings*, Cambridge University Press, Cambridge, 2008.

Miller, Carl, 'Taiwan is making democracy work again. It's time we paid attention', *Wired*, 26 November 2019 <https://www.wired.co.uk/article/taiwan-democracy-social-media> (accessed 20 August 2021).

Miller, Catherine, Jacob Ohrvik-Stott and Rachel Coldicutt, *'Regulating for Responsible Technology: Capacity, Evidence and Redress: a new system for a fairer future*, Doteveryone, London, 2018.

Miller, Paul B. and Andrew S. Gold, *Contract, Status, and Fiduciary Law*, Oxford University Press, Oxford, 2016.

Milton, John, *Areopagitica and Other Writings*, Penguin Classics, London, 2014.

—, *Political Writings* (ed. Martin Dzelzainis), Cambridge University Press, Cambridge, 1998.

MIS Integrity, *Not Fit-for-Purpose: The Grand Experiment of Multi-Stakeholder Initiatives in Corporate Accountability, Human Rights and Global Governance* (Summary Report), July 2020.

Mittelstadt, Brent, 'Principles alone cannot guarantee ethical AI', *Nature Machine Intelligence*, Vol. 1 (2019), 501–507.

Möller, Judith et al., 'Do Not Blame It on the Algorithm: An Empirical Assessment of Multiple Recommender Systems and Their Impact on Content Diversity', *Information, Communication & Society*, Vol. 21, No. 7 (2018), 959–977.

Montesquieu, *The Spirit of the Laws*, Cambridge University Press, Cambridge, 2018.

Moore, Martin and Damian Tambini (eds), *Digital Dominance: The Power of Google, Amazon, Facebook, and Apple*, Oxford University Press, Oxford, 2018.

Moore, Matthew, 'Facebook poaches social media regulator Tony Close from Ofcom, *The Times*, 29 April 2020 <https://www.thetimes.co.uk/article/facebook-poaches-tony-close-from-ofcom-mdrkv7t2w> (accessed 20 August 2021).

Morgan, Bronwen and Karen Yeung, *An Introduction to Law and Regulation: Texts and Materials*, Cambridge University Press, Cambridge, 2007.

Morozov, Evgeny, 'Capitalism's New Clothes', *The Baffler*, 4 February 2019 <https://thebaffler.com/latest/capitalisms-new-clothes-morozov> (accessed 19 August 2021).

Morris, Steven, 'Facebook apologises for flagging Plymouth Hoe as offensive term', *The Guardian*, 27 January 2021 <https://www.theg uardian.com/uk-news/2021/jan/27/facebook-apologises-flagging-plymouth-hoe-offensive-term> (accessed 20 August 2021).

Mostrous, Alexi and Peter Hoskin, 'Domestic policy: facing both ways', *Tortoise Media*, 8 January 2020 <https://www.tortoisemedia.com/2020/01/08/tech-states-apple-domestic-policy/> (accessed 20 August 2021).

—, 'Foreign policy: the great game', *Tortoise Media*, 7 January 2020 <https://www.tortoisemedia.com/2020/01/07/tech-states-apple-foreign-policy/> (accessed 20 August 2021).

—, 'Part II: The constitution', *Tortoise Media*, 7 January 2020 <https://www.tortoisemedia.com/2020/01/07/tech-states-apple-constitution/> (accessed 20 August 2021).

—, 'Welcome to Apple: A one-party state', *Tortoise Media*, 6 January 2020 <https://www.tortoisemedia.com/2020/01/06/day-1-apple-state-of-the-nation-2/> (accessed 20 August 2021).

Mulgan, Geoff, 'Anticipatory Regulation: 10 ways governments can better keep up with fast-changing industries', *Nesta*, 15 May 2017, <https://www.nesta.org.uk/blog/anticipatory-regulation-10-ways-governments-can-better-keep-up-with-fast-changing-industries/> (accessed 3 October 2021).

Mulhall, Stephen and Adam Swift, *Liberals & Communitarians*, 2nd edn, Wiley-Blackwell, Oxford, 1996.

Muris, Timothy J. and Jonathan E. Nuechterlein, 'Antitrust in the Internet Era: The Legacy of *United States v A&P*' (2018) George Mason Law & Economics Research Paper No. 18–15.

Murphy, Laura W. et al., *Facebook's Civil Rights Audit - Final Report*, Facebook, 2020.

Nadler, Anthony, Matthew Crain and Joan Donovan, 'Weaponizing the Digital Influence Machine: The Political Perils of Online Ad Tech', Data & Society Research Institute, 2018 <https://datasociety.net/library/weaponizing-the-digital-influence-machine/> (accessed 20 August 2021).

Napoli, Philip M., *Social Media and the Public Interest: Media Regulation in the Disinformation Age*, Columbia University Press, New York, 2019.

Nietzsche, Friedrich, *Thus Spoke Zarathustra*, Penguin Classics, London, 2003.

Nelson, Eric, *The Greek Tradition in Republican Thought*, Cambridge University Press, Cambridge, 2006.

Nesta, 'vTaiwan', 2020 <https://www.nesta.org.uk/feature/six-pione ers-digital-democracy/vtaiwan/> (accessed 20 August 2021).

Newell, Bryce Clayton, 'Privacy as Antipower: In Pursuit of Non-Domination', *European Data Protection Law Review*, Vol. 4, No. 1 (2018), 12–16

Newell, Bryce Clayton, Tjerk Timan and Bert-Jaap Koops (eds), *Surveillance, Privacy and Public Space*, Routledge, Abingdon, 2018.

Nisbett, Richard E., *The Geography of Thought*, Nicholas Brealey Publishing, London, 2005.

Nix, Naomi, 'Facebook Ran Multi-Year Charm Offensive to Woo State Prosecutors', *Bloomberg*, 27 May 2020 <https://www.bloomberg.com/news/articles/2020-05-27/facebook-ran-multi-year-charm-offensive-to-woo-state-prosecutors> (accessed 20 August 2021).

Noble, Safiya Umoja, *Algorithms of Oppression: How Search Engines Reinforce Racism*, New York University Press, New York, 2018.

Noordyke, Mitchell, 'US state comprehensive privacy law comparison', *International Association of Privacy Professionals*, 18 April 2019 <https://iapp.org/news/a/us-state-comprehensive-privacy-law-comparison/> (accessed 20 August 2021).

Nouwens, Midas et al., 'Dark Patterns after the GDPR: Scraping Consent Pop-ups and Demonstrating their Influence', *arXiv*: 2001.02479 (2020).

Ober, Josiah, *Demopolis: Democracy Before Liberalism in Theory and Practice*, Cambridge University Press, Cambridge, 2017.

Oberdiek, John (ed.), *Philosophical Foundations of the Law of Torts*, Oxford University Press, Oxford, 2018.

Obermeyer, Ziad et al., 'Dissecting racial bias in an algorithm used to manage the health of populations', *Science*, Vol. 366, No. 6464 (2019), 447–453.

Ochigame, Rodrigo, 'The Invention of "Ethical AI"', *The Intercept*, 20 December 2019 <https://theintercept.com/2019/12/20/mit-ethi cal-ai-artificial-intelligence/> (accessed 20 August 2021).

O'Connor, Sarah, 'When your boss is an algorithm', *Financial Times*, 8 September 2016 <https://www.ft.com/content/88fdc 58e-754f-11e6-b60a-de4532d5ea35> (accessed 20 August 2021).

OECD (ed.), 'Executive Summary', in OECD (ed.), *Innovative Citizen Participation and New Democratic Institutions*, OECD Publishing, Paris, 2020.

—, *Innovative Citizen Participation and New Democratic Institutions: Catching the Deliberative Wave: Highlights 2020*, OECD Publishing, Paris, 2020.

Office of Communications, *The Ofcom Broadcasting Code* (with the *Cross-promotion Code and the On Demand Programme Service Rules*), 2020.

O'Hara, Kieron and Nigel Shadbolt, *The Spy in the Coffee Machine: The End of Privacy as We Know it*, Oneworld Publications, London, 2008.

O'Hara, Kieron and Wendy Hall, 'Four Internets: The Geopolitics of Digital Governance' (December 2018) Centre for International Governance Innovation Paper No. 206.

Ohm, Paul, 'Broken Promises of Privacy: Responding to the Surprising Failure of Anonymization', *UCLA Law Review*, Vol. 57, No. 6 (2010), 1701–1777.

O'Neil, Cathy, *Weapons of Math Destruction: How Big Data Increases Inequality and Threatens Democracy*, Crown, New York, 2016.

Orts, Eric W. and Craig N. Smith, *The Moral Responsibility of Firms*, Oxford University Press, Oxford, 2017.

Osnos, Evan, 'Can Mark Zuckerberg Fix Facebook Before It Breaks Democracy?', *The New Yorker*, 17 September 2018 <https://www.newyorker.com/magazine/2018/09/17/can-mark-zuckerberg-fix-facebook-before-it-breaks-democracy> (accessed 19 August 2021).

O'Sullivan, Donie, 'The biggest Black Lives Matter page on Facebook is fake', CNN, 9 April 2018 <https://money.cnn.com/2018/04/09/technology/fake-black-lives-matter-facebook-page/index.html> (accessed 20 August 2021).

Oversight Board, 'Announcing the Oversight Board's first case decisions', January 2021 <https://oversightboard.com/news/16552323 5084273-announcing-the-oversight-board-s-first-case-decisions/> (accessed 20 August 2021).

Ovide, Shira, 'Facebook and Its Secret Policies', *The New York Times*, 28 May 2020 <https://www.nytimes.com/2020/05/28/technology/facebook-polarization.html> (accessed 20 August 2021).

Owen, David, 'Should We Be Worried About Computerized Facial Recognition?' *The New Yorker*, 10 December 2018 <https://www.newyorker.com/magazine/2018/12/17/should-we-be-worried-about-computerized-facial-recognition> (accessed 20 August 2021).

Pałka, Przemysław, 'Terms of Service Are Not Contracts - Beyond Contract Law in the Regulation of Online Platforms', in Stefan Grundmann et al. (eds), *European Contract Law in the Digital Age*, Intersentia, Cambridge, 2018.

Pardes, Arielle, 'How Facebook and Other Sites Manipulate Your Privacy Choices', *Wired*, 12 August 2020 <https://www.wired.com/story/

facebook-social-media-privacy-dark-patterns/> (accessed 20 August 2021).

Parfit, Derek, *Reasons and Persons*, Oxford University Press, Oxford, 1987.

Park, Seung-min et al., 'A mountable toilet system for personalized health monitoring via the analysis of excreta', *Nature Biomedical Engineering*, Vol. 4, No. 6 (2020), 624–635.

Pasquale, Frank, 'Data-Informed Duties in AI Development', *Columbia Law Review*, Vol. 119, No. 7 (2019), 1917–1940.

—, 'The Automated Public Sphere' (2017) University of Maryland Legal Studies Research Paper No. 2017-31.

—, *The Black Box Society: The Secret Algorithms that Control Money and Information*, Harvard University Press, Cambridge, MA, 2015.

Patel, Nilay, 'Facebook's $5 billion FTC fine is an embarrassing joke', *The Verge*, 12 July 2019 <https://www.theverge.com/2019/7/12/20692 524/facebook-five-billion-ftc-fine-embarrassing-joke> (accessed 20 August 2021).

Paton, Graeme, 'Admiral charges Hotmail users more for car insurance', *The Times*, 23 January 2018 <https://www.thetimes.co.uk/article/ admiral-charges-hotmail-users-more-for-car-insurance-hrzjxsslr> (accessed 20 August 2021).

Pennycook, Gordon and David G. Rand, 'Lazy, not biased: Susceptibility to partisan fake news is better explained by lack of reasoning than by motivated reasoning', *Cognition*, Vol. 188 (2019), 39–50.

Peppet, Scott R., 'Regulating the Internet of Things: First Steps Toward Managing Discrimination, Privacy, Security, and Consent', *Texas Law Review*, Vol. 93, No. 83 (2014), 85–176.

Perrin, William, 'Regulation, misinformation and COVID19', *Carnegie UK*, 30 April 2020 <https://www.carnegieuktrust.org.uk/blog-posts/reg ulation-misinformation-and-covid19/> (accessed 20 August 2021).

Persily, Nathaniel, *The Internet's Challenge to Democracy: Framing the Problem and Assessing Reforms*, Kofi Annan Foundation, Stanford, CA, 2019.

Persily, Nathaniel and Joshua A. Tucker (eds), *Social Media and Democracy: The State of the Field and Prospects for Reform*, Cambridge University Press, New York, 2020.

Pettit, Philip, 'Is Facebook Marking Us Less Free?' *The Institute of Art and Ideas*, 26 March 2018 <https://iai.tv/articles/the-big-brotherhood-of- digital-giants-is-taking-away-our-freedom-auid-884> (accessed 20 August 2021).

—, 'Two Concepts of Free Speech', in Jennifer Lackey (ed.), *Academic Freedom*, Oxford University Press, Oxford, 2018.

—, 'Political Realism Meets Civic Republicanism', *Critical Review of International Social and Political Philosophy*, Vol. 20, No. 3 (2017), 331–347.

—, 'A Brief History of Liberty - And Its Lessons', *Journal of Human Development and Capabilities*, Vol. 17, No. 1 (2016), 5–21.

—, *On the People's Terms: A Republican Theory and Model of Democracy*, Cambridge University Press, Cambridge, 2014.

—, *Just Freedom: A Moral Compass for a Complex World*, W. W. Norton and Company, New York, 2014.

—, *Republicanism: A Theory of Freedom and Government*, Oxford University Press, Oxford, 2010.

—, 'Freedom as Antipower', *Ethics*, Vol. 106, No. 3 (April 1996), 576–604.

—, 'On Corporate Governance: Comments on Ravi Kailas'.

Pickard, Victor, 'The Strange Life and Death of the Fairness Doctrine: Tracing the Decline of Positive Freedoms in American Policy Discourse', *International Journal of Communication*, Vol. 12 (2018), 3434–3453.

The Pillar, 'Location-based apps pose security risk for Holy See', *The Pillar*, 27 July 2021 <https://www.pillarcatholic.com/p/locat ion-based-apps-pose-security> (accessed 3 October 2021).

Pistor, Katharina, *The Code of Capital: How the Law Creates Wealth and Inequality*, Princeton University Press, Princeton, 2019.

Pocock, J. G. A., *The Machiavellian Moment: Florentine Political Thought and the Atlantic Republican Tradition*, Princeton University Press, Princeton, 1975.

Pomerantsev, Peter, *This Is Not Propaganda: Adventures in the War Against Reality*, Faber & Faber, London, 2019.

Popper, Nathaniel, 'Lost Passwords Lock Millionaires Out of Their Bitcoin Fortunes', *The New York Times*, 12 January 2021 <https://www.nytimes. com/2021/01/12/technology/bitcoin-passwords-wallets-fortunes. html> (accessed 19 August 2021).

Posner, Richard A., 'Natural Monopoly and Its Regulation', *Stanford Law Review*, Vol. 21, No. 3 (1968), 548–643.

Pozen, David E. (ed.), *The Perilous Public Square: Structural Threats to Free Expression Today*, Columbia University Press, New York, 2020.

Prewitt, Matt, 'A View Of The Future Of Our Data', *Noēma Magazine*, 23 February 2021 <https://www.noemamag.com/a-view-of-the-fut ure-of-our-data/> (accessed 20 August 2021).

Price II, W. Nicholson, 'Regulating Black-Box Medicine', *Michigan Law Review*, Vol. 116, No. 3 (2017), 421–474.

Radin, Margaret Jane, *Boilerplate: The Fine Print, Vanishing Rights, and the Rule of Law*, Princeton University Press, Princeton, 2013.

Rahman, K. Sabeel, 'Regulating Informational Infrastructure: Internet Platforms as the New Public Utilities', *Georgetown Law and Technology Review*, Vol. 2, No. 2 (2018), 234–251.

—, 'Monopoly Men', *Boston Review*, 11 October 2017 <https://bosto nreview.net/class-inequality/k-sabeel-rahman-monopoly-men> (accessed 3 October 2021).

—, *Democracy Against Domination*, Oxford University Press, Oxford, 2017.

Rahman, K. Sabeel and Zephyr Teachout, 'From Private Bads to Public Goods: Adapting Public Utility Regulation for Information Infrastructure: Dismantling surveillance-based business models', *Knight First Amendment Institute*, 4 February 2020 <https://knightcolum bia.org/content/from-private-bads-to-public-goods-adapting-pub lic-utility-regulation-for-informational-infrastructure> (accessed 20 August 2021).

Raj, Prateek, '"Antimonopoly Is as Old as the Republic"', *Promarket*, 22 May 2017 <https://promarket.org/2017/05/22/antimonop oly-old-republic/> (accessed 20 August 2021).

Rajkomar, Alvin et al., 'Ensuring Fairness in Machine Learning to Advance Health Equity', *Annals of Internal Medicine*, Vol. 169, No. 12 (2018), 866–872.

Randell, Charles, 'How can we ensure that Big Data does not make us prisoners of technology?' *Reuters Newsmaker Event*, Reuters News & Media, London, 11 July 2018 <https://www.fca.org.uk/news/speec hes/how-can-we-ensure-big-data-does-not-make-us-prisoners-tec hnology> (accessed 20 August 2021).

Rangan, Subramanian (ed.), *Performance & Progress: Essays on Capitalism, Business, and Society*, Oxford University Press, Oxford, 2017.

Ranking Digital Rights, '2019 Ranking Digital Rights Corporate Accountability Index', May 2019 <https://rankingdigitalrights.org/ index2019/assets/static/download/RDRindex2019report.pdf> (accessed 20 August 2021).

Raub, McKenzie, 'Bots, Bias, and Big Data: Artificial Intelligence, Algorithmic Bias and Disparate Impact Liability in Hiring Practices', *Arkansas Law Review*, Vol. 71, No. 2 (2018), 529–570.

Rauber, Jonas, Emily B. Fox and Leon A. Gatys, 'Modeling patterns of smartphone usage and their relationship to cognitive health', *arXiv*: 1911.05683 (2019).

Rawls, John, *Political Liberalism*, Columbia University Press, New York, 2015.

—, *A Theory of Justice*, Harvard University Press, Cambridge, MA, 1999.

Reality Check Team, 'Social media: How do other governments regulate it?' BBC, 12 February 2020 <https://www.bbc.co.uk/news/technology-47135058> (accessed 20 August 2021).

Regan, Priscilla M., 'Reviving the Public Trustee Concept and Applying it to Information Privacy Policy', Maryland Law Review, Vol. 76, No. 4 (2017), 1025–1043.

Reidenberg, Joel R., N. Cameron Russell, Alexander J. Callen, Sophia Casio and Thomas B. Norton, 'Privacy Harms and the Effectiveness of the Notice and Choice Framework', I/S: A Journal of Law and Policy, Vol. 11, No. 2 (2014), 485–524.

Reinsel, David, John Gantz and John Rydning, 'The Digitization of the World: From Edge to Core' (2019) IDC White Paper No. US44413318.

Reisman, Dillon et al., 'Algorithmic Impact Assessments: A Practical Framework for Public Agency Accountability', AI Now Institute, April 2018 <https://ainowinstitute.org/aiareport2018.pdf> (accessed 3 October 2021).

Rességuier, Anaïs and Rowena Rodrigues, 'AI ethics should not remain toothless! A call to bring back the teeth of ethics', Big Data & Society, Vol. 7, No. 2 (2020), 1–5.

Richards, Neil, Intellectual Privacy: Rethinking Civil Liberties in the Digital Age, Oxford University Press, Oxford, 2017.

Richards, Neil and Woodrow Hartzog, 'Privacy's Constitutional Moment and the Limits of Data Protection', Boston College Law Review, Vol. 61, No. 5 (2020), 1687–1761.

—, 'The Pathologies of Digital Consent', Washington University Law Review, Vol. 96 (2019), 1461–1503.

—, 'Trusting Big Data Research', DePaul Law Review, Vol. 66, No. 2 (2017), 579–590.

—, 'Privacy's Trust Gap: A Review', Yale Law Journal, Vol. 126, No. 4 (2017), 1180–1224.

—, 'Taking Trust Seriously in Privacy Law,' Stanford Technology Law Review, Vol. 19 (2016), 431–472.

Richter, Melvin, The Political Theory of Montesquieu, Cambridge University Press, New York, 1977.

Rigillo, Nichole, 'AI Must Explain Itself', Noēma Magazine, 16 June 2020 <https://www.noemamag.com/ai-must-explain-itself/> (accessed 20 August 2021).

Roberts, Sarah T., Behind the Screen: Content Moderation in the Shadows of Social Media, Yale University Press, New Haven and London, 2019.

Robertson, Adi, 'Facebook and Twitter are restricting a disputed New York Post story about Joe Biden's son', *The Verge*, 14 October 2020 <https://www.theverge.com/2020/10/14/21515972/faceb ook-new-york-post-hunter-biden-story-fact-checking-reduced-distr ibution-election-misinformation> (accessed 4 October 2021).

Robitzski, Dan, 'Ex-Googler: Company Has "Voodoo Doll, Avatar-Like Version of You"', *Futurism*, 2 May 2019 <https://futurism.com/goo gle-company-voodoo-doll-avatar/amp> (accessed 20 August 2021).

Romm, Tony, 'Amazon, Facebook and Google turn to deep network of political allies to battle back antitrust probes', *Washington Post*, 10 June 2020 <https://www.washingtonpost.com/technol ogy/2020/06/10/amazon-facebook-google-political-allies-antitr ust/> (accessed 20 August 2021).

Roose, Kevin, 'The Making of a YouTube Radical', *The New York Times*, 8 June 2019 <https://www.nytimes.com/interactive/2019/06/08/ technology/youtube-radical.html> (accessed 20 August 2021).

Rovatsos, Michael, Brent Mittelstadt and Ansgar Koene, 'Landscape Summary: Bias in Algorithmic Decision-Making', *Centre for Data Ethics and Innovation*, 19 July 2019 <https://assets.publishing.service.gov.uk/ government/uploads/system/uploads/attachment_data/file/819 055/Landscape_Summary_-_Bias_in_Algorithmic_Decision-Mak ing.pdf> (accessed 20 August 2021).

Rowbottom, Jacob, *Media Law*, Hart Publishing, Oxford, 2018.

Rubenstein, Ira S. and Nathaniel Good, 'Privacy by Design: A Counterfactual Analysis of Google and Facebook Privacy Incidents', *Berkeley Technology Law Journal*, Vol. 28, No. 2 (2013), 1333–1413.

Rudin, Cynthia, 'Stop explaining black box machine learning models for high stakes decisions and use interpretable models instead', *Nature Machine Intelligence*, Vol. 1 (May 2019), 206–215.

Ruhaak, Anouk, 'Data trusts: what are they and how do they work?', *The RSA*, 11 June 2020 <https://www.thersa.org/blog/2020/06/data-trusts-protection> (accessed 20 August 2021)

—, 'When One Affects Many: The Case For Collective Consent', *Mozilla*, 13 February 2020 <https://foundation.mozilla.org/en/blog/ when-one-affects-many-case-collective-consent/> (accessed 20 August 2021).

Russell, Stuart, *Human Compatible: AI and the Problem of Control*, Allen Lane, London, 2019.

Ryan, Johnny and Alan Toner, 'Europe's enforcement paralysis: ICCL's 2021 report on the enforcement capacity of data protection authorities',

Irish Council for Civil Liberties, 2021 <https://www.iccl.ie/digi tal-data/2021-gdpr-report/> (accessed 3 October 2021).

—, 'Europe's governments are failing the GDPR', Brave, 2020 <https:// brave.com/wp-content/uploads/2020/04/Brave-2020-DPA-Rep ort.pdf> (accessed 20 August 2021).

Sales, Philip, 'Algorithms, Artificial Intelligence and the Law', Sir Henry Brooke Lecture for BAILII, Freshfields Bruckhaus Deringer LLP, London, 12 November 2019.

Sallust, Catiline's War, The Jugurthine War, Histories, Penguin Classics, London, 2007.

Sandel, Michael (ed.), Democracy's Discontent: America in Search of a Public Philosophy, Harvard University Press, Cambridge, MA, 1998.

—, Liberalism and its Critics, New York University Press, New York, 1984.

Satariano, Adam, 'Europe's Privacy Law Hasn't Shown Its Teeth, Frustrating Advocates', The New York Times, 27 April 2020 <https:// www.nytimes.com/2020/04/27/technology/GDPR-privacy-law- europe.html> (accessed 20 August 2021).

—, 'Europe Is Toughest on Big Tech, Yet Big Tech Still Reigns', The New York Times, 11 November 2019 <https://www.nytimes. com/2019/11/11/business/europe-technology-antitrust-regulat ion.html> (accessed 20 August 2021).

—, 'Google Fined $1.7 Billion by E.U. for Unfair Advertising Rules', The New York Times, 20 March 2019 <https://www.nytimes. com/2019/03/20/business/google-fine-advertising.html> (accessed 20 August 2021).

Satariano, Adam and Mike Isaac, 'The Silent Partner Cleaning Up Facebook for $500 Million a Year', The New York Times, 31 August 2021 <https:// www.nytimes.com/2021/08/31/technology/facebook-accenture- content-moderation.html> (accessed 3 October 2021).

Scharff, Robert C. and Val Dusel (eds), Philosophy of Technology: The Technological Condition: An Anthology, 2nd edn, John Wiley & Sons, Maldon, 2014.

Schauer, Frederick, Profiles, Probabilities, and Stereotypes, Harvard University Press, Cambridge, MA, 2003.

Schick, Nina, Deepfakes: The Coming Infocalypse, Monoray, London, 2020.

Schmitt, Carl, Political Theology: Four Chapters on the Concept of Sovereignty, University of Chicago Press, Chicago, 2005.

Schneier, Bruce, 'We're Banning Facial Recognition. We're Missing the Point', The New York Times, 20 January 2020, <https://www.nytimes. com/2020/01/20/opinion/facial-recognition-ban-privacy.html> (accessed 20 August 2021).

—, *Click Here to Kill Everybody: Security and Survival in a Hyper-Connected World*, W. W. Norton & Company, New York, 2018.

Sciencewise, 'Supporting socially informed policy making', 2021 <https://sciencewise.org.uk> (accessed 20 August 2021).

Scott, Mark, Laurens Cerulus and Steven Overly, 'How Silicon Valley gamed Europe's privacy rules', *Politico*, 22 May 2019 <https://www.politico.eu/article/europe-data-protection-gdpr-general-data-protection-regulation-facebook-google/> (accessed 20 August 2021).

Scruton, Roger, *England: An Elegy*, Bloomsbury, London, 2006.

Seddon, Max and Madhumita Murgia, 'Apple and Google drop Navalny app after Kremlin piles on pressure', *Financial Times*, 17 September 2021 <https://www.ft.com/content/faaada81-73d6-428c-8d74-88d273adbad3> (accessed 3 October 2021).

Sedley, Stephen, *Lions Under the Throne: Essays on the History of English Public Law*, Cambridge University Press, Cambridge, 2015.

Sejnowski, Terrence J., *The Deep Learning Revolution*, MIT Press, Cambridge, MA, 2018.

Selbst, Andrew D. and Julia Powles, 'Meaningful information and the right to explanation', *International Data Privacy Law*, Vol. 7, No. 4 (2017), 233–242.

Selbst, Andrew D. and Solon Barocas, 'The Intuitive Appeal of Explainable Machines', *Fordham Law Review*, Vol. 87, No. 3 (2018), 1085–1140.

Senate Democrats, 'A Better Deal: Cracking Down on Corporate Monopolies', June 2017 <https://www.democrats.senate.gov/imo/media/doc/2017/07/A-Better-Deal-on-Competition-and-Costs-1.pdf> (accessed 20 August 2021).

Shapiro, Carl, 'Protecting Competition in the American Economy: Merger Control, Tech Titans, Labor Markets', *Journal of Economic Perspectives*, Vol. 33, No. 3 (Summer 2019), 69–93.

Sherman, Justin, 'Data Brokers and Sensitive Data on U.S. Individuals: Threats to American Civil Rights, National Security, and Democracy', Duke University Sanford Cyber Policy Program, 2021 <https://sites.sanford.duke.edu/techpolicy/report-data-brokers-and-sensitive-data-on-u-s-individuals/> (accessed 3 October 2021).

—, 'Data Brokers Are a Threat to Democracy', *Wired*, 13 April 2021 <https://www.wired.com/story/opinion-data-brokers-are-a-threat-to-democracy/> (accessed 3 October 2021).

Silverman, Craig and Jane Lytvynenko, 'Amazon Is Pushing Readers Down A "Rabbit Hole" Of Conspiracy Theories About The Coronavirus',

BuzzFeed News, 15 March 2021 <https://www.buzzfeednews.com/article/craigsilverman/amazon-covid-conspiracy-books> (accessed 3 October 2021).

Simon, Matt, 'This Robot Can Guess How You're Feeling by the Way You Walk', *Wired*, 18 May 2020 <https://www.wired.com/story/proxemo-robot-guesses-emotion-from-walking/> (accessed 20 August 2021).

Simonite, Tom, 'The AI Text Generator That's Too Dangerous to Make Public', *Wired*, 14 February 2019 <https://www.wired.com/story/ai-text-generator-too-dangerous-to-make-public/> (accessed 20 August 2021).

Singer, Natasha and Cade Metz, 'Many Facial-Recognition Systems Are Biased, Says U.S. Study', *The New York Times*, 19 December 2019 <https://www.nytimes.com/2019/12/19/technology/facial-recognition-bias.html> (accessed 20 August 2021).

Skinner, Quentin, *Hobbes and Republican Liberty*, Cambridge University Press, Cambridge, 2008.

—, *Liberty before Liberalism*, Cambridge University Press, Cambridge, 1998.

Slack, Megan, 'From the Archives: President Teddy Roosevelt's New Nationalism Speech', *The White House: President Barack Obama*, 6 December 2011 <https://obamawhitehouse.archives.gov/blog/2011/12/06/archives-president-teddy-roosevelts-new-nationalism-speech> (accessed 20 August 2021).

Smith, Brad and Carol Ann Browne, *Tools and Weapons: The Promise and the Peril of the Digital Age*, Hodder & Stoughton, London, 2019.

Smith, Robert Elliott, *Rage Inside the Machine: The Prejudice of Algorithms, and How to Stop the Internet Making Bigots of Us All*, Bloomsbury Business, London, 2019.

Solove, Daniel J., 'Introduction: Privacy Self-Management and the Consent Dilemma', *Harvard Law Review*, Vol. 126, No. 7 (2013), 1880–1903.

Solove, Daniel J. and Woodrow Hartzog, 'The FTC and the New Common Law of Privacy', *Columbia Law Review*, Vol. 114:583 (2014).

Statt, Nick, 'Google expands AI calling service Duplex to Australia, Canada, and the UK', *The Verge*, 8 April 2020 <https://www.theverge.com/2020/4/8/21214321/google-duplex-ai-automated-calling-australia-canada-uk-expansion> (accessed 20 August 2021).

Stewart, Fenner L., 'Dominium and the Empire of Laws', *Windsor Yearbook of Access to Justice*, Vol. 35 (2019), 36–62.

Stigler Committee on Digital Platforms, 'Final Report', Stigler Centre for the Study of the Economy and the State, 2019 <https://www.chica

gobooth.edu/~/media/research/stigler/pdfs/digital-platforms---committee-report---stigler-center.pdf> (accessed 20 August 2021).

Stilgoe, Jack, *Who's Driving Innovation? New Technologies and the Collaborative State*, Palgrave Macmillan, Cham, 2020.

Stokel-Walker, Chris, 'If You're a Remote Worker, You're Going to Be Surveilled. A Lot', *OneZero*, 23 April 2020 <https://onezero.medium.com/if-youre-a-remote-worker-you-re-going-to-be-surveilled-a-lot-f3f8d4308ee> (accessed 20 August 2021).

Stolton, Samuel, 'Vestager distances Commission from option of 'Big Tech Breakups'', *Euractiv*, 27 October 2020 <https://www.euractiv.com/section/digital/news/vestager-distances-commission-from-option-of-big-tech-breakups/> (accessed 20 August 2021).

Storbeck, Olaf, Madhumita Murgia and Rochelle Toplensky, 'Germany blocks Facebook from pooling user data without consent', *The Financial Times*, 7 February 2019 <https://www.ft.com/content/3a035 1b6-2ab9-11e9-88a4-c32129756dd8> (accessed 20 August 2021).

Storr, Will, *The Science of Storytelling*, William Collins, London, 2019.

Sullivan, Charles A., 'Employing AI', *Villanova Law Review*, Vol. 63, No. 3 (2018), 395–430.

Sumption, Jonathan, *Trials of the State: Law and the Decline of Politics*, Profile Books, London, 2019.

Sunstein, Cass, *Free Markets and Social Justice*, Oxford University Press, New York, 1997.

—, *Democracy and the Problem of Free Speech*, Free Press, New York, 1995.

—, *After the Rights Revolution: Reconceiving the Regulatory State*, Harvard University Press, Cambridge, MA, 1993.

Susarla, Anjana, 'Biases in algorithms hurt those looking for information on health', *The Conversation*, 14 July 2020 <https://theconversation.com/biases-in-algorithms-hurt-those-looking-for-information-on-health-140616> (accessed 20 August 2021).

Susskind, Jamie, 'Chatbots Are a Danger to Democracy', *The New York Times*, 4 December 2018 <https://www.nytimes.com/2018/12/04/opinion/chatbots-ai-democracy-free-speech.html> (accessed 20 August 2021)

—, *Future Politics: Living Together in a World Transformed by Tech*, Oxford University Press, Oxford, 2018.

—, 'What we need from social media is transparency, not apologies', *The New Statesman*, 6 September 2018 <https://www.newstatesman.

com/science-tech/2018/09/what-we-need-social-media-transpare
ncy-not-apologies> (accessed 24 September 2021).

Susskind, Richard, 'The Future of Courts', The Practice, Vol. 6, Issue 5, July/
August 2020.

—, Online Courts and the Future of Justice, Oxford University Press, Oxford, 2019.

—, The Future of Law: Facing the Challenges of Legal Technology, Oxford University
Press, Oxford, 1996.

Susskind, Richard and Daniel Susskind, The Future of the Professions: How
Technology will Transform the Work of Human Experts, Oxford University Press,
Oxford, 2015.

Sutton, Richard S. and Andrew G. Barto, Reinforcement Learning: An Introduction,
2nd edn, MIT Press, Cambridge, MA, 2018.

Suzor, Nicolas P., Lawless: The Secret Rules that Govern Our Digital Lives, Cambridge
University Press, Cambridge, 2019.

Suzor, Nicolas P. et al., 'What Do We Mean When We Talk About
Transparency? Toward Meaningful Transparency in Commercial
Content Moderation', International Journal of Communication, Vol. 13 (2019),
1526–1543.

Swedloff, Rick, 'The New Regulatory Imperative for Insurance', Boston
College Law Review, Vol. 61, No. 6 (2020), 2031–2084.

Tambini, Damian, Danilo Leonardi and Chris Marsden, 'The privatisation
of censorship: self regulation and freedom of expression', in Damian
Tambini, Danilo Leonardi and Chris Marsden (eds), Codifying
cyberspace: communications self-regulation in the age of internet convergence,
Routledge, Abingdon, 2008.

Tamò-Larrieux, Aurelia, Designing for Privacy and its Legal Framework: Data
Protection by Design and Default for the Internet of Things, Springer, Cham, 2018.

Taylor, Charles, Philosophy and the Human Sciences: Philosophical Papers 2,
Cambridge University Press, Cambridge, 1999.

Taylor, Charlie, 'Data Protection Commission criticised as WhatsApp
decision nears', Irish Times, 15 January 2020 <https://www.irishti
mes.com/business/technology/data-protection-commission-cri
ticised-as-whatsapp-decision-nears-1.4139804> (accessed 20
August 2021).

Taylor, Robert S., Exit Left: Markets and Mobility in Republican Thought, Oxford
University Press, Oxford, 2017.

Teachout, Zephyr, Break 'Em Up: Recovering our Freedom from Big Ag, Big Tech, and
Big Money, All Points Books, New York, 2020.

Tegmark, Max, Life 3.0: Being human in the age of Artificial Intelligence, Allen Lane,
London, 2017.

Thaler, Richard H. and Cass Sunstein, *Nudge: Improving Decisions about Health, Wealth and Happiness*, 1st edn, Penguin Books, London, 2009.

Thierer, Adam, 'The Perils of Classifying Social Media Platforms as Public Utilities', *CommLaw Conspectus - Journal of Communications Law and Policy*, Vol. 21, No. 2 (2013), 249–297.

Thompson, Clive, *Coders: Who They Are, What They Think And How They Are Changing Our World*, Picador, London, 2019.

Thompson, Stuart A. and Charlie Warzel, 'How to Track President Trump', *The New York Times*, 20 December 2019 <https://www.nytimes.com/interactive/2019/12/20/opinion/location-data-national-security.html> (accessed 20 August 2021).

—, 'Twelve Million Phones, One Dataset, Zero Privacy', *The New York Times*, 19 December 2019 <https://www.nytimes.com/interactive/2019/12/19/opinion/location-tracking-cell-phone.html> (accessed 20 August 2021).

Tidy, Joe, 'Twitter apologises for letting ads target neo-Nazis and bigots', BBC, 16 January 2020 <https://www.bbc.co.uk/news/technology-51112238> (accessed 20 August 2021).

Tiffany, Kaitlyn, 'No, the Internet Is Not Good Again', *The Atlantic*, 16 April 2020 <https://www.theatlantic.com/technology/archive/2020/04/zoom-facebook-moderation-ai-coronavirus-internet/610099/> (accessed 20 August 2021).

Timmins, Nicholas, *The Five Giants: A Biography of the Welfare State*, William Collins, London, 2017.

Townsend, Mark, 'Facebook algorithm found to "actively promote" Holocaust denial', *The Guardian*, 16 August 2020 <https://www.theguardian.com/world/2020/aug/16/facebook-algorithm-found-to-actively-promote-holocaust-denial> (accessed 20 August 2021).

Townsend, Tess, 'Keith Ellison and the New "Antitrust Caucus" Want to Know Exactly How Bad Mergers Have Been for the American Public', *Intelligencer*, 4 December 2017 <https://nymag.com/intelligencer/2017/12/antitrust-bill-from-keith-ellison-seek-info-on-mergers.html> (accessed 20 August 2021).

Transatlantic High Level Working Group on Content Moderation, 'Freedom and Accountability: A Transatlantic Framework for Moderating Speech Online', Annenberg Public Policy Center, Philadelphia, June 2020.

Tucker, Ian, 'Yaël Eisenstat: "Facebook is ripe for manipulation and viral misinformation"', *The Guardian*, 26 July 2020 <https://www.theguardian.com/technology/2020/jul/26/yael-eisenstat-faceb

ook-is-ripe-for-manipulation-and-viral-misinformation> (accessed 20 August 2021).

Tucker, Paul, *Unelected Power: The Quest for Legitimacy in Central Banking and the Regulatory State*, Princeton University Press, Princeton, 2018.

Tufekci, Zeynep, 'The Latest Data Privacy Debacle', *The New York Times*, 30 January 2018 <https://www.nytimes.com/2018/01/30/opinion/strava-privacy.html> (accessed 20 August 2021).

Turner, Fred, *From Counterculture to Cyberculture: Stewart Brand, the Whole Earth Network, and the Rise of Digital Utopianism*, University of Chicago Press, Chicago and London, 2008.

Turner, Jacob, *Robot Rules: Regulating Artificial Intelligence*, Palgrave Macmillan, London, 2019.

Turvill, William, 'Apple may be forced to disclose censorship requests from China', *The Guardian*, 25 February 2020 <https://www.theg uardian.com/technology/2020/feb/25/apple-censorship-reque sts-china-shareholder-groups-proposal> (accessed 20 August 2021).

Tusikov, Natasha, *Chokepoints: Global Private Regulation on the Internet*, University of California Press, Oakland, 2017.

Vaidhyanathan, Siva, *Antisocial Media: How Facebook Disconnects Us And Undermines Democracy*, Oxford University Press, New York, 2018.

—, *The Googlization of Everything (and Why We Should Worry)*, University of California Press, Berkeley, 2011.

Van Gelderen, Martin and Quentin Skinner (eds), *Republicanism: A Shared European Heritage (Volume I)*, Cambridge University Press, Cambridge, 2006.

—, *Republicanism: A Shared European Heritage (Volume II)*, Cambridge University Press, Cambridge, 2006.

Van Loo, Rory, 'The New Gatekeepers: Private Firms as Public Enforcers', *Virginia Law Review*, Vol. 106, No. 2 (2020), 467–522.

—, 'The Missing Regulatory State: Monitoring Businesses in an Age of Surveillance', *Vanderbilt Law Review*, Vol. 72, No. 5 (2019), 1563–1631.

Veale, Michael, 'A Critical Take on the Policy Recommendations of the EU High-Level Expert Group on Artificial Intelligence', *European Journal of Risk Regulation* (2020), 1–10.

Veale, Michael, Reuben Binns and Lilian Edwards, 'Algorithms that remember: model inversion attacks and data protection law', *Philosophical Transactions of the Royal Society of London: Series A*, Vol. 376, No. 2133 (2018), 20180083–20180098.

Véliz, Carissa, *Privacy is Power: Why and How you Should Take Back Control of Your Data*, Transworld Publishers, London, 2020.

—, 'Three things digital ethics can learn from medical ethics', *Nature Electronics*, Vol. 2, No. 8 (2019), 1–3.

Vetterli, Richard and Gary Bryner, *In Search of the Republic: Public Virtue and the Roots of American Government*, Rowman & Littlefield Publishers, Lanham, MD, 1996.

Vibert, Frank, *The Rise of the Unelected: Democracy and the New Separation of Powers*, Cambridge University Press, Cambridge, 2007.

Vigen, Tyler, *Spurious Correlations*, Hachette Books, New York, 2015.

Vinocur, Nicholas, '"We have a huge problem": European tech regulator despairs over lack of enforcement', *Politico*, 27 December 2019 <https://www.politico.com/news/2019/12/27/europe-gdpr-technology-regulation-089605> (accessed 20 August 2021).

Wachter, Sandra and Brent Mittelstadt, 'A Right to Reasonable Inferences: Re-Thinking Data Protection Law in the Age of Big Data and AI', *Columbia Business Law Review*, Vol. 2019, No. 2 (2019), 494–620.

Wachter, Sandra, Brent Mittelstadt and Chris Russell, 'Why fairness cannot be automated: Bridging the gap between EU non-discrimination law and AI', *Computer Law & Security Review*, Vol. 41 (2021), 105567–105597.

—, 'Counterfactual Explanations Without Opening the Black Box: Automated Decisions and the GDPR', *Harvard Journal of Law & Technology*, Vol. 31, No. 2 (2018), 841–888.

Wachter, Sandra and Luciano Floridi, 'Why a right to explanation of automated decision-making does not exist in the General Data Protection Regulation', *International Data Privacy Law*, Vol. 7, No. 2 (2017), 76–99.

Wagner, Ben, 'Ethics as an escape from regulation: From "ethics-washing" to ethics-shopping?' in Emre Bayamlıoğlu et al. (eds), *Being Profiled: Cogitas Ergo Sum: 10 Years of Profiling the European Citizen*, Amsterdam University Press, Amsterdam, 2018.

Waldron, Jeremy, *Political Political Theory: Essays on Institutions*, Harvard University Press, Cambridge, MA, 2016.

Walker, Robert, 'The First Amendment and Article Ten: Sisters Under The Skin?' *Holdsworth Club Presidential Address*, Birmingham Law School, Birmingham, 24 October 2008.

Wallor, Shannon, *Technology and the Virtues: A Philosophical Guide to a Future Worth Wanting*, Oxford University Press, Oxford, 2016.

Walsh, Toby, *Android Dreams: The Past, Present and Future of Artificial Intelligence*, C. Hurst & Co (Publishers), London, 2017.

Walzer, Michael, 'The Communitarian Critique of Liberalism', *Political Theory*, Vol. 18, No. 1 (Feb 1990), 6–23.

—, *Exodus and Revolution*, Basic Books, New York, 1985.

—, *Spheres of Justice: A Defense of Pluralism and Equality*, Basic Books, New York, 1983.

Warren, Elizabeth, 'Here's how we can break up Big Tech', *Team Warren*, 8 March 2019 <https://medium.com/@teamwarren/here s-how-we-can-break-up-big-tech-9ad9e0da324c> (accessed 20 August 2021).

Warren, Samuel D. and Louis D. Brandeis, 'The Right to Privacy', *Harvard Law Review*, Vol. 4, No. 5 (1980), 193–220.

Waters, Richard, 'Google scraps ethics council for artificial intelligence', *Financial Times* <https://www.ft.com/content/6e291 2f8-573e-11e9-91f9-b6515a54c5b1> (accessed 20 August 2021).

Watts, Edward J., *Mortal Republic: How Rome Fell Into Tyranny*, Basic Books, New York, 2018.

Webb, Amy, *The Big Nine: How the Tech Titans & Their Thinking Machines Could Warp Humanity*, Hachette Book Group, New York, 2019.

Weinberger, David, 'How Machine Learning Pushes Us to Define Fairness', *Harvard Business Review*, 6 November 2019 <https://hbr.org/2019/11/how-machine-learning-pushes-us-to-define-fairness> (accessed 20 August 2021).

—, *Everyday Chaos: Technology, Complexity, and How We're Thriving in a New World of Possibility*, Harvard Business Review Press, Boston, MA, 2019.

Weller, Adrian, 'Challenges for Transparency', *arXiv*: 1708.01870v1 (2017).

Wendell Holmes Jr., Oliver, *The Common Law*, Dover, New York, 1991.

Werbach, Kevin, *The Blockchain and the New Architecture of Trust*, MIT Press, Cambridge, MA, 2018.

West, Darrell M., '10 actions that will protect people from facial recognition software', *The Brookings Institution*, 31 October 2019 <https://www.brookings.edu/research/10-actions-that-will-protect-peo ple-from-facial-recognition-software/> (accessed 20 August 2021).

Wichowksi, Alexis, *The Information Trade: How Big Tech Conquers Countries, Challenges our Rights, and Transforms our World*, HarperOne, New York, 2020.

Wikipedia, 'Draco (lawgiver)', 30 June 2021 <https://en.wikipedia.org/wiki/Draco_(lawgiver)> (accessed 20 August 2021).

Williams, James, *Stand Out of Our Light: Freedom and Resistance in the Attention Economy*, Cambridge University Press, Cambridge, 2018.

Wilson, Tom and Kate Starbird, 'Cross-platform disinformation campaigns: Lessons learned and next steps', *The Harvard Kennedy School Misinformation Review*, Vol. 1, No. 1 (2020).

Wilson, Woodrow, 'The Study of Administration', *Political Science Quarterly*, Vol. 2, No. 2 (June 1887), 197–222.

Wollstonecraft, Mary, *A Vindication of the Rights of Women*, Penguin Classics, London, 2004.

Wong, Julia Carrie and Hannah Ellis-Petersen, 'Facebook planned to remove fake accounts in India – until it realized a BJP politician was involved', *The Guardian*, 15 April 2021 <https://www.theguardian.com/technology/2021/apr/15/facebook-india-bjp-fake-accounts> (accessed 3 October 2021).

Wu, Tim, 'A TikTok Ban Is Overdue', *The New York Times*, 18 August 2020 <https://www.nytimes.com/2020/08/18/opinion/tiktok-wechat-ban-trump. html> (accessed 20 August 2021).

——, *The Curse of Bigness: Antitrust in the New Gilded Age*, Columbia Global Reports, New York, 2018.

The Attention Merchants: The Epic Scramble to Get Inside our Heads, Alfred A. Knopf, New York, 2016.

——, *The Master Switch: The Rise and Fall of Information Empires*, Atlantic, London, 2010.

Yeung, Karen and Martin Lodge, *Algorithmic Regulation*, Oxford University Press, Oxford, 2019.

Zittrain, Jonathan, 'A Jury of Random People Can Do Wonders for Facebook', *The Atlantic*, 14 November 2019 <https://www.theatlantic.com/ideas/archive/2019/11/let-juries-review-facebook-ads/601996/> (accessed 20 August 2021).

——, 'How to Exercise the Power You Didn't Ask For', *Harvard Business Review*, 19 September 2018 <https://hbr.org/2018/09/how-to-exercise-the-power-you-didnt-ask-for> (accessed 20 August 2021).

Zuboff, Shoshana, *The Age of Surveillance Capitalism: The Fight for a Human Future at the New Frontier of Power*, Profile Books, London, 2019.

Zuckerberg, Mark, 'Mark Zuckerberg: Big Tech needs more regulation', *Financial Times*, 16 February 2020 <https://www.ft.com/content/602ec7ec-4f18-11ea-95a0-43d18ec715f5> (accessed 20 August 2021).Index

Index

Ada Lovelace Institute, 159
addictions, 73
Air Mail Act (1934), 241
alcohol consumption, 53
Alexander VI, Pope, 277
algorithmic discrimination, 260
Allan, Richard, 296
Allen, Tamsin, 192
Amazon, 38, 87, 101, 271
    abuse of market power, 235–8,
        241
    Amazon Prime, 36, 237
    recruitment system, 67–8
    Ring doorbell camera, 142
    and surveillance, 43–4
American Civil War, 116
American Revolution, 10, 22, 305
Analytical Engine, 69
anonymity, 42
anti-discrimination law, 140–1, 185,
    345, 373–4
antitrust laws, 8, 116, 120,
    124, 235–47
Apple, 67, 87, 94, 141, 236
Approved Persons, 200
Arendt, Hannah, 28, 33, 72
Aristotle, 160, 262
artificial intelligence (AI), 68–9, 99,
    171, 204, 244, 345, 364
    AI ethics, 6, 99, 101–2, 308

draft legislation, 179, 230
    see also machine learning
Association of Computing Machinery
    (ACM), 201
AT&T, 242
Athens, ancient, see Greece, ancient
Attenborough, David, 164
Australia, 124, 142, 172, 186, 188,
    192, 250, 274

Bagehot, Walter, 116
Banking Act, 241
Battle of the Standard, 170
Belfield, Haydn, 244
Benjamin, Ruha, 66
Bentham, Jeremy, 30
Bickert, Monika, 269
Biden, Joe, 4–5, 55, 145
Bills of Rights, 117, 239, 278
Bimber, Bruce, 149
Bing, 49
Black Lives Matter, 41, 274
Blackstone, William, 128
Blair, Tony, 153
body shape, 47
body-mass index, 74
Brandeis, Louis, 106, 239–40,
    289
Brazil, 54
Brexit, 206

British Columbia Civil Resolution Tribunal, 182
British Standards Institute (BSI), 170–1
broadcasting regulation, *see* media regulation
Broussard, Meredith, 66
Brownsword, Roger, 80, 82
Buolamwini, Joy, 66

Canada, 124, 142, 156, 159, 250, 287
cancer, 75
Catholic Church, 41
certification, 191–5
chatbots, 58–9, 120, 219, 229
Chauvin, Derek, 23
Chicago School, 241
child pornography, 295
China, 26, 120, 204, 207–8, 236
Cicero, 7, 20, 81, 83, 160, 262, 313
citizen juries, 156–7
'citizen welfare', 246
citizens' assemblies, 156
Citizens' Biometrics Council, 159
citizens' councils, 157
citizens' dialogues, 157
Citron, Danielle, 293
Civil Rights Act (1965), 90
Clearview AI, 142
Clegg, Nick, 14
Clinton, Bill, 130
Clinton, Hillary, 48, 274
Cohen, Julie E., 223, 304
'commercial society', rise of, 91
Communications Decency Act (Section 230), 122–3, 127, 295, 300
compensation, 163, 166, 199–200, 215
computational ideology, 72–7, 261
conditioning, 46
consensus conferences, 157
consent, 105–12, 125, 335
'consumer welfare', 241, 246
contract law, 129–30, 165–7

Controlled Functions, 200
corporate social responsibility, 102
corporations, and law, 128–30, 165, 167
counterpower, 177–8, 189, 209, 243, 245, 290, 353
couriers, 43
Covid-19 pandemic, 38, 41, 51, 54, 182, 271, 274
    and exam fiasco, 71–2
Coyle, Diane, 302
credit (and credit scores), 4, 37, 43, 72, 74, 90, 108, 120, 177, 179, 183, 219, 222, 228, 241, 254–5, 258, 261
credit cards, 35, 74, 108, 222
credit rating agencies, 215
criminal law, 187–8
Cummings, Dominic, 51–2
CVs, 38, 67–8, 100
Cyrus, Miley, 54

'dark patterns', 219
data portability right, 244
data protection impact assessments, 194
Deep Patient, 220
deepfakes, 59, 229, 262
defamation, law of, 122
'Delaware effect', 204
deliberative democracy, 151, 153–61, 247, 259, 289, 300
democracy principle, 137–9, 141, 205, 258
Denmark, 236
Department for Education, 71
deregulation, 8, 130
Didi Chuxing, 37
digital harms, 159, 169–70
digital literacy, 43
digital payment platforms, 87
digital personal assistants, 43, 49
digital republicanism, principles of, 136–43

digital rights management (DRM),
    140
digital transparency, 101–2, 108, 171,
    210, 218, 220, 222–6, 230
dikastēria, 159
Director of National Intelligence, 150
DiResta, Renee, 293
distributed machine learning
    analytics, 253
District Court of the Hague, 164
diversity, lack of, 99–100
domination principle, 137–8
Dorsey, Jack, 50, 55
Doteveryone, 150
Douglass, Frederick, 23
Draco, 217
Dunkirk evacuation, 73
Dutch Republic, 22

eating disorders, 290
eBay, 208
'echo chambers', 157, 378
'edge computing', 253
Edwards, Lilian, 244, 254
Einstein, Albert, 1
Eisenstat, Yaël, 97
electric scooters, 35
Emerson, Thomas, 265
Engels, Friedrich, 86
English Revolution, 10, 21, 43
Enlightenment, 91, 106
epikeia, 262
Espionage Act (1917), 278
ethics, 99–103
European Union, 118, 124–5, 204,
    206–7, 250, 295, 340–1
    draft Artificial Intelligence Act,
        179, 230
    draft Digital Services Act,
        179, 182–3
    see also GDPR
European Commission, 48, 236, 341
European Convention on Human
    Rights (Article 10), 287–8, 300

Fabricius, 91
FaceApp, 89
Facebook, 4–5, 14, 16, 38, 48, 52–3,
    74, 123, 204, 243–4, 269, 274
    abuse of market power, 235–8, 241
    advertising revenue, 270
    democracy experiment, 145–6
    and elections integrity, 97
    fined by FTC, 121–2
    groups, 295
    and industrial relations, 139
    and media regulation, 281–2
    online speech policy, 269–72,
        374–5
    Oversight Board, 180–1
    and political advertising, 55
    privacy policy, 107
    recruitment, 100
    and surveillance, 41–3
    'whitelisting', 55
    'world's largest censorship
        body', 294
facial recognition systems, 42, 68, 163,
    193, 260
Fair Information Practices (FIPs),
    250–2, 255
fairness doctrine, 279, 289
Federal Deposit Insurance
    Corporation, 117
Federal Reserve Board, 116
Federal Trade Commission (FTC), 116,
    120–2, 268
Ferguson, Adam, 24, 91
fiduciary law, 167
Five Eyes, 142
Flaubert, Gustave, 25
Floyd, George, 23
Food and Drug Administration
    (FDA), 268
football, 345
Ford, 37, 148
Foucault, Michel, 43, 214
Founding Fathers, 91, 239
France, 205, 209, 288

Franklin, Benjamin, 111
free speech, 122, 164, 267–9,
 275, 289–90
French Revolution, 22, 214
Friedman, Milton, 94

Galileo, 277
GDPR (General Data Protection
 Regulation), 12, 87, 124–5, 186,
 204, 207, 230, 244, 249–51,
 253–4, 336–7, 369, 372–3
Germany, 208, 288, 294–5, 299
gig economy, 150
Gillespie, Tarleton, 146
global governance, 13, 118, 205–6
Goodwin, Morag, 82
Google, 37, 42, 48–9, 108, 146, 205,
 208, 223, 229, 238, 244, 270
 abuse of market power, 235–6
 and ethics, 94, 97, 101
 and neutrality fallacy, 65–6
 and political advertising, 54–5
 and property laws, 127–8
 recruitment, 100
 and Russia, 141–2
GPT-3, 59
Great Depression, 117
Greece, ancient, 20, 155, 217, 253,
 277, 313
Grossman, Vasily, 63

Hachette, 237
Hadfield, Gillian, 194–5
Hamilton, Alexander, 22, 239,
 272, 304
Harm Reduction Plans, 296
Harrington, James, 30
Harris, Kamala, 49
Hart, Herbert, 197
hate speech, regulation of, 280, 295
Hayek, Friedrich, 82
Hebrews, ancient, 22
high-risk algorithms, 179, 181, 183,
 192–3, 227, 235, 259–61

Hill, Jonah, 122
'hinterland of naughtiness', 139–40
Hitachi, 87
Hobbes, Thomas, 80
Hoffman, Reid, 93–4
Holmes, Oliver Wendell, 269, 287
Holocaust denial, 270, 288
home-workers, 43
homosexuality, 53
Hong Kong, 67
Hotmail, 74, 261
Hua, Shin-Shin, 244
human rights, 164–8
Hume, David, 304

IBM, 59, 101
ICANN, 119
IETF, 119
industrial relations, 138–9
information overload, 222
Innocent IV, Pope, 128
Instagram, 241, 244, 269–70
 ban on nipples, 54
Institute of Electrical and Electronics
 Engineers (IEEE), 119, 171, 201
insurance, 37, 72, 75, 90, 138, 192,
 200, 219, 228, 254–5, 258,
 261, 358
Intel, 236
intellectual property, 120, 214
International League Against Racism
 and Anti-Semitism, 209
International Organization for
 Standardization (ISO), 171
'internet of things', 36
interoperability, 244
Interstate Commerce Commission,
 116
inuria, 69
Ireland, 125, 156
isothymia, 69

Japan, 250
Jefferson, Thomas, 113, 128, 239

job interviews, 38, 75–6, 100
juries, 76, 160, 187

Kant, Immanuel, 107
Keller, Daphne, 291, 293
Khamenei, Ayatollah, 54
Khashoggi, Jamal, 58
Klonick, Kate, 293
Kundera, Milan, 286

LaJeunesse, Ross, 97
Learned Hand, Billings, 110, 211
Lincoln, Abraham, 302
LinkedIn, 93
Livy, 20
Locke, John, 106
Lovelace, Ada, 37
Luxembourg, 125
Lyft, 150

Machiavelli, Niccolò, 133, 239
machine learning, 68–9, 73, 99, 109,
    122, 166, 213, 220–1, 227, 231,
    238, 251–4, 262, 270
Madison, James, 9–10, 30, 111, 142,
    161, 239, 289
Magaziner, Ira, 130
mailing lists, 45
market individualism, 11–13, 15–16,
    25–31, 131, 136, 251, 269,
    302, 304
markets, 85–91, 111–12, 135–6
  abuse of market power, 235–42
  and choice, 87–90, 103, 107
  and self-regulation, 91, 95
Markle, Meghan, 271
Marx, Groucho, 101
Marx, Karl, 86
Mazzucato, Mariana, 146
media regulation, 259, 278–83
medical ethics, 101
Microsoft, 87, 101, 236, 244
Mieklejohn, Alexander, 289–90
migrants, 58

Mill, John Stuart, 275
Milton, John, 21, 111, 275, 293
Mohammed bin Salman, Crown
    Prince, 58
Montesquieu, 175, 179, 239, 301
mortgages, 73–5, 89, 153, 163, 166,
    226, 257, 261

names, 69, 75
National Algorithms Centre
    (proposed), 150
National Artificial Intelligence Centre
    (proposed), 150
National Cyber office (proposed),
    150
National Labor Relations Board, 117
National Robotics Centre
    (proposed), 150
National Security Agency, 142
Navalny, Alexei, 142
Nazi memorabilia, 209
negligence, law of, 297
Netflix, 46
Netzwerkdurchsetzungsgesetz law, 124
neutrality fallacy, 65–6, 272
New Brandeis Movement, 236, 241
New York Post, 4
New York Times, 110, 281
New Yorker, 5, 273
New Zealand, 192, 287
Neymar, 55
Nietzsche, Friedrich, 141
Noble, Safiya, 66
nursing homes, 172

Ofcom, 14, 280–1
Office for Responsible Technology
    (proposed), 150
Office of Technology Assessment
    (OTA), 149
oil rigs, 298
online dispute resolution, 182
online harassment, 138
Ostbelgien Citizens' Council, 157

Page, Larry, 127–8
Pakistan, 53
Parfit, Derek, 79
parsimony principle, 137, 141, 188, 245, 259, 347
Pascal, Blaise, 233
PayPal, 93
Pelosi, Nancy, 5
Perrin, Will, 293
Persily, Nathaniel, 275, 293
personal data, market in, 39–41, 71–2, 87, 109–10, 136
  see also privacy
'personal data containers', 253
Physiocrats, 86
Pinker, Steven, 40
Plato, 313
Poland, 21
Polis social media platform, 158
political advertising, 54–5, 273
Polybius, 239
precautionary principle, 193
preservation principle, 137
Priestley, Joseph, 22
privacy, 40, 87–9, 100–2, 105–8, 164–5, 171, 186, 238, 249–55
  'privacy by design', 193–4
  privacy complaints, 121–2
  privacy policies, 105, 107
  tendency to overestimate, 88
  and transparency, 102, 222–3, 230
professionalism, 198–201
Progressives, 116
property law, 127–8, 130
public service broadcasting, 279
Putin, Vladimir, 204
Pyrrhus, 90–1

Qualcomm, 236

racism, 65–6, 90, 148, 209, 272
Ranking Digital Rights, 223
rape, 39, 55, 280

Ravelry, 53
Reagan, Ronald, 118
recruitment, 38, 67–8, 100, 166, 199, 214, 219, 229–30, 254, 257
  see also job interviews
Reddit, 57, 66, 295
referendums, 150
regulatory access points, 208–9
regulatory arbitrage, 204, 208
'regulatory state', 118
republicanism (the term), 9–10
revenge porn, 123, 299
'revolving door', 14–15
Rihanna, 54
Roman Republic, 10, 20, 90–1, 239, 262, 313–14
Romans, 14, 20–1, 23, 28, 50, 69
Roosevelt, Franklin, 117
Roosevelt, Theodore, 81, 239, 241–3
Rousseau, Jean-Jacques, 214
Russia, 89, 141–2, 274

Sabeel, K. Rahman, 14
Sales, Lord, 262
Sallust, 20, 314
Saudi Arabia, 58, 67
Schauer, Frederick, 262
Schenier, Bruce, 36
Schmitt, Carl, 55
Schneier, Bruce, 150
'Scunthorpe Problem', 52
search engines, 4, 47–9, 159, 223, 340–1
Securities and Exchange Commission (SEC), 117, 120, 268
Sedition Acts, 278, 278, 288
self-driving vehicles, 37, 147–8, 159, 166, 228–9, 358
self-regulation, 6, 91, 95–7, 103, 201
  'by engineers through standards', 120
sexual harassment, 295

Sherman Act, 116
Singapore, 188–9, 250
slavery, 23, 26
slot machines, 192–3
smart toilets, 36
smartphone apps, 66–7
Smith, Adam, 86
Snapchat, 54
Snowden, Edward, 142
social democracy (the term),
    147
Socrates, 277
Solon, 253, 277
South Africa, 287
speed limits, 75, 80
sport, professional, 139
Stamp Act, 22
Standard Oil, 241
standards (the term), 170
streaming platforms, 45–6
suicide, 57–8, 269
surveillance, 40–4, 87, 142, 206,
    223, 253
'surveillance capitalism', 270
Susskind, Richard, 181, 215
Suzor, Nicholas, 209
Switzerland, 22

Tacitus, 20
Taiwan, 158
tea consumption, 74
Teachout, Zephyr, 237, 241
Teigen, Chrissy, 54
Tennyson, Alfred Lord, 73
Tesla, 148
text-generation systems, 228,
    258
Thailand, 53
thegentlemetre, 58–9
Thermopylae, 73
Tiananmen Square, 53
Tibet, 53
TikTok, 47–8, 53–4
torts, law of, 121–2, 166–7

Traffic Penalty Appeal Tribunal
    (TPT), 181–2
transparency paradox, 108
Trump, Donald, 53–5, 110, 271,
    274
Turing, Alan, 328
Turing Red Flag Law, 230
Turkey, 53
Tushnet, Rebecca, 123
Twitter, 4, 50, 52–5, 244, 270, 294

Uber, 38, 150, 158
UK Supreme Court, 21, 262
Ungern-Sterneberg, Antje von, 148
United Nations, 164
Universal Declaration of Human
    Rights, 164
US Capitol, storming of, 270
US Congress, 50, 116, 122–3, 149–50,
    165, 185, 235, 238, 241, 286
US Constitution, 113, 117, 154, 287
    First Amendment, 128, 278–9, 281,
        286–91, 300, 384
US healthcare system, 68
US Justice Department, 120
US Steel, 241
US Supreme Court, 128, 278, 288–9,
    291

Véliz, Carissa, 252
Venezuela, 54
Venice, fifteenth-century, 343
venture capitalism, 93, 100, 237

W3C, 119
Waldron, Jeremy, 81
Walzer, Michael, 17
Warren, Elizabeth, 235
Warren, Samuel, 106
Waze app, 167
Weber, Max, 79
WhatsApp, 241
Wolf of Wall Street, The, 122
Wollstonecraft, Mary, 23, 198

Woods, Lorna, 293
Wu, Tim, 206, 241

Yahoo!, 48, 209
YouTube, 223, 271–2, 274, 282

Zittrain, Jonathan, 167
Zoom, 41, 319
Zuboff, Shoshana, 107, 270
Zuckerberg, Mark, 5, 7, 10, 53, 55, 60,
    91, 145–6, 268, 272

A Note on the Author

Jamie Susskind is the author of the award-winning bestseller *Future Politics*, which received the Estoril Global Issues Distinguished Book Prize, and was an *Evening Standard* and *Prospect* Book of the Year. It was described as 'a formidable achievement' (*Guardian*), 'crucial reading ... superb and necessary' (*Observer*), and 'as convincing as it is shocking' (*Sunday Times*). A practising barrister, Jamie is a graduate of the University of Oxford and has held fellowships at Harvard University and the University of Cambridge. His writing has appeared in the *New York Times*, *The Times*, *New Statesman*, *Wired* and elsewhere. The *Evening Standard* has written that Jamie 'could be one of the great public intellectual rock stars of our time'.